# Lecture Notes in Intelligent Transportation and Infrastructure

**Series Editor**

The series "Lecture Notes in Intelligent Transportation and Infrastructure" (LNITI) publishes new developments and advances in the various areas of intelligent transportation and infrastructure. The intent is to cover the theory, applications, and perspectives on the state-of-the-art and future developments relevant to topics such as intelligent transportation systems, smart mobility, urban logistics, smart grids, critical infrastructure, smart architecture, smart citizens, intelligent governance, smart architecture and construction design, as well as green and sustainable urban structures. The series contains monographs, conference proceedings, edited volumes, lecture notes and textbooks. Of particular value to both the contributors and the readership are the short publication timeframe and the world-wide distribution, which enable wide and rapid dissemination of high-quality research output.

More information about this series at http://www.springer.com/series/15991

Joanna Soszyńska-Budny

# Safety Analysis of Critical Infrastructure

## Impact of Operation Processes and Climate Change

Springer

Joanna Soszyńska-Budny
Sopot University of Applied Sciences
Sopot, Poland

ISSN 2523-3440 ISSN 2523-3459 (electronic)
Lecture Notes in Intelligent Transportation and Infrastructure
ISBN 978-3-030-65338-5 ISBN 978-3-030-65339-2 (eBook)
https://doi.org/10.1007/978-3-030-65339-2



This Springer imprint is published by the registered company Springer Nature Switzerland AG
The registered company address is: Gewerbestrasse 11, 6330 Cham, Switzerland

# Preface

The book is focused on the critical infrastructure safety analysis based on reliability modelling of multistate ageing system. This modelling is developed by the assumption that changes of the operation process and the climate-weather process, in the operating area of the system, have influence on the safety parameters of system components. Above assumption, consequently, influences the system safety characteristics degradation. The book is also dedicated to critical infrastructure safety mathematical modelling and prediction, and due to that the critical infrastructure is defined as a complex multistate ageing system composed of multistate ageing components/assets in its operating environment. In practical usage, the critical infrastructure significant features are its inside and outside dependencies; i.e. infrastructure degradation has significant destructive influence on health, safety and security, economics and social conditions of large human communities and/or territories.

Many multistate methods, used in analysis of complex system reliability and safety, are difficult to apply practically with acceptable accuracy. Necessity of considering a large number of transitions between the reliability and safety states precludes obtaining of exact solutions, and approximate solution is often not precise enough. The proposed approach eliminates this problem and gives exact values of basic critical infrastructure safety indicators, regardless of the number of assets and the number of their safety states. More exactly, in case when critical infrastructure is free of outside impacts, the large number of its assets and their safety states do not restrict usage of the proposed approach, which gives exact solutions. Another, practically important, idea of the approach to the multistate ageing critical infrastructure safety analysis presented in this book is considering influence of the operation process and climate-weather change process on critical infrastructure assets' safety states. It means that degradation of assets' safety states causes decreasing of safety level of the whole critical infrastructure. And, in order to realize this novel idea, exact results from the safety model of the critical infrastructure, which is free of outside impacts, are combined with approximate results obtained from the operation process model and the climate-weather change process model. What's important, we are forced to use approximate characteristics of those two processes, what cause the final results approximate as well.

The book's main principle of the critical infrastructure safety modelling is simultaneous usage of the multistate complex critical infrastructure safety model and semi-Markov models for its operation process and for the climate-weather change process in the infrastructure's operating area. Above combination is used to construct the joint general safety model of the critical infrastructure related to its operation process and the climate-weather change process in its operating area. This principle allows to create useful and practical tool in safety examination of real critical infrastructures. The joint model which links:

- the safety model of the multistate critical infrastructure,
- varying in time operation process model and
- the climate-weather change process in its operating area model,

is constructed and applied in safety analysis of real critical infrastructure. Proposed in the book model and methods of finding critical infrastructure safety, risk and resilience indicators are practically applied. For port oil terminal critical infrastructure analysis, identification and prediction of safety parameters in relation to varying in time its operation process is presented. Influence of the climate-weather change process in operating area on, changing in time, safety structure and assets' safety parameters is taken into account in calculations.

The book delivers a complete and up-to-date elaboration on the newest mathematical methods of safety identification, evaluation and prediction for a wide range of critical infrastructures. Pointing out practical application of these methods in the operating processes of critical infrastructures and climate-weather change processes is another important reason for this book. The book contains complete solutions of the formulated problems for the considered critical infrastructures, with the assumption of the piecewise exponential safety functions of their assets. Such assumption is necessary in the analytical approach to consider subject, in order to obtain significant results, when ageing and degradation of critical infrastructures caused by outside operation and weather conditions is taken into account. The notation "climate-weather change process" corresponds to the weather change process in a short-term impact analysis and to the climate change process in long-term impact analysis.

The book consists of 7 parts, including this Preface and Chaps. 1–6.

In Preface, the book introduction and its contents are given.

In Chap. 1, the reliability approach to critical infrastructure examination is presented. The critical infrastructure reliability backgrounds are introduced. Two-state and multistate approaches to critical infrastructure reliability analysis are considered. The below listed critical infrastructure reliability indicators (RelIs) are defined in both considered approaches:

- critical infrastructure reliability function (RelI1),
- critical infrastructure risk function (RelI2),
- graph of the critical infrastructure risk function/the fragility curve (RelI3),

- mean value of the critical infrastructure lifetime up to exceeding critical reliability state (RelI4),
- standard deviation of the critical infrastructure lifetime up to the exceeding critical reliability state (RelI5),
- moment of exceeding acceptable value of critical infrastructure risk function level (RelI6),
- mean values of the critical infrastructure lifetime in the reliability state subsets (RelI7),
- standard deviations of the critical infrastructure lifetime in the reliability state subsets (RelI8),
- mean values of the critical infrastructure lifetime in particular reliability states (RelI9),
- intensities of degradation (ageing) of the critical infrastructure, i.e. intensities of critical infrastructure departure from the reliability state subsets (RelI10).

Moreover, applications of the proposed reliability models to the evaluation and prediction of the reliability indicators for an exemplary system and the port oil terminal critical infrastructure are presented as well.

In Chap. 2, the methodology and general approach to critical infrastructure safety and resilience analysis is proposed. Principles of multistate approach to critical infrastructure safety analysis are presented. There are introduced notions of critical infrastructure basic safety indicators, for example, critical infrastructure safety function or critical infrastructure risk function and fragility curve. Basic critical infrastructure safety structures are defined, and the general formulae for their safety functions and their particular forms are given in case of assets' piecewise exponential safety functions. The critical infrastructure safety and resilience indicators are obtained by using probabilistic approach to modelling of operation threats and extreme weather hazard impacts on its assets' safety.

The following, important from practical side, safety and resilience indicators (SafIs, ResIs) are proposed:

- critical infrastructure safety function (SafI1),
- critical infrastructure risk function (SafI2),
- critical infrastructure fragility curve (SafI3),
- mean value of the critical infrastructure lifetime up to exceeding critical safety state (SafI4),
- standard deviation of the critical infrastructure lifetime up to the exceeding critical safety state (SafI5),
- moment of exceeding acceptable value of critical infrastructure risk function level (SafI6),
- mean values of the critical infrastructure lifetime in the safety state subsets (SafI7),
- standard deviations of the critical infrastructure lifetime in the safety state subsets (SafI8),
- mean values of the critical infrastructure lifetime in particular safety states (SafI9),

- intensities of degradation (ageing) of the critical infrastructure, i.e. the intensities of critical infrastructure departure from the safety state subsets (SafI10),
- coefficients of operation process and climate-weather change process impact on the critical infrastructure intensities of degradation (ResI1),
- the indicator of critical infrastructure resilience to operation process and climate-weather change process impact (ResI2).

Chapter 3 is dedicated to modelling of critical infrastructure safety without considering outside impacts. Safety parameters of the critical infrastructure assets are defined. The following safety indicators (SafIs) of the critical infrastructure free of any outside impacts are introduced:

- critical infrastructure safety function (SafI1),
- critical infrastructure risk function (SafI2),
- critical infrastructure fragility curve (SafI3),
- mean value of the critical infrastructure lifetime up to exceeding critical safety state (SafI4),
- standard deviation of the critical infrastructure lifetime up to the exceeding critical safety state (SafI5),
- moment of exceeding acceptable value of critical infrastructure risk function level (SafI6),
- mean values of the critical infrastructure lifetime in the safety state subsets (SafI7),
- standard deviations of the critical infrastructure lifetime in the safety state subsets (SafI8),
- mean values of the critical infrastructure lifetime in particular safety states (SafI9),
- intensities of degradation (ageing) of the critical infrastructure/intensities of critical infrastructure departure from the safety state subsets (SafI10).

A real technical application of the proposed safety model is presented as well. Namely, the port oil terminal critical infrastructure, placed at the coast of Gulf of Gdansk, free of any outside impacts is examined. Safety analysis of this oil terminal is performed, and its safety indicators are evaluated.

Chapter 4 is concerned with modelling of safety of a critical infrastructure impacted by its operation process. The semi-Markov process is used to construct probabilistic model of the critical infrastructure operation process. The critical infrastructure operation process and its parameters are defined. The operation characteristics of the critical infrastructure operation process are determined. Thereafter, the integrated general safety model is created for a critical infrastructure impacted by its operation process. That general safety model links its multistate safety model constructed in Chap. 3 and its operation process model. The model considers also variable, at the different operation states, its safety structure and its assets' safety parameter. Definitions of safety parameters of the critical infrastructure assets, impacted by its operation process, are given. The following safety and resilience indicators (SafIs, ResIs) of the critical infrastructure impacted by its operation process are introduced:

- critical infrastructure safety function (SafI1),
- critical infrastructure risk function (SafI2),
- critical infrastructure fragility curve (SafI3),
- mean value of the critical infrastructure lifetime up to exceeding critical safety state (SafI4),
- standard deviation of the critical infrastructure lifetime up to the exceeding critical safety state (SafI5),
- moment of exceeding acceptable value of critical infrastructure risk function level (SafI6),
- mean values of the critical infrastructure lifetime in the safety state subsets (SafI7),
- standard deviations of the critical infrastructure lifetime in the safety state subsets (SafI8),
- mean values of the critical infrastructure lifetime in particular safety states (SafI9),
- intensities of degradation (ageing) of the critical infrastructure/the intensities of critical infrastructure departure from the safety state subset (SafI10),
- coefficients of operation process impact on the critical infrastructure intensities of degradation/the coefficients of operation process impact on critical infrastructure intensities of departure from the safety state subset (ResI1),
- indicator of critical infrastructure resilience to operation process impact (ResI2).

Application for the proposed safety model is also presented on the example of port oil terminal critical infrastructure impacted by its operation process. Definition of operation process for the considered port oil terminal critical infrastructure is given. The parameters of the port oil terminal operation process are fixed, and its operation characteristics are evaluated. Safety analysis of the port oil terminal critical infrastructure impacted by its operation process is examined. Safety and resilience indicators are evaluated and compared with those obtained for the same critical infrastructure, but free of operation process influence, considered in Chap. 3.

Chapter 5 is a dedicated model of safety of a critical infrastructure impacted by the climate-weather change process. The semi-Markov process is used to construct a probabilistic model of the climate-weather change process which affects critical infrastructure at its operation area. Definition of climate-weather change process at the critical infrastructure operating area is given, as well as parameters of that process. The characteristics of this climate-weather change process are also determined. Then, the integrated general safety model of a critical infrastructure impacted by the climate-weather change process at its operating area is presented. That general model links its multistate safety model from Chap. 3 and the model of the climate-weather change process at its operation area. Additionally, it considers also variable safety structure and its assets' safety parameters at the different climate-weather states. The critical infrastructure assets' safety parameters are defined, in situation when they are impacted by the climate-weather change process. The following safety and resilience indicators (SafIs, ResIs) of the critical infrastructure impacted by the climate-weather change process at its operating area are introduced:

- critical infrastructure safety function (SafI1),
- critical infrastructure risk function (SafI2),
- critical infrastructure fragility curve (SafI3),
- mean value of the critical infrastructure lifetime up to exceeding critical safety state (SafI4),
- standard deviation of the critical infrastructure lifetime up to the exceeding critical safety state (SafI5),
- moment of exceeding acceptable value of critical infrastructure risk function level (SafI6),
- mean values of the critical infrastructure lifetime in the safety state subsets (SafI7),
- standard deviations of the critical infrastructure lifetime in the safety state subsets (SafI8),
- mean values of the critical infrastructure lifetime in particular safety states (SafI9),
- intensities of degradation (ageing) of the critical infrastructure/the intensities of critical infrastructure departure from the safety state subset (SafI10),
- coefficients of climate-weather change process impact on the critical infrastructure intensities of degradation/the coefficients of climate-weather change process impact on critical infrastructure intensities of departure from the safety state subset (ResI1),
- indicator of critical infrastructure resilience to climate-weather change process impact (ResI2).

Application of the proposed safety model is presented as well. The port oil terminal critical infrastructure impacted by the climate-weather change process at its operating area is again taken into consideration. The climate-weather change process, impacting the considered port oil terminal critical infrastructure, is defined. The parameters of the climate-weather change process at the port oil terminal operating area are fixed, and its characteristics are evaluated. Safety analysis of considered port oil terminal critical infrastructure, impacted by the climate-weather change process at its operating area, is examined. Its safety and resilience indicators are evaluated and compared with their values in case without climate-weather influences from Chap. 3. Those indicators are also compared with their values from the case of this critical infrastructure impacted by its operation process considered in Chap. 4.

In Chap. 6, two previously presented models are integrated. Namely, safety model of the critical infrastructure impacted by its operation process from Chap. 4 and safety model of the critical infrastructure impacted by the climate-weather change process at its operating area are introduced in Chap. 5. Combined, they create general joint safety model of critical infrastructure safety influenced, simultaneously, by its operation process and by the climate-weather change process.

This general joint safety model is proposed in two cases, when the critical infrastructure operation process and climate-weather change process are independent and when they are dependent. The critical infrastructure operation process related to climate-weather change process is defined as well as its parameters. Characteristics of that process are also determined. Safety parameters for assets of the critical infrastructure, impacted by its operation process related to the climate-weather change at its operating area, are defined. Definition is given for the following safety and resilience indicators (SafIs, ResIs) of the critical infrastructure impacted by its operation process related to the climate-weather change at its operating area:

- critical infrastructure safety function (SafI1),
- critical infrastructure risk function (SafI2),
- critical infrastructure fragility curve (SafI3),
- mean value of the critical infrastructure lifetime up to exceeding critical safety state (SafI4),
- standard deviation of the critical infrastructure lifetime up to the exceeding critical safety state (SafI5),
- moment of exceeding acceptable value of critical infrastructure risk function level (SafI6),
- mean values of the critical infrastructure lifetime in the safety state subsets (SafI7),
- standard deviations of the critical infrastructure lifetime in the safety state subsets (SafI8),
- mean values of the critical infrastructure lifetime in particular safety states (SafI9),
- intensities of degradation (ageing) of the critical infrastructure/the intensities of critical infrastructure departure from the safety state subset (SafI10),
- coefficients of its operation process related to climate-weather change process impact on the critical infrastructure intensities of degradation/the coefficients of its operation process related to climate-weather change process impact on critical infrastructure intensities of departure from the safety state subset (ResI1),
- indicator of critical infrastructure resilience to its operation process related to climate-weather change process impact (ResI2).

Application of created general joint safety model of the critical infrastructure is presented. The port oil terminal critical infrastructure, impacted by its operation process related to the climate-weather change at its operating area, is considered. Definition of the operation process, related to climate-weather change of considered port oil terminal critical infrastructure, is given. Further, the parameters of the joint operation and climate-weather change process of this critical infrastructure at its operating area are fixed. Characteristics of these parameters are evaluated as well. Safety analysis of the considered oil terminal critical infrastructure, impacted by its operation process related to climate-weather change, is conducted. Safety and resilience indicators for the oil terminal are determined. Values of safety and resilience indicators of the port oil terminal critical infrastructure are discussed and compared with their values obtained in the previous three cases, i.e. without considering outside impact, with considering operation process impact and, separately, with considering climate-weather change process impact.

The possibility of other, future, theoretical developments and practical applications of safety analysis, presented in the book, for multistate ageing systems and critical infrastructures impacted by outside conditions are suggested in conclusion.

The book is a review of the author's previous works and summarizes in a coherent and comprehensive way the author's research on the safety of critical infrastructures. The models presented in the book have been modified to make them more understandable to practitioners dealing with critical infrastructures. A completely new way of numbering the climate-weather states of the joint climate-weather change process is proposed. The new procedures for determining the limit values of the joint climate-weather change process transient probabilities are also presented.

Sopot, Poland Joanna Soszyńska-Budny

# Acknowledgements

I would like to thank Prof. Krzysztof Kołowrocki for his substantive remarks that had valuable impact on the content of this book.

# Contents

# Chapter 1
# Reliability of Critical Infrastructure

The backgrounds for critical infrastructure reliability analysis are introduced. Two-state approach and multistate approach to critical infrastructure reliability examination are presented. The critical infrastructure reliability indicators are defined in both considered approaches. Applications of the proposed reliability approaches to the evaluation and prediction of the reliability indicators of an exemplary critical infrastructure and the port oil terminal critical infrastructure are presented as well.

## 1.1 Two-State Approach to Critical Infrastructure Reliability Analysis

In the two-state reliability analysis to define the critical infrastructure, we assume that:

- the critical infrastructure has the reliability state set $\{0, 1\}$;
- the critical infrastructure is in the reliability state 0 if it is failed;
- the critical infrastructure is in the reliability state 1 if it is not failed;
- $T$ is a random variable representing the lifetime of the critical infrastructure in the reliability state 1, while it was in the reliability state 1 at the moment $t = 0$ (a time up to its failure);
- $s(t)$ is the critical infrastructure reliability state at the moment $t$, $t \in < 0, \infty)$, given that it was in the reliability state 1 at the moment $t = 0$.

As we denoted the critical infrastructure unconditional lifetime in the reliability state 1 (time to critical infrastructure failure) by $T$, then we define the critical infrastructure reliability function, the first reliability indicator (RelI1)

$$\boldsymbol{R}(t) = P(s(t) = 1 | s(0) = 1) = P(T > t) \text{ for } t \in < 0, \infty), \tag{1.1}$$

 
J. Soszyńska-Budny, *Safety Analysis of Critical Infrastructure*,
Lecture Notes in Intelligent Transportation and Infrastructure,
https://doi.org/10.1007/978-3-030-65339-2_1

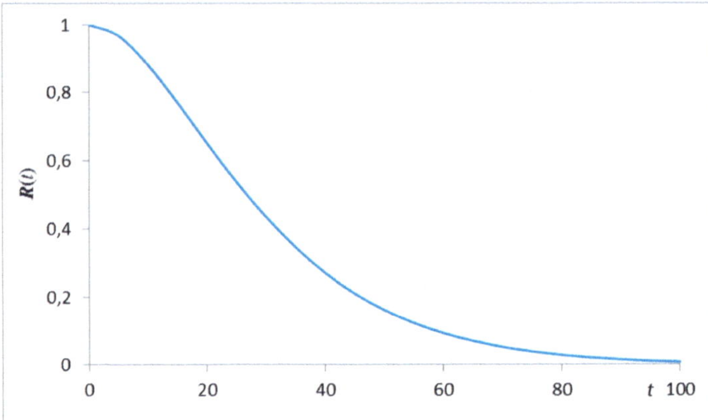


**Fig. 1.1** The graph of a critical infrastructure reliability function $\boldsymbol{R}(t)$

as a probability that the critical infrastructure is in the reliability state 1 at the moment $t, t \in < 0, \infty)$, while it was in the reliability state 1 at the moment $t = 0$.

The exemplary graph of a critical infrastructure reliability function $\boldsymbol{R}(t)$ for $t \in < 0, \infty)$, is shown in Fig. 1.1.

From that assumption it follows that between a critical infrastructure reliability function $\boldsymbol{R}(t)$ and a critical infrastructure time to failure $T$ distribution function

$$\boldsymbol{F}(t) = P(T \leq t), t \in < 0, \infty), \tag{1.2}$$

there exists a relationship given by

$$\boldsymbol{R}(t) = 1 - \boldsymbol{F}(t) \text{ for } t \in < 0, \infty). \tag{1.3}$$

Thus, if we define the critical infrastructure risk function, the second reliability indicator (RelI2),

$$\boldsymbol{r}(t) = P(s(t) = 0 | s(0) = 1) = P(T \leq t), t \in < 0, \infty), \tag{1.4}$$

as a probability that the critical infrastructure is in the reliability state 0 at the moment $t, t \in < 0, \infty)$, while it was in the reliability state 1 at the moment $t = 0$, then

$$\boldsymbol{r}(t) = \boldsymbol{F}(t) = 1 - \boldsymbol{R}(t), t \in < 0, \infty), \tag{1.5}$$

where $\boldsymbol{F}(t)$ is the distribution function given by (1.2) and $\boldsymbol{R}(t)$ is the reliability function given by (1.1).

The graph of the critical infrastructure risk function, the third reliability indicator (RelI3), is called the fragility curve [6].

The graph of the exemplary fragility curve critical infrastructure is presented in Fig. 1.2.

The critical infrastructure reliability function (RelI1), the critical infrastructure risk function (RelI2) and the critical infrastructure fragility curve (RelI3) are proposed as main basic critical infrastructure reliability indicators.

Other practically useful critical infrastructure reliability indicators are:

– the mean value of the critical infrastructure lifetime $T$ (RelI4) given by

$$\mu = \int_0^\infty \boldsymbol{R}(t)dt, \tag{1.6}$$

where $\boldsymbol{R}(t)$ is defined by (1.1);

– the standard deviation of the critical infrastructure lifetime $T$ (RelI5), given by

$$\sigma = \sqrt{\boldsymbol{n} - [\mu]^2}, \tag{1.7}$$

where

$$\boldsymbol{n} = 2\int_0^\infty t\boldsymbol{R}(t)dt, \tag{1.8}$$

and $\boldsymbol{R}(t)$ is given by (1.1) and $\mu$ is given by (1.6);

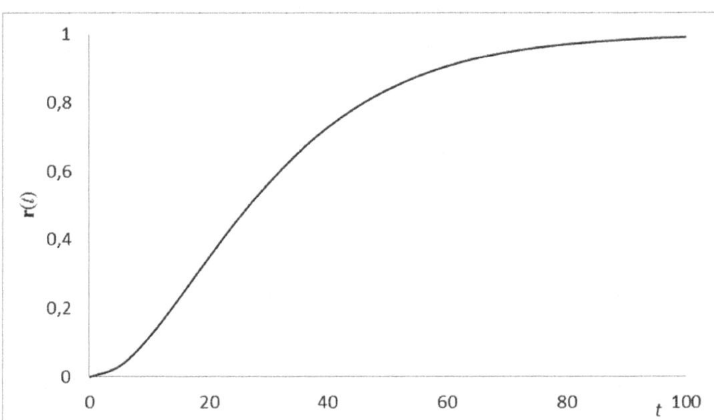


**Fig. 1.2** The graph of the exemplary critical infrastructure risk function $\boldsymbol{r}(t)$

- the moment $\tau$ of exceeding acceptable value of critical infrastructure risk function level $\delta$ (RelI6), given by

$$\boldsymbol{\tau} = \boldsymbol{r}^{-1}(\delta), \tag{1.9}$$

and illustrated in Fig. 1.2, where $\boldsymbol{r}^{-1}(t)$ is the inverse function of the risk function $\boldsymbol{r}(t)$ given by (1.5);

- the failure rate/the intensity of failure of the critical infrastructure (RelI7) given by

$$\lambda(t) = \frac{-\frac{d\boldsymbol{R}(t)}{dt}}{\boldsymbol{R}(t)}, t \in < 0, +\infty \tag{1.10}$$

In the particular case, when the critical infrastructure has the exponential reliability function (RelI1), i.e.

$$\boldsymbol{R}(t) = \exp[-\boldsymbol{\lambda}t], t \in < 0, \infty), \geq 0, \tag{1.11}$$

the failure rate of the critical infrastructure/intensity of failure of the critical infrastructure (RelI7), is constant, as according to (1.10), we have

$$\boldsymbol{\lambda}(t) = \frac{-\frac{d}{dt}\exp[-\boldsymbol{\lambda}t]}{\exp[-\boldsymbol{\lambda}t]} = \frac{\lambda\exp[-\boldsymbol{\lambda}t]}{\exp[-\boldsymbol{\lambda}t]} = \boldsymbol{\lambda}, t \in < 0, +\infty). \tag{1.12}$$

Moreover, at this case, the mean value $\boldsymbol{\mu}$ of the critical infrastructure lifetime $T$ defined by (1.6) is

$$\boldsymbol{\mu} = \int_0^\infty \boldsymbol{R}(t)dt = \int_0^\infty \exp[-\boldsymbol{\lambda}t]dt = -\frac{1}{\boldsymbol{\lambda}}\exp[-\boldsymbol{\lambda}t]_0^\infty = \frac{1}{\boldsymbol{\lambda}}. \tag{1.13}$$

Thus, if the critical infrastructure has the exponential reliability function, given by (1.11), then between the failure rate $\boldsymbol{\lambda}$ and the mean value $\boldsymbol{\mu}$ the following relationship holds

$$\boldsymbol{\lambda} = \frac{1}{\boldsymbol{\mu}}. \tag{1.14}$$

We will need in the next applique Sect. 1.2 the definitions of selected two-state critical infrastructure reliability structures and their reliability functions. To introduce them, we assume that the critical infrastructure is composed of $n$ two-state assets/components

$$E_i, i = 1, 2, \ldots, n, n \in N,$$

having reliability functions

$$R_i(t) = P(T_i > t), t \in < 0, +\infty), i = 1, 2, \ldots, n,$$

where

$$T_i, i = 1, 2, \ldots, n,$$

are independent random variables representing the lifetimes of components $E_i$, $i = 1, 2, \ldots, n$, with distribution functions

$$F_i(t) = P(T_i \leq t), t \in < 0, +\infty), i = 1, 2, \ldots, n.$$

The simplest two-state reliability structures are series and parallel. We define these structures first.

**Definition 1.1** A two-state critical infrastructure is called series if its lifetime $T$ is given by

$$T = \min_{1 \leq i \leq n} \{T_i\}.$$

The number $n$ is called the critical infrastructure reliability structure shape parameter.

The scheme of a series critical infrastructure reliability structure is given in Fig. 1.3.

The above definition means that the series critical infrastructure is not failed if and only if all its components are not failed, and therefore its reliability function is given by

$$\boldsymbol{R}(t) = \prod_{i=1}^{n} R_i(t), t \in < 0, +\infty). \tag{1.15}$$

**Definition 1.2** A two-state series critical infrastructure is called homogeneous if its component lifetimes $T_i$ have an identical distribution function

$$F(t) = P(T_i \leq t), t \in, 0, +\infty), i = 1, 2, \ldots, n,$$

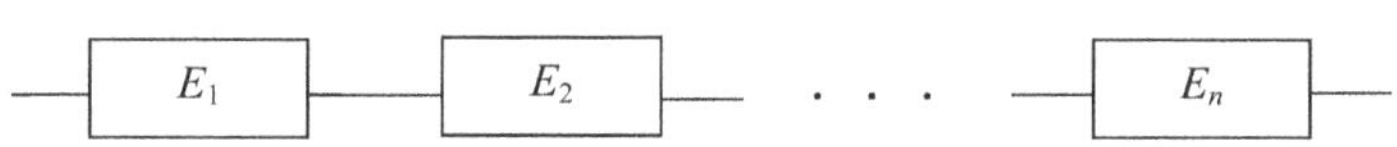


**Fig. 1.3** The scheme of a series critical infrastructure reliability structure

i.e. if its components $E_i$, $t \in <0, +\infty)$, have the same reliability function

$$R(t) = 1 - F(t), t \in <0, +\infty).$$

The above definition results in the following simplified formula

$$\boldsymbol{R}(t) = [R(t)]^n, t \in <0, +\infty), \tag{1.16}$$

for the reliability function of the homogeneous two-state series critical infrastructure.

**Definition 1.3** A two-state critical infrastructure is called parallel if its lifetime $T$ is given by

$$T = \max_{1 \le i \le n} \{T_i\}.$$

The above definition means that the parallel critical infrastructure is failed if and only if all its components are failed and therefore its reliability function is given by

$$\boldsymbol{R}(t) = 1 - \prod_{i=1}^{n} F_i(t) = 1 - \prod_{i=1}^{n} [1 - R_i(t)], t \in <0, +\infty). \tag{1.17}$$

The number $n$ is called the critical infrastructure reliability structure shape parameter.

The scheme of a parallel critical infrastructure reliability structure is given in Fig. 1.4.

**Definition 1.4** A two-state parallel critical infrastructure is called homogeneous if its component lifetimes $T_i$, $i = 1, 2, \ldots, n$, have an identical distribution function

$$F(t) = P(T_i \le t), t \in <0, +\infty), i = 1, 2, \ldots, n$$

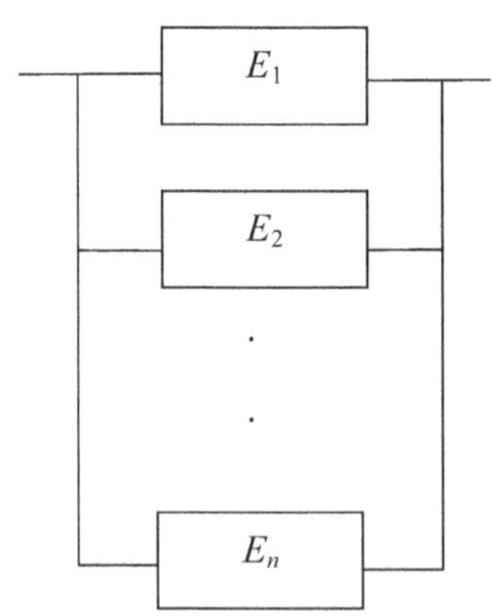


**Fig. 1.4** The scheme of a parallel critical infrastructure reliability structure

i.e. if its components $E_i$, $i = 1, 2, \ldots, n$, have the same reliability function

$$R(t) = 1 - F(t), t \in < 0, +\infty).$$

Under this definition we get the following simplified formula

$$\boldsymbol{R}(t) = 1 - [F(t)]^n = 1 - [1 - R(t)]^n, t \in < 0, +\infty), \tag{1.18}$$

for the reliability function of the homogeneous two-state parallel critical infrastructure.

Another basic, a bit more complex, two-state reliability structure is parallel-series critical infrastructure. To define it, we assume that

$$E_{ij}, i = 1, 2, \ldots, k, j = 1, 2, \ldots, l_i, k, l_1, l_2, \ldots, l_k \in N,$$

are two-state components of the critical infrastructure reliability functions

$$R_{ij}(t) = P\big(T_{ij} > t\big), t \in < 0, +\infty), i = 1, 2, \ldots, k, j = 1, 2, \ldots, l_i,$$

where

$$T_{ij}, i = 1, 2, \ldots, k, j = 1, 2, \ldots, l_i,$$

are independent random variables representing the lifetimes of components $E_{ij}$, $i = 1, 2, \ldots, k$, $j = 1, 2, \ldots, l_i$, with distribution functions

$$F_{ij}(t) = P\big(T_{ij} \le t\big), t \in < 0, +\infty), i = 1, 2, \ldots, k, j = 1, 2, \ldots, l_i,$$

**Definition 1.5** A two-state critical infrastructure is called parallel-series if its lifetime $T$ is given by

$$T = \min_{1 \le i \le k} \left\{ \max_{1 \le j \le l_i} \{T_{ij}\} \right\}.$$

The numbers $k, l_1, l_2, \ldots, l_k$, are called the parallel-series critical infrastructure reliability structure shape parameters.

The scheme of a parallel-series critical infrastructure reliability structure is given in Fig. 1.5.

By superposition of the formulae for the reliability functions of two-state parallel and series critical infrastructures,

it is easy to conclude that the reliability function of the two-state parallel-series critical infrastructures is given by

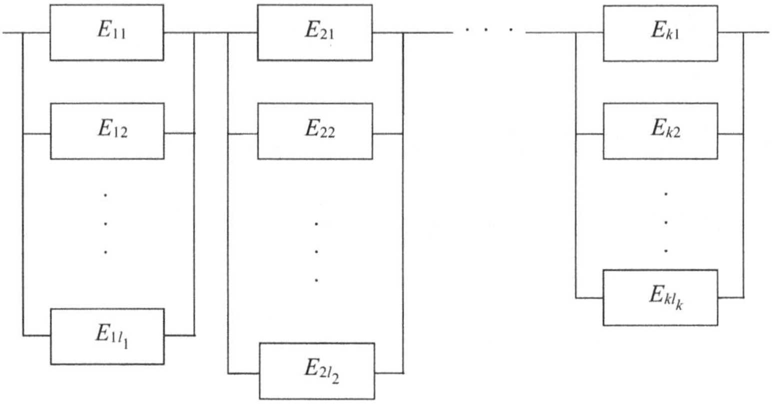


**Fig. 1.5** The scheme of a parallel-series critical infrastructure reliability structure

$$\boldsymbol{R}(t) = \prod_{i=1}^{k}\left[1 - \prod_{j=1}^{l_i} F_{ij}(t)\right] = \prod_{i=1}^{k}\left[1 - \prod_{j=1}^{l_i}[1 - R_{ij}(t)]\right], t \in < 0, +\infty), \tag{1.19}$$

where $k$ is the number of parallel subsystems linked in series and $l_i$, $i = 1, 2, \ldots, k$, are the numbers of components in the parallel subsystems.

**Definition 1.6** A two-state parallel-series critical infrastructure is called homogeneous if its component lifetimes $T_{ij}$ have an identical distribution function

$$F(t) = P\big(T_{ij} \le t\big), t \in < 0, +\infty), i = 1, 2, \ldots, k, j = 1, 2, \ldots, l_i,$$

i.e. if its components $E_{ij}$ have the same reliability function

$$R(t) = 1 - F(t), t \in < 0, +\infty).$$

**Definition 1.7** A two-state parallel-series critical infrastructure is called regular if

$$l_1 = l_2 = \cdots = l_k = l, l \in N,$$

i.e. if the numbers of components in its parallel subsystems are equal.

The numbers $k$, $l$, are called the regular parallel-series critical infrastructure reliability structure shape parameters.

The scheme of a regular parallel-series critical infrastructure reliability structure is given in Fig. 1.6.

The reliability function of the homogeneous regular two-state parallel-series critical infrastructure is given by

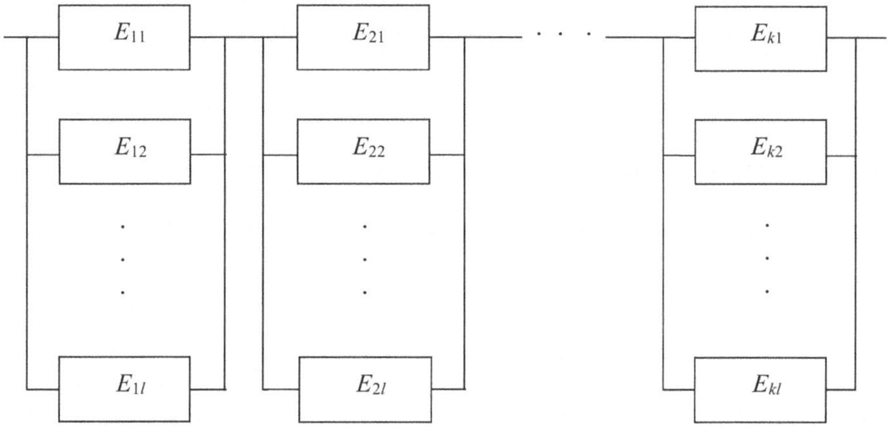


**Fig. 1.6** The scheme of a regular parallel-series critical infrastructure reliability structure

$$\boldsymbol{R}(t) = [1 - [F(t)]^l]^k = 1 - [1 - R(t)]^l]^k, \ t \in < 0, +\infty), \tag{1.20}$$

where $k$ is the number of parallel subsystems linked in series and $l$ is the number of components in the parallel subsystems.

## 1.2 Application

We consider an exemplary two-state critical infrastructure composed of four assets/components with the reliability structure presented in Fig. 1.7.

**Fig. 1.7** The scheme of the exemplary critical infrastructure reliability structure

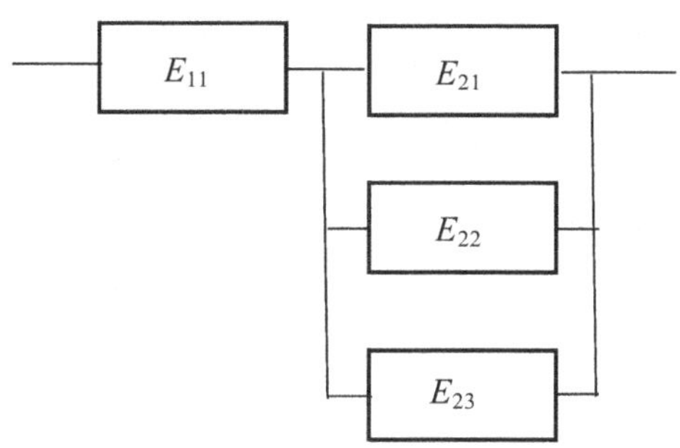

### 1.2.1 Exemplary Critical Infrastructure Reliability Parameters

All four components of the critical infrastructure have identical Weibull (Rayleigh) reliability functions given by

$$R_{ij}(t) = e^{-0.04t^2}, t \in < 0, +\infty), i = 1, j = 1; i = 2, j = 1, 2, 3. \tag{1.21}$$

### 1.2.2 Exemplary Critical Infrastructure Reliability Indicators

The considered critical infrastructure is a homogeneous parallel-series critical infrastructure with Weibull reliability function of its components.

According to the formula for parallel-series critical infrastructure the structure shape parameters $k = 2$, $l_1 = 1$, $l_2 = 3$, according to (1.19), its reliability function, the first reliability indicator, (RelI1) is given by

$$\begin{aligned} \boldsymbol{R}(t) &= \prod_{i=1}^{2}\left[1 - \prod_{j=1}^{l_i}[1 - R_{ij}(t)]\right] = e^{-0.04t^2}\left[1 - [1 - e^{-0.04t^2}]^3\right] \\ &= e^{-0.04t^2}[3e^{-0.04t^2} - 3e^{-0.08t^2} + e^{-0.12t^2}] \\ &= 3e^{-0.08t^2} - 3e^{-0.12t^2} + e^{-0.16t^2} \text{ for } t \in < 0, +\infty). \end{aligned} \tag{1.22}$$

The graph of the exemplary critical infrastructure reliability function is shown in Fig. 1.8.

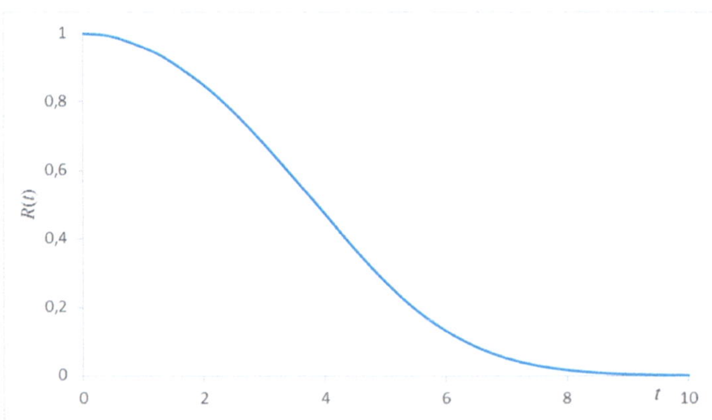


**Fig. 1.8** The graph of the exemplary critical infrastructure reliability function

The critical infrastructure risk function, the second reliability indicator (RelI2), according to (1.5), is given by

$$\boldsymbol{r}(t) = 1 - \boldsymbol{R}(t) = 1 - 3e^{-0.08t^2} + 3e^{-0.12t^2} - e^{-0.16r^2}] \text{ for } t \in < 0, +\infty). \quad (1.23)$$

The graph of the risk function $\boldsymbol{r}(t)$ of the considered exemplary critical infrastructure, the fragility curve (RelI3), is shown in Fig. 1.9.

From (1.22), after considering (1.6)–(1.8) and applying the approximate trapezium method for integration and using the values of reliability function $\boldsymbol{R}(t)$ and the values of $2t\boldsymbol{R}(t)$ calculated with the step $h = 0.5$ given in Table 1.1, the mean value and standard deviation of the exemplary critical infrastructure lifetime (RelI4-5), can be calculated as follows.

$$\begin{aligned}\mu &= \int_0^\infty \boldsymbol{R}(t)dt \cong \frac{h}{2}[\boldsymbol{R}(0) + 2[\boldsymbol{R}(1h) + \boldsymbol{R}(2h) + \cdots + \boldsymbol{R}((n-1)h)] + \boldsymbol{R}(nh)] \\ &= \frac{0.5}{2}[1 + 2[\boldsymbol{R}(0.5) + \boldsymbol{R}(1) + \cdots + \boldsymbol{R}(9.5)] + \boldsymbol{R}(10)] \\ &= 0.25 \cdot [1 + 2[0.990049 + 0.960732 + \cdots + 0.002137] + 0.000] \\ &= 0.25 \cdot [1 + 2 \cdot 7.379219] = 0.25 \cdot 15.758438 \cong 3.94 \qquad (1.24)\end{aligned}$$

Next considering (1.8), using the approximate trapezium method for integration and the values of $2t\boldsymbol{R}(t)$ from Table 1.1, we have

$$\boldsymbol{n} = 2\int_0^\infty t\boldsymbol{R}(t)dt \cong \frac{h}{2}[2 \cdot 0\boldsymbol{R}(0h) + 2[2(1h)\boldsymbol{R}(1h) + 2(2h)\boldsymbol{R}(2h)$$

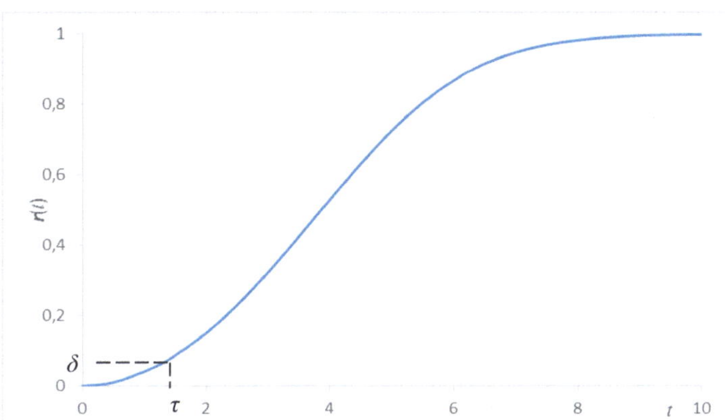


**Fig. 1.9** The graph of the risk function $\boldsymbol{r}(t)$ (the fragility curve) of the exemplary critical infrastructure

**Table 1.1** The values of the $\boldsymbol{R}(t)$ and $2t\boldsymbol{R}(t)$ of exemplary critical infrastructure

| $t$ | $\boldsymbol{R}(t)$ | $2t\boldsymbol{R}(t)$ |
|---|---|---|
| 0 | 1 | 0 |
| 0.5 | 0.990049 | 0.990049 |
| 1.0 | 0.960732 | 1.921463 |
| 1.5 | 0.913348 | 2.740045 |
| 2.0 | 0.849389 | 3.397557 |
| 2.5 | 0.770372 | 3.851859 |
| 3.0 | 0.678398 | 4.070388 |
| 3.5 | 0.577015 | 4.039107 |
| 4.0 | 0.471596 | 3.772766 |
| 4.5 | 0.368749 | 3.318745 |
| 5.0 | 0.27496 | 2.749603 |
| 5.5 | 0.195123 | 2.146357 |
| 6.0 | 0.131656 | 1.579869 |
| 6.5 | 0.084454 | 1.097906 |
| 7.0 | 0.051533 | 0.721456 |
| 7.5 | 0.029938 | 0.449066 |
| 8.0 | 0.016578 | 0.265246 |
| 8.5 | 0.008761 | 0.148932 |
| 9.0 | 0.004424 | 0.079624 |
| 9.5 | 0.002137 | 0.040595 |
| 10.0 | $\approx 0.000$ | $\approx 0000$ |

$$+ \cdots + 2(n-1)h\boldsymbol{R}((n-1)h)] + 2nh\boldsymbol{R}(nh)]$$
$$= \frac{0.5}{2}[0 + 2[0.990049 + 1.921463 + \cdots + 0.040595] + 0]$$
$$= 0.25 \cdot [2 \cdot 37.38063] = 0.25 \cdot 74.76126 \cong 18.69. \tag{1.25}$$

Applying (1.7), after considering (1.24), (1.25) the standard deviation of the exemplary critical infrastructure lifetime is

$$\boldsymbol{\sigma} = \sqrt{\boldsymbol{n} - [\boldsymbol{\mu}]^2} = \sqrt{18.69 - (3.94)^2} = \sqrt{18.69 - 15.52} \cong 1.78. \tag{1.26}$$

Considering (1.9) and the values of the risk function given in Table 1.2, we can determine the moment of exceeding the acceptable value of exemplary critical infrastructure risk function level $\delta = 0.05$ (RelI6), that amounts

$$\tau \cong 1.13. \tag{1.27}$$

**Table 1.2** The values of the risk function $\boldsymbol{r}(t)$ of the exemplary critical infrastructure

| $t$ | $\boldsymbol{r}(t) = 1 - \boldsymbol{R}(t)$ |
|---|---|
| 1.0 | 0.039268 |
| 1.01 | 0.040044 |
| 1.02 | 0.040827 |
| 1.03 | 0.041617 |
| 1.04 | 0.042414 |
| 1.05 | 0.043219 |
| 1.06 | 0.04403 |
| 1.07 | 0.044849 |
| 1.08 | 0.045675 |
| 1.09 | 0.046508 |
| 1.10 | 0.047348 |
| 1.11 | 0.048195 |
| 1.12 | 0.049049 |
| **1.13** | **0.049911** |
| 1.14 | 0.050779 |
| 1.15 | 0.051655 |
| 1.16 | 0.052537 |
| 1.17 | 0.053427 |
| 1.18 | 0.054324 |

From (1.10) and (1.22) it follows that the intensity of failure (RelI7), of the considered critical infrastructure is

$$\boldsymbol{\lambda}(t) = \frac{0.48te^{-0.08t^2} - 0.72te^{-0.12t^2} + 0.32te^{-0.16t^2}}{3e^{-0.08t^2} - 3e^{-.12t^2} + e^{-0.16t^2}}, \text{ for } t \in < 0, +\infty). \quad (1.28)$$

The graphs of the intensity of failure of the considered critical infrastructure is given in Fig. 1.10.

The limit value of this critical infrastructure intensity of failure, considering (1.28) is

$$\boldsymbol{\lambda} = \lim_{t\to\infty} \boldsymbol{\lambda}(t) = \lim_{t\to\infty} \frac{t(1 + 0(t))}{(1 + 0(t))} = +\infty, \quad (1.29)$$

and considering (1.13) and (1.24) the approximate mean value of this critical infrastructure intensity of failure is

$$\boldsymbol{\lambda} \cong \frac{1}{\boldsymbol{\mu}} = \frac{1}{3.94} \cong 0.25. \quad (1.30)$$

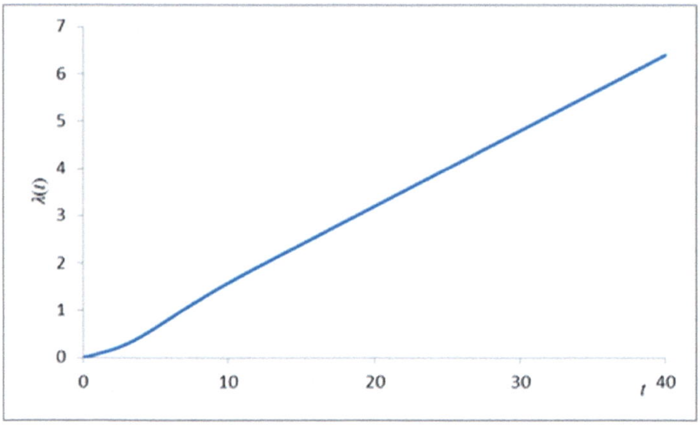


**Fig. 1.10** The graph of the intensity of failure of the critical infrastructure

## 1.3 Multistate Approach to Critical Infrastructure Reliability Analysis

In the multistate reliability analysis to define the critical infrastructure, after considering approaches in [1–5], [7–18], [20–23],we assume that [13, 14], [20–23]:

- the critical infrastructure have the reliability state set $\{0, 1, \ldots, z\}$, $z \geq 1$;
- the reliability states are ordered, the reliability state 0 is the worst and the reliability state $z$ is the best,
- $r$, $r \in \{1, 2, \ldots, z\}$, is the critical reliability state;
- $T(u)$ is a random variable representing the lifetime of the critical infrastructure in the reliability state subset $\{u, u+1, \ldots, z\}$, $u = 1, 2, \ldots, z$, while it was in the reliability state $z$ at the moment $t = 0$;
- the critical infrastructure states degrade with time $t$;
- $s(t)$ is the critical infrastructure reliability state at the moment $t$, $t \in < 0, \infty)$, given that it was in the reliability state $z$ at the moment $t = 0$.

The critical reliability state $r$ means that the critical infrastructure staying in the reliability states less than this reliability state is highly dangerous for the critical infrastructure and for its operating environment.

The above assumptions mean that the reliability states of the critical infrastructure may be changed in time only from better to worse [12–14]. The way in which the critical infrastructure reliability states are changing is illustrated in Figs. 1.11 and 1.12.

As we denoted by $T(u)$ the critical infrastructure lifetime in the reliability state subset $\{u, u+1, \ldots, z\}$, $u = 1, 2, \ldots, z$, then we define the critical infrastructure reliability function, the first reliability indicator (RelI1), by the vector [14]

$$\boldsymbol{R}(t, \cdot) = [\boldsymbol{R}(t, 0), \boldsymbol{R}(t, 1), \ldots, \boldsymbol{R}(t, z)], t \in < 0, \infty), \qquad (1.31)$$

Fig. 1.11 Illustration of a critical infrastructure reliability states changing

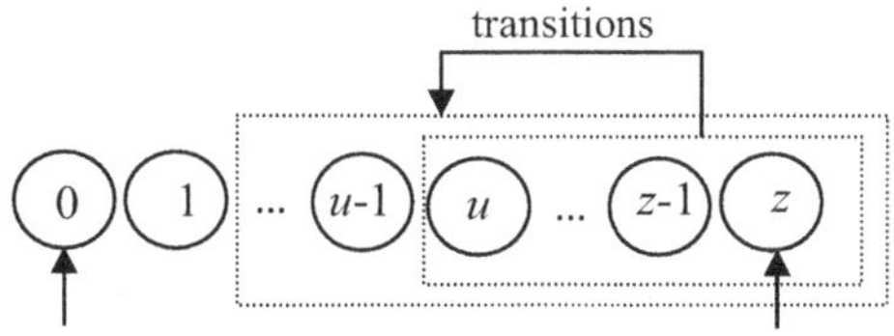


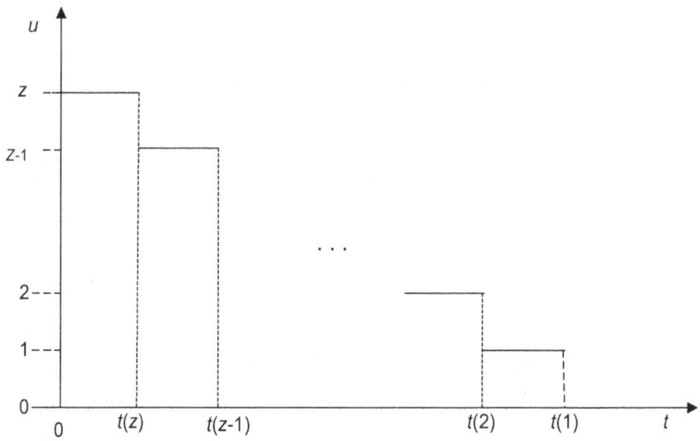


**Fig. 1.12** The relationship between the realizations $t(u)$, $u = 1, 2, \ldots, z$, of the critical infrastructure lifetime $T(u)$, in the reliability state subsets $\{u, u+1, \ldots, z\}$, $u = 1, 2, \ldots, z$

with the coordinates defined by

$$\boldsymbol{R}(t, 0) = P(T(0) > t) = P(s(t) \geq 0 | s(0) = z) = 1 \text{ for } t \in <0, \infty), u = 0, \tag{1.32}$$

$$\boldsymbol{R}(t, u) = P(T(u) > t) = P(s(t) \geq u | s(0) = z) \text{ for } t \in <0, \infty), u = 1, 2, \ldots, z. \tag{1.33}$$

The exemplary graph of a four-state ($z = 3$) critical infrastructure reliability function (RelI1)

$$\boldsymbol{R}(t, \cdot) = [1, \boldsymbol{R}(t, 1), \boldsymbol{R}(t, 2), \boldsymbol{S}(t, 3)], t \in <0, \infty),$$

is shown in Fig. 1.13.

If $r$ is the critical reliability state, then the second reliability indicator (RelI2), the critical infrastructure risk function

$$\boldsymbol{r}(t) = P(s(t) < r | s(0) = z) = P(T(r) \leq t), t \in <0, \infty), \tag{1.34}$$

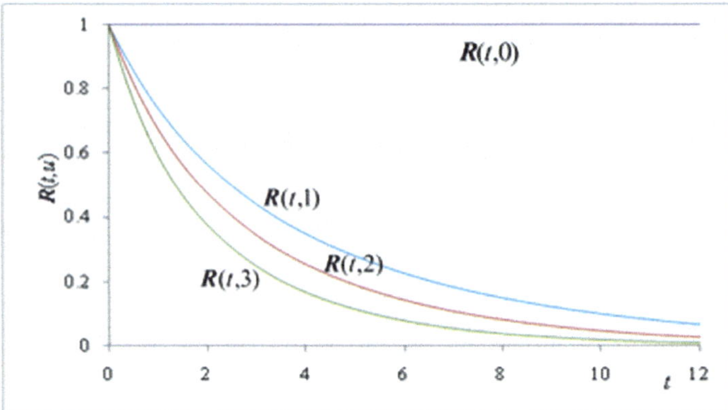


**Fig. 1.13** The graphs of a four-state critical infrastructure reliability function $\boldsymbol{R}(t, \cdot)$ coordinates

is defined as a probability that the critical infrastructure is in the subset of reliability states worse than the critical reliability state $r$, $r \in \{1,\ldots,z\}$ while it was in the best reliability state $z$ at the moment $t = 0$ and given by [12–14]

$$\boldsymbol{r}(t) = 1 - \boldsymbol{R}(t, r), t \in < 0, \infty), \tag{1.35}$$

where $\boldsymbol{R}(t, r)$ is the coordinate of the critical infrastructure reliability function given by (1.31)–(1.33) for $u = r$.

The graph of the critical infrastructure risk function, the third reliability indicator (RelI3), is called the fragility curve [6].

The graph of the exemplary critical infrastructure fragility curve is presented in Fig. 1.14.

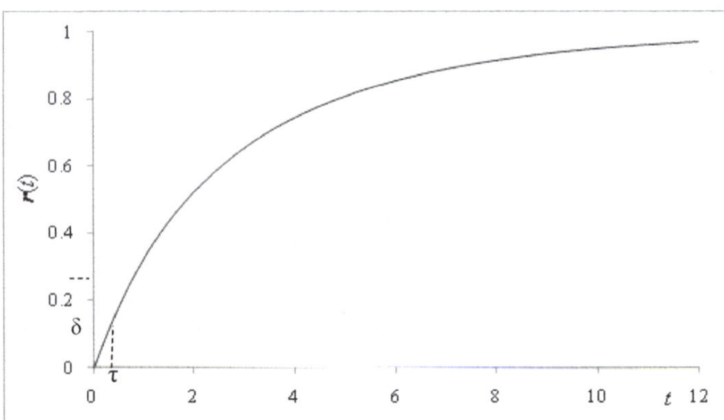


**Fig. 1.14** The graph of the exemplary critical infrastructure risk function $\boldsymbol{r}(t)$

The critical infrastructure reliability function (RelI1), the critical infrastructure risk function (RelI2) and the critical infrastructure fragility curve (RelI3) are proposed as main basic critical infrastructure reliability indicators.

Other practically useful critical infrastructure reliability indicators are:

- the mean value of the critical infrastructure lifetime $T(r)$ up to exceeding critical reliability state $r$ (RelI4) given by

$$\boldsymbol{\mu}(r) = \int_0^\infty \boldsymbol{R}(t, r)dt, \tag{1.36}$$

where $\boldsymbol{R}(t, r)$ is defined by (1.33) for $u = r$;

- the standard deviation of the critical infrastructure lifetime $T(r)$ up to the exceeding the critical reliability state $r$ (RelI5) given by

$$\boldsymbol{\sigma}(r) = \sqrt{\boldsymbol{n}(r) - [\boldsymbol{\mu}(r)]^2}, \tag{1.37}$$

where

$$\boldsymbol{n}(r) = 2\int_0^\infty t R(t, r)dt, \tag{1.38}$$

and $\boldsymbol{R}(t, r)$ is given by (1.31) for $u = r$ and $\mu(r)$ is given by (1.36);

- the moment $\tau$ of exceeding acceptable value of critical infrastructure risk function level $\delta$ (RelI6) given by

$$\boldsymbol{\tau} = \boldsymbol{r}^{-1}(\delta), \tag{1.39}$$

and illustrated in Fig. 1.14, where $\boldsymbol{r}^{-1}(t)$ is the inverse function of the risk function $\boldsymbol{r}(t)$ given by (1.34);

- the mean lifetimes of the critical infrastructure in the reliability state subsets $\{u, u+1, \ldots, z\}$, $u = 1, 2, \ldots, z$, (RelI7) given by

$$\boldsymbol{\mu}(u) = \int_0^\infty [\boldsymbol{R}(t, u)]dt, u = 1, 2, \ldots, z, i = 1, 2, \ldots, n; \tag{1.40}$$

- the standard deviations of the critical infrastructure lifetimes in the reliability state subsets $\{u, u+1, \ldots, z\}$, $u = 1, 2, \ldots, z$, (RelI8) given by

$$\sigma(u) = \sqrt{\boldsymbol{n}(u) - [\boldsymbol{\mu}(u)]^2},\ u = 1, 2, \ldots, z, \tag{1.41}$$

where

$$\boldsymbol{n}(u) = 2\int_0^\infty t\boldsymbol{R}(t, u)dt,\ u = 1, 2, \ldots, z, i = 1, 2, \ldots, n; \tag{1.42}$$

- the mean lifetimes $\bar{\boldsymbol{\mu}}(u)$, $u = 1, 2, \ldots, z$, of the critical infrastructure in the particular reliability states (RelI9)

$$\bar{\mu}(u) = \mu(u) - \mu(u+1),\ u = 0, 1, \ldots, z-1,\ \bar{\mu}(z) = \mu(z); \tag{1.43}$$

- the intensities of degradation (ageing) of the critical infrastructure/the intensities of critical infrastructure departure from the reliability state subset $\{u, u+1, \ldots, z\}$, $u = 1, 2, \ldots, z$, (RelI10), i.e. the coordinates of the vector

$$\boldsymbol{\lambda}(t, \cdot) = [0, \boldsymbol{\lambda}(t, 1), \ldots, \boldsymbol{\lambda}(t, z)],\ t \in <0, +\infty), \tag{1.44}$$

where

$$\boldsymbol{\lambda}(t, u) = \frac{-\frac{d\mathbf{R}(t,u)}{dt}}{\mathbf{R}(t, u)},\ t \in <0, +\infty),\ u = 1, 2, \ldots, z. \tag{1.45}$$

With the above definition of the critical infrastructure intensities of degradation correspond:

- the limit intensities of degradation of the critical infrastructure given by

$$\boldsymbol{\lambda}(u) = \lim_{t\to\infty} \boldsymbol{\lambda}(t, u),\ u = 1, 2, \ldots, z, \tag{1.46}$$

where $\boldsymbol{\lambda}(t, u)$, $t \in <0, +\infty)$, $u = 1, 2, \ldots, z$, are defined by (1.45);

- the approximate mean of intensities of critical infrastructure departure from the reliability state subset $\{u, u+1, \ldots, z\}$, $u = 1, 2, \ldots, z$,

$$\boldsymbol{\lambda}(u) = \frac{1}{\boldsymbol{\mu}(u)}, u = 1, 2, \ldots, z, \tag{1.47}$$

where $\boldsymbol{\mu}(u)$, $u = 1, 2, \ldots, z$, are given by (1.40).

In the particular case, when the critical infrastructure has the piecewise exponential reliability function (RelI1), i.e.

$$\boldsymbol{R}(t, \cdot) = [1, \boldsymbol{R}(t, 1), \ldots, \boldsymbol{R}(t, z)], t \in < 0, +\infty), \tag{1.48}$$

where

$$\boldsymbol{R}(t, u) = \exp[-\lambda(u)t], t \in < 0, +\infty), \boldsymbol{\lambda}(u) \geq 0, u = 1, 2, \ldots z, \tag{1.49}$$

the intensities of degradation of the critical infrastructure/the intensities of critical infrastructure departure from the reliability state subset $\{u, u+1, \ldots, z\}$, $u = 1, 2, \ldots, z$, (RelI10), i.e. the coordinates of the vector

$$\boldsymbol{\lambda}(\cdot) = [0, \boldsymbol{\lambda}(1), \ldots, \boldsymbol{\lambda}(1)], \tag{1.50}$$

are constant and

$$\boldsymbol{\lambda}(u) = \frac{1}{\boldsymbol{\mu}(u)}, \tag{1.51}$$

where $\boldsymbol{\mu}(u)$, $u = 1, 2, \ldots z$, is the mean value of the critical infrastructure lifetime $T(u)$,$u = 1, 2, \ldots z$, in the reliability state subset $\{u, u+1, \ldots, z\}$, $u = 1, 2, \ldots z$, given by (1.36) and $\boldsymbol{R}(t, u)$ is defined by (1.49) for $t \in < 0, +\infty)$, $u = 1, 2, \ldots z$.

In the next applique Sect. 1.4, we will need the definitions of selected multistate critical infrastructure reliability structures and their reliability functions. To introduce them, we assume that:

- the critical infrastructure is composed of $n$ multistate assets/components,
- $T_i(u)$, $i = 1, 2, \ldots, n$, are independent random variables representing the lifetimes of components $E_i$ in the reliability state subset $\{u, u+1, \ldots, z\}$, while they were in the reliability state $z$ at the moment $t = 0$,
- $s_i(t)$ is a component $E_i$ reliability state at the moment $t$, $t \in < 0, \infty)$, while it was in the reliability state $z$ at the moment $t = 0$.

**Definition 1.8** A vector

$$R_i(t, \cdot) = [1, R_i(t, 1), \ldots, R_i(t, z)], t \in < 0, \infty), i = 1, 2, \ldots, n, \tag{1.52}$$

where

$$\begin{aligned} R_i(t, u) &= P(s_i(t) \geq u | s_i(0) = z) \\ &= P(T_i(u) > t), t \in < 0, \infty), u = 1, 2, \ldots, z, i = 1, 2, \ldots, n, \end{aligned} \tag{1.53}$$

is the probability that the component $E_i$, $i = 1, 2, \ldots, n$, is in the reliability state subset $\{u, u + 1, \ldots, z\}$ at the moment $t$, $t \in< 0, \infty)$, while it was in the reliability state $z$ at the moment $t = 0$, is called the reliability function of a multistate component $E_i$, $i = 1, 2, \ldots, n$.

The reliability functions $R_i(t, u)$, $t \in< 0, \infty)$, $u = 1, 2, \ldots z$, $i = 1, 2, \ldots, n$, defined by (1.53) are called the coordinates of the component $E_i$, $i = 1, 2, \ldots, n$, multistate reliability function $R_i(t, \cdot)$ given by (1.52). Thus, the relationship between the distribution function $F_i(t, u)$ of the component $E_i$, $i = 1, 2, \ldots, n$, lifetime $T_i(u)$ in the reliability state subset $\{u, u + 1, \ldots, z\}$ and the coordinate $R_i(t, u)$ of its multistate reliability function is given by

$$\begin{aligned} F_i(t, u) &= P(T_i(u) \leq t) = 1 - P(T_i(u) > t) \\ &= 1 - R_i(t, u), t \in< 0, \infty), u = 1, 2, \ldots, z, i = 1, 2, \ldots, n. \end{aligned}$$

Now, we may define selected basic multi-state critical infrastructure reliability structures.

**Definition 1.9** A multistate critical infrastructure is called series if its lifetime $T(u)$ in the reliability state subset $\{u, u + 1, \ldots, z\}$ is given by

$$T(u) = \min_{1 \leq i \leq n} \{T_i(u)\}, u = 1, 2, \ldots, z.$$

The number $n$ is called the critical infrastructure reliability structure shape parameter.

The above definition means that a multi-state series critical infrastructure is in the reliability state subset $\{u, u + 1, \ldots, z\}$ if and only if all its $n$ components are in this subset of reliability states. That meaning is very close to the definition of a two-state series critical infrastructure that is not failed if all its components are not failed. This fact can justify the same reliability structure scheme presented in Fig. 1.3 for a two-state series critical infrastructure and for a multistate series critical infrastructure.

It is easy to work out that the reliability function of the multi-state series critical infrastructure is given by the vector

$$\boldsymbol{R}(t, \cdot) = [1, \boldsymbol{R}(t, 1), \ldots, \boldsymbol{R}(t, z)], t \in< 0, \infty), \tag{1.54}$$

with the coordinates

$$\boldsymbol{R}(t, u) = \prod_{i=1}^{n} R_i(t, u), t \in< 0, \infty), u = 1, 2, \ldots, z. \tag{1.55}$$

**Definition 1.10** A multistate series critical infrastructure is called homogeneous if its component lifetimes $T_i(u)$, $u = 1, 2, \ldots, z$, $i = 1, 2, \ldots, n$, have an identical distribution function, i.e.

$$F(t, u) = P(T_i(u) \leq t), t \in< 0, +\infty), u = 1, 2, \ldots, z, i = 1, 2, \ldots, n,$$

and its components $E_i$, $t \in< 0, +\infty)$, have the same reliability function, i.e.

$$R(t, u) = 1 - F(t, u), t \in< 0, +\infty), u = 1, 2, \ldots, z.$$

The above definition results in the following simplified formula

$$\boldsymbol{R}(t, \cdot) = [1, \boldsymbol{R}(t, 1), \ldots, \boldsymbol{R}(t, z)], t \in< 0, \infty), \quad (1.56)$$

where

$$\boldsymbol{R}(t, u) = [R(t, u]^n, t \in< 0, \infty), u = 1, 2, \ldots, z. \quad (1.57)$$

for the reliability function of the homogeneous multistate series critical infrastructure.

**Definition 1.11** A multistate critical infrastructure is called parallel if its lifetime $T(u)$ in the reliability state subset $\{u, u + 1, \ldots, z\}$ is given by

$$T(u) = \max_{1 \leq 15n} \{T_i(u)\}, u = 1, 2, \ldots, z.$$

The number $n$ is called the critical infrastructure reliability structure shape parameter.

The above definition means that the multistate parallel critical infrastructure is in the reliability state subset $\{u, u + 1, \ldots, z\}$ if and only if at least one of its $n$ components is in this subset of reliability states. That meaning is very close to the definition of a two-state parallel critical infrastructure that is not failed if at least one of its components is not failed what can justify the same reliability structure scheme presented in Fig. 1.4 for a two-state parallel critical infrastructure and for a multistate parallel critical infrastructure.

The reliability function of the multi-state parallel critical infrastructure is given by the vector

$$\boldsymbol{R}(t, \cdot) = [1, \boldsymbol{R}(t, 1), \ldots, \boldsymbol{R}(t, z)], t \in< 0, \infty), \quad (1.58)$$

with the coordinates

$$\boldsymbol{R}(t, u) = 1 - \prod_{i=1}^{n} F_i(t, u) = 1 - \prod_{i=1}^{n} [1 - R_i(t, u)], t \in< 0, \infty), u = 1, 2, \ldots, z. \quad (1.59)$$

**Definition 1.12** A multistate parallel critical infrastructure is called homogeneous if its component lifetimes $T_i(u)$, $u = 1, 2, \ldots, z$, $i = 1, 2, \ldots, n$, have an identical distribution function

$$F(t,u) = P(T_i(u) \le t), t \in <0, +\infty), u = 1, 2, \ldots, z, i = 1, 2, \ldots, n,$$

i.e. if its components $E_i$, $i = 1, 2, \ldots, n$, have the same reliability function

$$R(t,u) = 1 - F(t,u), t \in <0, +\infty), u = 1, 2, \ldots, z.$$

Under this definition we get the following simplified formula

$$\boldsymbol{R}(t,\cdot) = [1, \boldsymbol{R}(t,1), \ldots, \boldsymbol{R}(t,z)], t \in <0, \infty), \tag{1.60}$$

where

$$\boldsymbol{R}(t,u) = 1 - [F(t,u)]^n = 1 - [1 - R(t,u)]^n, t \in <0, +\infty), u = 1, 2, \ldots, z, \tag{1.61}$$

for the reliability function of the homogeneous multistate parallel critical infrastructure.

**Definition 1.13** A multistate critical infrastructure is called an "$m$ out of $n$" critical infrastructure if its lifetime $T(u)$ in the reliability state subset $\{u, u+1, \ldots, z\}$ is given by

$$T(u) = T_{(n-m+1)}(u), m = 1, 2, \ldots, n, u = 1, 2, \ldots, z,$$

where $T_{(n-m+1)}(u)$ is the $(n - m + 1)$-th order statistic in the sequence of the component lifetimes $T_1(u), T_2(u), \ldots, T_n(u)$, $u = 1, 2, \ldots, z$.

The above definition means that the multistate „$m$ out of $n$" critical infrastructure is in the reliability state subset $\{u, u+1, \ldots, z\}$ if and only if at least $m$ out of its $n$ components are in this reliability state subset. The multistate „$m$ out of $n$" critical infrastructure is a multistate parallel critical infrastructure if $m = 1$ and it is a multistate series critical infrastructure if $m = n$. The numbers $m$ and $n$ are called the critical infrastructure reliability structure shape parameters. The scheme of an "$m$ out of $n$" multistate critical infrastructure reliability structure, is given in Fig. 1.15, where $i_1, i_2, \ldots, i_n \in \{1, 2, \ldots, n\}$ and $i_a \ne i_b$ for $a \ne b$.

It can be simply shown that the reliability function of the multistate "$m$ out of $n$" system is given either by the vector

$$\boldsymbol{R}(t,\cdot) = [1, \boldsymbol{R}(t,1), \ldots, \boldsymbol{R}(t,z)], t \in <0, \infty), \tag{1.62}$$

with the coordinates

$$\boldsymbol{R}(t,u) = \sum_{\substack{r_1, r_2, \ldots, r_n \in \{0,1|\} \\ m \le r_1 + r_2 + \cdots + r_n \le n}} [R_i(t,u)]^{r_i} [F_i(t,u)]^{1-r_i}, t \in <0, \infty), u = 1, 2, \ldots, z, \tag{1.63}$$

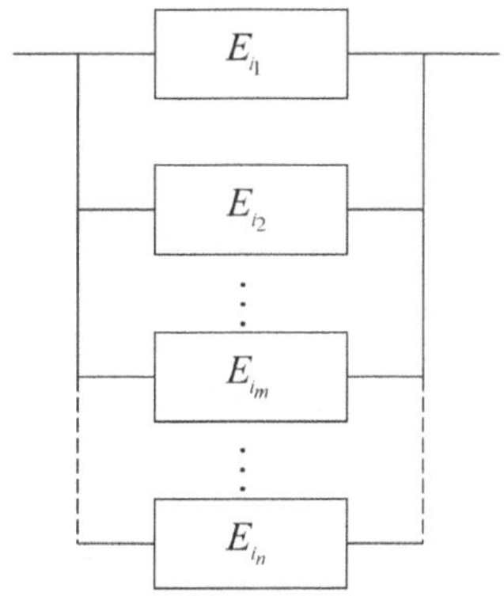


**Fig. 1.15** The scheme of an "$m$ out of $n$" critical infrastructure reliability structure

or with the coordinates

$$\boldsymbol{R}(t,u) = 1 - \sum_{\substack{r_1, r_2, \ldots, r_n \in \{0,1\} \\ 0 \le r_1 + r_2 + \cdots + r_n \le m-1}} [R_i(t,u)]^{r_i} [F_i(t,u)]^{1-r_i}, \; t \in <0, \infty), \; u = 1, 2, \ldots, z. \quad (1.64)$$

**Definition 1.14** A multistate "$m$ out of $n$" critical infrastructure is called homogeneous if its component lifetimes $T_i\ (u)$, $u = 1, 2, \ldots, z$, $i = 1, 2, \ldots, n$, have an identical distribution function

$$F(t,u) = P(T_i(u) \le t), \; t \in <0, +\infty), \; u = 1, 2, \ldots, z, \; i = 1, 2, \ldots, n,$$

i.e. if its components $E_i$, $i = 1, 2, \ldots, n$, have the same reliability function

$$R(t,u) = 1 - F(t,u), \; t \in <0, +\infty), \; u = 1, 2, \ldots, z.$$

Under this definition we get the following simplified formula

$$\boldsymbol{R}(t,\cdot) = [1, \boldsymbol{R}(t,1), \ldots, \boldsymbol{R}(t,z)], \; t \in <0, \infty), \quad (1.65)$$

where

$$\boldsymbol{R}(t,u) = \sum_{i=m}^{n} \binom{n}{i} [R(t,u)]^i [F(t,u)]^{n-i}$$
$$= \sum_{i=m}^{n} \binom{n}{i} [R(t,u)]^i [1 - R(t,u)]^{n-i} t \in <0, \infty), \; u = 1, 2, \ldots, z], \quad (1.66)$$

or

$$\boldsymbol{R}(t,u) = 1 - \sum_{i=0}^{m-1} \binom{n}{i} \left[ R(t,u) \right]^j [F(t,u)]^{n-i}$$
$$= 1 - \sum_{i=0}^{m-1} \binom{n}{i} \left[ R(t,u) \right]^i \left[ 1 - R(t,u) \right]^{n-i} t \in < 0, \infty), u = 1, 2, \ldots, z, \tag{1.67}$$

for the reliability function of the homogeneous multistate "$m$ out of $n$" critical infrastructure.

Another basic multistate reliability structure with components degrading in time is an "$m_i$ out of $l_i$"-series critical infrastructure. To define it, we assume that:

- $k$ is the number of the critical infrastructure subsystems;
- $l_i$, $i = 1, 2, \ldots, k$, are the numbers of the subsystem components/assets;
- $E_{ij}$, $i = 1, 2, \ldots, k$, $j = 1, 2, \ldots, l_i$, $k, l_1, l_2, \ldots, l_k \in N$, are components of a critical infrastructure;
- all components $E_{ij}$ have the same reliability state set as before $\{0, 1, \ldots, z\}$;
- $T_{ij}(u)$, $i = 1, 2, \ldots, k$, $j = 1, 2, \ldots, l_i$, $k, l_1, l_2, \ldots, l_k \in N$, are independent random variables representing the lifetimes of components $E_{ij}$ in the reliability state subset $\{u, u + 1, \ldots, z\}$ while they were in the state $z$ at the moment $t = 0$;
- $s_{ij}(t)$ is a component $E_{ij}$ reliability state at the moment $t$, $t \in < 0, \infty)$, while they were in the state $z$ at the moment $t = 0$.

**Definition 1.15** A vector

$$R_{ij}(t, \cdot) = \left[ R_{ij}(t, 0), R_{ij}(t, 1), \ldots, R_{jj}(t, z) \right],$$
$$t \in < 0, \infty), i = 1, 2, \ldots, k, j = 1, 2, \ldots, l_i, \tag{1.68}$$

where

$$R_{ij}(t,u) = P\left(s_{ij}(t) \geq u | s_{ij}(0) = z\right)$$
$$= P\left(T_{ij}(u) > t\right), t \in < 0, \infty), u = 1, 2, \ldots, z, \tag{1.69}$$

is the probability that the component $E_{ij}$ is in the reliability state subset $\{u, u + 1, \ldots, z\}$, $u = 1, 2, \ldots, z$, at the moment $t$, $t \in < 0, \infty)$, while it was in the reliability state $z$ at the moment $t = 0$, is called the multistate reliability function of a component $E_{ij}$.

The reliability functions $R_{ij}(t, u)$, $t \in < 0, \infty)$, $u = 1, 2, \ldots, z$, defined by (1.69) are called the coordinates of the component $E_{ij}$, $i = 1, 2, \ldots, k$, $j = 1, 2, \ldots, l_i$, multistate reliability function $R_{ij}(t, \cdot)$, $t \in < 0, \infty)$, $i = 1, 2, \ldots, k$, $j = 1, 2, \ldots, l_i$, given by (1.68). Thus, the relationship between the distribution function $F_{ij}(t,u)$, $t \in < 0, \infty)$, $u = 1, 2, \ldots, z$, $i = 1, 2, \ldots, k$, $j = 1, 2, \ldots, l_i$, of the component/asset $E_{ij}$, $i = 1, 2, \ldots, k$, $j = 1, 2, \ldots, l_i$, lifetime $T_{ij}(u)$, $u = 1, 2, \ldots, z$, $i = 1, 2, \ldots, k$, $j = 1, 2, \ldots, l_i$, in the reliability state subset $\{u, u + 1, \ldots, z\}$, $u = 1, 2, \ldots, z$, and the coordinate $R_{ij}(t, u)$, $t \in< 0, \infty)$, $u = 1, 2, \ldots, z$, $i = 1, 2, \ldots, k$, $j = 1, 2, \ldots, l_i$, of its multistate reliability function is given by

$$F_{ij}(t, u) = P\big(T_{ij}(u) \le t\big) = 1 - P\big(T_{ij}(u) > t\big)$$
$$= 1 - R_{ij}(t, u), t \in< 0, \infty), u = 1, 2, \ldots, z.$$

**Definition 1.16** A multistate system is called an "$m_i$ out of $l_i$"-series system if its lifetime $T(u)$ in the state subset $\{u, u+1, \ldots, z\}$ is given by

$$T(u) = \min_{1 \le i \le k} T_{(l_i - m_i + 1)}(u), u = 1, 2, \ldots, \mathrm{z}, m_i = 1, 2, \ldots, l_i,$$

where $T_{(l_i - m_i + 1)}(u)$, $u = 1, 2, \ldots, \mathrm{z}$, $m_i = 1, 2, \ldots, l_i$, is the $(l_i - m_i + 1)$-th order statistic in the set of random variables

$$T_{i1}(u), T_{i2}(u), \ldots, T_{il_i}(u), i = 1, 2, \ldots, k, u = 1, 2, \ldots, \mathrm{z}.$$

The above definition means that the multistate "$m_i$ out of $l_i$"-series critical infrastructure is composed of $k$ subsystems that are multistate "$m_i$ out of $l_i$" systems and it is in the reliability state subset $\{u, u+1, \ldots, z\}$, $u = 1, 2, \ldots, z$, if all its "$m_i$ out of $l_i$" subsystems are in this reliability state subset. In this definition $l_i$, $i = 1, 2, \ldots, k$, denote the numbers of components in the "$m_i$ out of $l_i$" subsystems. The numbers $k$, $m_1, m_2, \ldots, m_k$ and $l_1, l_2, \ldots, l_k$ are called the critical infrastructure shape parameters. Joining the justification for the reliability structure scheme of the multistate "$m_i$ out of $l_i$" critical infrastructure presented in Fig. 1.15 and the multistate series critical infrastructure reliability structure scheme presented in Fig. 1.3 leads to the scheme of a multistate of an "$m_i$ out of $l_i$"-series critical infrastructure reliability structure given in Fig. 1.16, where $j_1, j_2, \ldots, j_{l_i} \in \{1, 2, \ldots, l_i\}$, for $i = 1, 2, \ldots, k$ and $i_a \ne i_b$ for $a \ne b$.

The reliability function of the multi-state "$m_i$ out of $l_i$"-series critical infrastructure is given by the vector [14]

$$\boldsymbol{R}(t, \cdot) = [1, \boldsymbol{R}(t, 1), \ldots, \boldsymbol{R}(t, z)], t \in< 0, \infty) \tag{1.70}$$

where

$$\boldsymbol{R}(t, u) = \prod_{i=1}^{k} \left[ \sum_{\substack{\sum l r_1, r_2, \ldots, r_{l_i} \in \{0, 1\} \\ m_i \le r_1 + r_2 + \cdots + r_{l_i} \le l_i}} \left[ \prod_{j=1}^{l_i} R_{ij}(t, u) \right]^{r_i} \left[ 1 - \prod_{j=1}^{l_i} R_{ij}(t, u) \right]^{1-r_i} \right],$$
$$t \in< 0, \infty), u = 1, 2, \ldots, z, \tag{1.71}$$

or

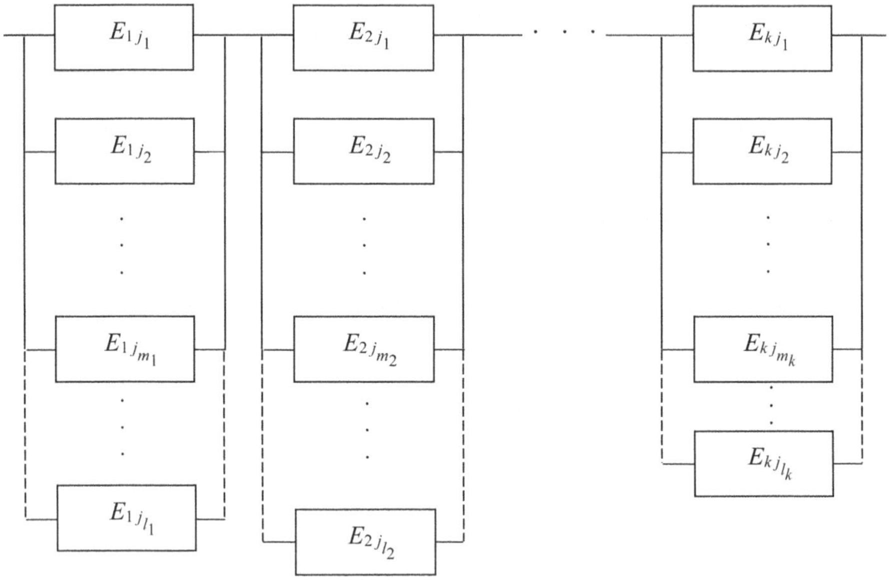


**Fig. 1.16** The scheme of an "$m_i$ out of $l_i$"-series critical infrastructure reliability structure

$$\boldsymbol{R}(t,u) = \prod_{i=1}^{k}\left[\sum_{\substack{\sum l r_1, r_2, \ldots, r_{l_i} \in \{0,1\} \\ 0 \le r_1 + r_2 + \cdots + r_{l_i} \le m_i - 1}} \left[\prod_{j=1}^{l_i} R_{ij}(t,u)\right]^{r_i}\left[1 - \prod_{j=1}^{l_i} R_{ij}(t,u)\right]^{1-r_i}\right],$$
$$t \in <0, \infty), u = 1, 2, \ldots, z, \tag{1.72}$$

## 1.4 Application

We consider the port oil terminal critical infrastructure. The considered port oil terminal placed at the Baltic seaside is designated for receiving oil products from ships, storage and sending them by carriages or trucks. The terminal can operate in reverse way as well.

### *1.4.1 Port Oil Terminal Critical Infrastructure Assets*

The considered terminal is composed of three parts *A*, *B* and *C*, linked by the piping transportation system with the pier. The neighborhood area of the port oil piping transportation system is presented in Figs. 1.17 and 1.18. The approximate length of the port oil piping transportation system is equal to around 25 km.

The main technical assets of the port oil terminal critical infrastructure are:

$A_1$ port oil piping transportation system;
$A_2$ internal pipeline technological system;
$A_3$ supporting pump station;
$A_4$ internal pump system;
$A_5$ port oil tanker shipment terminal;
$A_6$ loading railway carriage station;
$A_7$ loading road carriage station;
$A_8$ unloading railway carriage station;
$A_9$ oil storage reservoir system.

The functional scheme of the asset $A_1$, the port oil piping transportation system, is presented in Fig. 1.19.

The asset $A_1$, the port oil piping transportation system operating as a part of the port oil terminal critical infrastructure consists of three subsystems:

- the subsystem $S_1$ composed of two pipelines, each composed of 176 pipe segments and 2 valves;
- the subsystem $S_2$ composed of two pipelines, each composed of 717 pipe segments and 2 valves;
- the subsystem $S_3$ composed of three pipelines, each composed of 360 pipe segments and 2 valves.

The asset $A_1$, operation is the main activity of the port oil terminal critical infrastructure involving the remaining assets $A_2 - A_9$.

The port oil piping transportation system, the asset $A_1$, presented in Fig. 1.19, is a series system composed of series-parallel subsystems $S_1$, containing two pipelines (assets $A_{11}$, $A_{12}$), a series-parallel subsystems $S_2$, containing two pipelines (assets $A_{21}$, $A_{22}$), and one series-"2 out of 3" subsystem $S_3$ containing 3 pipelines (assets $A_{31}, A_{32}, A_{33}$) [14].

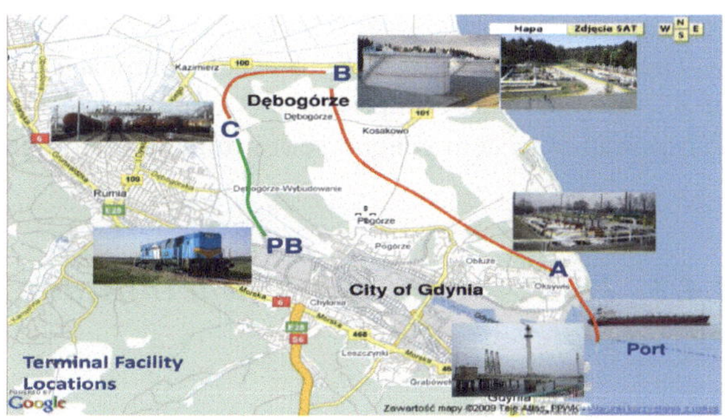


**Fig. 1.17** The port oil piping transportation system operating between the Port of Gdynia and the terminal in Dębogórze

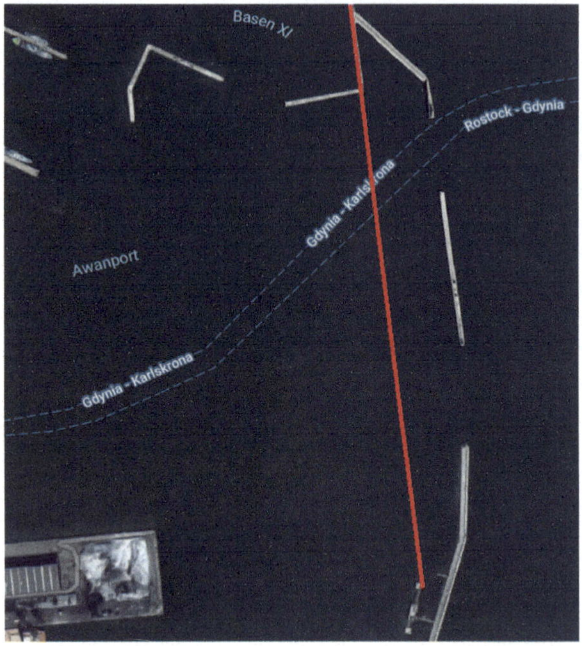


**Fig. 1.18** The port oil piping transportation system alignment in the Port of Gdynia

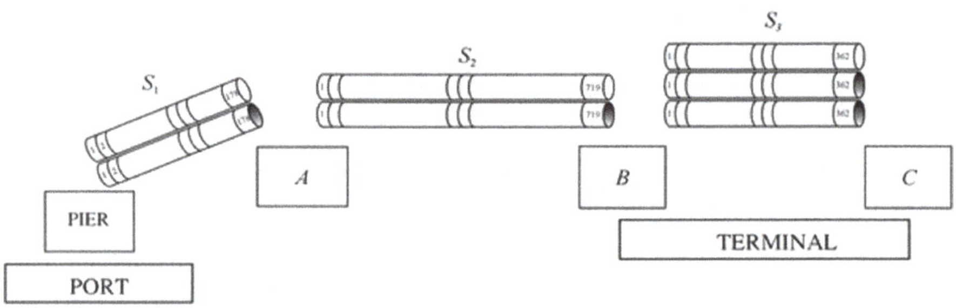


**Fig. 1.19** The functional scheme of the port oil piping transportation system

The parallel subsystems $S_1$, $S_2$ each composed of 2 components and the "2 out of 3" subsystem $S_3$ composed of 3 components are forming the port oil piping transportation system general series reliability structure presented in Fig. 1.20.

### 1.4.2 Port Oil Terminal Critical Infrastructure Reliability States

After considering the comments and opinions coming from experts, taking into account the effectiveness and safety aspects of the operation of the port oil terminal

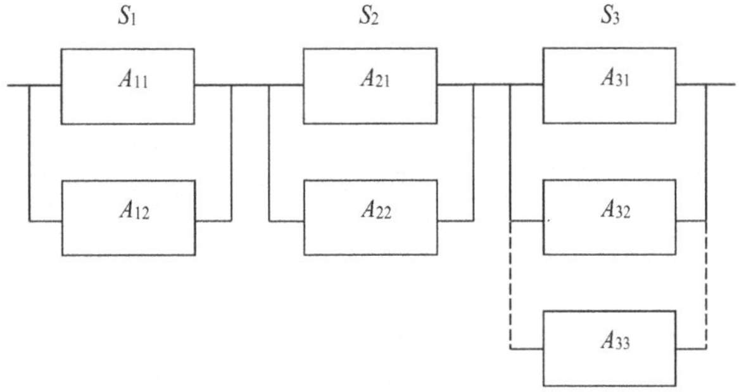


**Fig. 1.20** General scheme of the port oil piping transportation system reliability structure

critical infrastructure and its assets, we fix for all of them the following reliability parameters [19]:

- the number of reliability states (excluding reliability state 0) $z = 2$;
- three $(z + 1)$ reliability states:
  - a reliability state 2—an asset and the port oil terminal critical infrastructure operation is fully safe,
  - a reliability state 1—an asset and the port oil terminal critical infrastructure is less safe and more dangerous because of the possibility of environment pollution,
  - a reliability state 0—an asset and the port oil terminal critical infrastructure is destroyed;

and, we assume that:

- there are possible the transitions between the assets and the port oil terminal critical infrastructure reliability states only from better to worse ones,
- the critical reliability state of an asset and the port oil terminal critical infrastructure is $r = 1$,
- the port oil terminal critical infrastructure risk function permitted level $\delta = 0.05$.

### *1.4.3 Port Oil Piping Transportation System Reliability Parameters*

Now, as an illustration, we will perform detailed reliability analysis of the asset $A_1$, the port oil piping transportation system. The reliability parameters of its assets are as follows:

- the mean values of the asset $A_{11}$ of the port oil piping transportation system lifetimes in the reliability state subsets $\{1, 2\}$, $\{2\}$, calculated on the basis of reliability data of its components coming from experts, are as follows:

- for reliability state subset {1, 2}

$$\mu_{11}(1) = 276 \text{ years}, \tag{1.73}$$

- for reliability state subset {2}

$$\mu_{11}(2) = 185 \text{ years}; \tag{1.74}$$

- the mean values of the asset $A_{12}$ of the port oil piping transportation system lifetimes in the reliability state subsets {1, 2}, {2}, calculated on the basis of reliability data of its components coming from experts, are as follows:
  - for reliability state subset {1, 2}

$$\mu_{12}(1) = 276 \text{ years}, \tag{1.75}$$

  - for reliability state subset {2}

$$\mu_{12}(2) = 185 \text{ years}; \tag{1.76}$$

- the mean values of the asset $A_{21}$ of the port oil piping transportation system lifetimes in the reliability state subsets {1, 2}, {2}, calculated on the basis of reliability data of its components coming from experts, are as follows:
  - for reliability state subset {1, 2}

$$\mu_{21}(1) = 69 \text{ years}, \tag{1.77}$$

  - for reliability state subset {2}

$$\mu_{21}(2) = 46 \text{ years}; \tag{1.78}$$

- the mean values of the asset $A_{22}$ of the port oil piping transportation system lifetimes in the reliability state subsets {1, 2}, {2}, calculated on the basis of reliability data of its components coming from experts, are as follows:
  - for reliability state subset {1, 2}

$$\mu_{22}(1) = 69 \text{ years}, \tag{1.79}$$

  - for reliability state subset {2}

$$\mu_{22}(2) = 46 \text{ years}; \tag{1.80}$$

- the mean values of the asset $A_{31}$ of the port oil piping transportation system lifetimes in the reliability state subsets {1, 2}, {2}, calculated on the basis of reliability data of its components coming from experts, are as follows:
  - for reliability state subset {1, 2}

$$\mu_{31}(1) = 137 \text{ years}, \tag{1.81}$$

  - for reliability state subset {2}

$$\mu_{31}(2) = 110 \text{ years}; \tag{1.82}$$

- the mean values of the asset $A_{32}$ of the port oil piping transportation system lifetimes in the reliability state subsets {1, 2}, {2}, calculated on the basis of reliability data of its components coming from experts, are as follows:
  - for reliability state subset {1, 2}

$$\mu_{32}(1) = 137 \text{ years}, \tag{1.83}$$

  - for reliability state subset {2}

$$\mu_{32}(2) = 110 \text{ years}; \tag{1.84}$$

- the mean values of the asset $A_{33}$ of the port oil piping transportation system lifetimes in the reliability state subsets {1, 2}, {2}, calculated on the basis of reliability data of its components coming from experts, are as follows:
  - for reliability state subset {1, 2}

$$\mu_{33}(1) = 114 \text{ years}, \tag{1.85}$$

  - for reliability state subset {2}

$$\mu_{33}(2) = 102 \text{ years}. \tag{1.86}$$

Applying (1.73)–(1.86) and (1.51), we get the intensities of ageing of the critical infrastructure assets $A_{ij}$, $i = 1, 2, j = 1, 2, i = 3, j = 1, 2, 3$,/the intensities of critical infrastructure assets $A_{ij}$, $i = 1, 2, j = 1, 2, i = 3, j = 1, 2, 3$, departure from the reliability state subset {1, 2} and {2}, i.e. the coordinates of the vectors

$$\lambda_{ij}(\cdot) = [0, \lambda_{ij}(1), \lambda_{ij}(2)], i = 1, 2, j = 1, 2, i = 3, j = 1, 2, 3. \tag{1.87}$$

They are as follows:

- the intensities of departure of the asset $A_{11}$

- for reliability state subset {1, 2}

$$\lambda_{11}(1) = \frac{1}{\mu_{11}(1)} = \frac{1}{276} = 0.00362, \tag{1.88}$$

- for reliability state subset {2}

$$\lambda_{11}(2) = \frac{1}{\mu_{11}(2)} = \frac{1}{185} = 0.00540; \tag{1.89}$$

– the intensities of departure of the asset $A_{12}$

- for reliability state subset {1, 2}

$$\lambda_{12}(1) = \frac{1}{\mu_{12}(1)} = \frac{1}{276} = 0.00362, \tag{1.90}$$

- for reliability state subset {2}

$$\lambda_{12}(2) = \frac{1}{\mu_{12}(2)} = \frac{1}{185} = 0.00540; \tag{1.91}$$

– the intensities of departure of the asset $A_{21}$

- for reliability state subset {1, 2}

$$\lambda_{21}(1) = \frac{1}{\mu_{21}(1)} = \frac{1}{69} = 0.01444, \tag{1.92}$$

– for reliability state subset {2}

$$\lambda_{21}(2) = \frac{1}{\mu_{21}(2)} = \frac{1}{46} = 0.02163; \tag{1.93}$$

– the intensities of departure of the asset $A_{22}$

- for reliability state subset {1, 2}

$$\lambda_{22}(1) = \frac{1}{\mu_{22}(1)} = \frac{1}{69} = 0.01444, \tag{1.94}$$

– for reliability state subset {2}

$$\lambda_{22}(2) = \frac{1}{\mu_{22}(2)} = \frac{1}{46} = 0.02163; \tag{1.95}$$

– the intensities of departure of the asset $A_{31}$

- for reliability state subset {1, 2}

$$\lambda_{31}(1) = \frac{1}{\mu_{31}(1)} = \frac{1}{137} = 0.00730, \tag{1.96}$$

- for reliability state subset {2}

$$\lambda_{31}(2) = \frac{1}{\mu_{31}(2)} = \frac{1}{110} = 0.00912; \tag{1.97}$$

– the intensities of departure of the asset $A_{32}$
  - for reliability state subset {1, 2}

$$\lambda_{32}(1) = \frac{1}{\mu_{32}(1)} = \frac{1}{137} = 0.00730, \tag{1.98}$$

  - for reliability state subset {2}

$$\lambda_{32}(2) = \frac{1}{\mu_{32}(2)} = \frac{1}{110} = 0.00912; \tag{1.99}$$

– the intensities of departure of the asset $A_{33}$
  - for reliability state subset {1, 2}

$$\lambda_{33}(1) = \frac{1}{\mu_{33}(1)} = \frac{1}{114} = 0.00874, \tag{1.100}$$

  - for reliability state subset {2}

$$\lambda_{33}(2) = \frac{1}{\mu_{33}(2)} = \frac{1}{102} = 0.00984. \tag{1.101}$$

Assuming that the assets $A_{ij}$, $i = 1, 2, j = 1, 2, i = 3, j = 1, 2, 3$, of the asset $A_1$, the port oil piping transportation system have piecewise exponential reliability functions of the form

$$R_{ij}(t, \cdot) = \left[1, R_{ij}(t, 1), R_{ij}(t, 2)\right], t \in< 0, \infty), i = 1, 2, j = 1, 2, i = 3, j = 1, 2, 3,$$

with the coordinates

$$R_{ij}(t, u) = \exp\left[-\lambda_{ij}(u)t\right] \text{for} t \in< 0, \infty),$$
$$u = 1, 2, i = 1, 2, j = 1, 2, i = 3, j = 1, 2, 3,$$

where the intensities of degradation

$$\lambda_{ij}(u), u = 1, 2, i = 1, 2, j = 1, 2, i = 3, j = 1, 2, 3,$$

are respectively given by (1.88)–(1.101), we get their reliability functions of the following forms:

- the reliability function of the asset $A_{11}$

$$R_{11}(t, \cdot) = [1, R_{11}(t, 1), R_{11}(t, 2], t \in< 0, \infty), \quad (1.102)$$

where

$$R_{11}(t, 1) = \exp[-0.00362t], R_{11}(t, 2) = \exp[-0.00540t], t \in< 0, \infty); \quad (1.103)$$

- the reliability function of the asset $A_{12}$

$$R_{12}(t, \cdot) = [1, R_{12}(t, 1), R_{12}(t, 2], t \in< 0, \infty), \quad (1.104)$$

where

$$R_{12}(t, 1) = \exp[-0.00362t], R_{12}(t, 2) = \exp[-0.00540t], t \in< 0, \infty); \quad (1.105)$$

- the reliability function of the asset $A_{21}$

$$R_{21}(t, \cdot) = [1, R_{21}(t, 1), R_{21}(t, 2], t \in< 0, \infty), \quad (1.106)$$

where

$$R_{21}(t, 1) = \exp[-0.01444t], R_{21}(t, 2) = \exp[-0.02163t], t \in< 0, \infty); \quad (1.107)$$

- the reliability function of the asset $A_{22}$

$$R_{22}(t, \cdot) = [1, R_{22}(t, 1), R_{22}(t, 2], t \in< 0, \infty), \quad (1.108)$$

where

$$R_{22}(t, 1) = \exp[-0.01444t], R_{22}(t, 2) = \exp[-0.02163t], t \in< 0, \infty); \quad (1.109)$$

- the reliability function of the asset $A_{31}$

$$R_{31}(t,\cdot) = [1, R_{31}(t,1), R_{31}(t,2], t \in <0, \infty), \tag{1.110}$$

where

$$R_{31}(t,1) = \exp[-0.00730t], R_{31}(t,2) = \exp[-0.00912t], t \in <0, \infty); \tag{1.111}$$

– the reliability function of the asset $A_{32}$

$$R_{32}(t,\cdot) = [1, R_{32}(t,1), R_{32}(t,2], t \in <0, \infty), \tag{1.112}$$

where

$$R_{32}(t,1) = \exp[-0.00730t], R_{32}(t,2) = \exp[-0.00912t], t \in <0, \infty); \tag{1.113}$$

– the reliability function of the asset $A_{33}$

$$R_{33}(t,\cdot) = [1, R_{33}(t,1), R_{33}(t,2], t \in <0, \infty), \tag{1.114}$$

where

$$R_{33}(t,1) = \exp[-0.00874t], R_{33}(t,2) = \exp[-0.00984t], t \in <0, \infty). \tag{1.115}$$

### 1.4.4 Port Oil Piping Transportation System Reliability Characteristics

Using the reliability functions (1.102)–(1.105) of the assets $A_{11}$ and $A_{12}$ of the subsystem $S_1$ and the formula for reliability of parallel critical infrastructure (1.60), (1.61), we get its reliability function

$$\boldsymbol{R}_1^1(t,\cdot) = [1, \boldsymbol{R}_1^1(t,1), \boldsymbol{R}_1^1(t,2)] \text{ for } t \in <0, \infty), \tag{1.116}$$

where

$$\begin{aligned} R_1^1(t,1) &= 1 - [[1 - R_{11}(t,1)][1 - R_{12}(t,1)]] \\ &= 1 - [1 - \exp[-0.00362t]]^2 \\ &= 1 - [1 - 2\exp[-0.00362t] + \exp[-0.00724t]] \\ &= 2\exp[-0.00362t] - \exp[-0.00724t] \text{for} t \in <0, \infty), \end{aligned} \tag{1.117}$$

$$
\begin{aligned}
R_1^1(t,2) &= 1 - [[1 - R_{11}(t,2)][1 - R_{12}(t,2)]] \\
&= 1 - [1 - \exp[-0.00540t]]^2 \\
&= 1 - [1 - 2\exp[-0.00540t] + \exp[-0.0108t]] \\
&= 2\exp[-0.00540t] - \exp[-0.0108t] \text{ for } t \in< 0, \infty).
\end{aligned}
\tag{1.118}
$$

Using the reliability functions (1.106)–(1.109) of the assets $A_{21}$ and $A_{22}$ of the subsystem $S_2$ and the formula for reliability of parallel critical infrastructure (1.60), (1.61), we get its reliability function

$$
\boldsymbol{R}_1^2(t,\cdot) = [1, \boldsymbol{R}_1^2(t,1), \boldsymbol{R}_1^2(t,2)] \text{ for } t \in< 0, \infty),
\tag{1.119}
$$

where

$$
\begin{aligned}
R_1^2(t,1) &= 1 - [(1 - R_{21}(t,1))(1 - R_{22}(t,1))] \\
&= 1 - [1 - \exp[-0.01444t]]^2 \\
&= 1 - [1 - 2\exp[-0.01444t] + \exp[-0.02888t]] \\
&= 2\exp[-0.01444t] - \exp[-0.02888t] \text{ for } t \in< 0, \infty),
\end{aligned}
\tag{1.120}
$$

$$
\begin{aligned}
R_1^2(t,2) &= 1 - [(1 - R_{21}(t,2))(1 - R_{22}(t,2))] \\
&= 1 - [1 - \exp[-0.02163t]]^2 \\
&= 1 - [1 - 2\exp[-0.02163t] + \exp[-0.04326t]] \\
&= 2\exp[-0.02163t] - \exp[-0.04326t] \text{ for } t \in< 0, \infty).
\end{aligned}
\tag{1.121}
$$

Using the reliability functions (1.110)–(1.115) of the assets $A_{31}, A_{32}$ and $A_{33}$ of the subsystem $S_3$ and the formula for reliability of an "$m$ out of $n$" critical infrastructure (1.62), (1.63), we get its reliability function

$$
\boldsymbol{R}_1^3(t,\cdot) = [1, \boldsymbol{R}_1^3(t,1), \boldsymbol{R}_1^3(t,2)] \text{ for } t \in< 0, \infty),
\tag{1.122}
$$

where

$$
\begin{aligned}
\boldsymbol{R}_1^3(t,1) &= [1 - R_{31}(t,1)]R_{32}(t,1)R_{33}(t,1) + R_{31}(t,1)[1 - R_{32}(t,1)]R_{33}(t,1) \\
&\quad + R_{31}(t,1)R_{32}(t,1)[1 - R_{33}(t,1)] + R_{31}(t,1)[R_{32}(t,1)R_{33}(t,1) \\
&= [1 - \exp[-0.0073t]]\exp[-0.0073t]\exp[-0.00874t] \\
&\quad + \exp[-0.0073t][1 - \exp[-0.0073t]]\exp[-0.00874t] \\
&\quad + \exp[-0.0073t]\exp[-0.0073t][1 - \exp[-0.008744t] \\
&\quad + \exp[-0.0073t]\exp[-0.0073t]\exp[-0.00874t] \\
&= \exp[-0.0146t] + 2\exp[-0.01604t] - 2\exp[-0.02334t] \\
&\quad + [\exp[-0.073t]]^2\exp[-0.00874t]
\end{aligned}
\tag{1.123}
$$

for $t \in< 0, \infty)$,

$$\begin{aligned}\boldsymbol{R}_1^3(t,2) &= [1 - R_{31}(t,2)]R_{32}(t,2)R_{33}(t,2) + R_{31}(t,2)[1 - R_{32}(t,2)]R_{33}(t,2)\\ &\quad + R_{31}(t,2)R_{32}(t,2)[1 - R_{33}(t,2)] + R_{31}(t,2)[R_{32}(t,2)R_{33}(t,2)\\ &= [1 - \exp[-0.00912t]\exp[-0.00912t]\exp[-0.00984t]\\ &\quad + \exp[-0.00912t][1 - \exp[-0.00912t]\exp[-0.00984t]\\ &\quad + \exp[-0.00912t]\exp[-0.00912t][1 - \exp[-0.00984t]\\ &\quad + \exp[-0.00912t]\exp[-0.00912t]\exp[-0.00984t]\\ &= \exp[-0.01824t] + 2\exp[-0.01896t] - 2\exp[-0.02808t] \text{ for } t \in< 0, \infty)\end{aligned} \tag{1.124}$$

Next assuming that the port oil piping transportation system is series system composed from subsystem $S_1$, $S_2$, $S_3$ and using formula for reliability of series critical infrastructure (1.54), (1.55), we get the reliability function of the asset $A_1$,

$$\boldsymbol{R}_1(t,\cdot) = [1, \boldsymbol{R}_1(t,1), \boldsymbol{R}_1(\mathrm{t},2) \text{ for } t \in< 0, \infty), \tag{1.125}$$

where

$$\begin{aligned}\boldsymbol{R}_1(t,1) = \boldsymbol{R}_1^1(t,1)\cdot\boldsymbol{R}_1^2(t,1)\cdot\boldsymbol{R}_1^3(t,1) &= 8\exp[-0.0341t]\\ &- 8\exp[-0.0414t] + 4\exp[-0.03266t]\\ &- 4\exp[-0.04854t] + 4\exp[-0.05584t] - 4\exp[-0.03772t]\\ &+ 4\exp[-0.04502t] - 2\exp[-0.0471t]\\ &- 2\exp[-0.03628t] + 2\exp[-0.05216t] - 2\exp[-0.05946t]\\ &+ \exp[-0.05072t] \text{ for } t \in< 0, \infty),\end{aligned} \tag{1.126}$$

$$\begin{aligned}\boldsymbol{R}_1(t,2) &= \boldsymbol{R}_1^1(t,2)\cdot\boldsymbol{R}_1^2(t,2)\cdot\boldsymbol{R}_1^3(t,2)\\ &= 8\exp[-0.04598t] - 8\exp[-0.05511t] + 4\exp[-0.04527t]\\ &\quad - 4\exp[-0.06762t] + 4\exp[-0.07674t]\\ &\quad - 4\exp[-0.05139t] + 4\exp[-0.06051t]\\ &\quad - 2\exp[-0.0669t] - 2\exp[-0.08214t]\\ &\quad + 2\exp[-0.07302t] - 2\exp[-0.0506t]\\ &\quad + \exp[-0.0723t] \text{for} t \in< 0, \infty).\end{aligned} \tag{1.127}$$

Equivalently, using the reliability functions of subsystem $S_1$ given by (1.102)–(1.105), the reliability functions of subsystem $S_2$ given by (1.106)–(1.109), the reliability functions of subsystem $S_3$ given by (1.110)–(1.115) and applying formulae Eqs. (1.70), (1.71), for the reliability function of the "$m_i$ out of $l_i$"-series critical infrastructure with the shape parameters $k = 2$, $m_1 = 1$, $l_1 = 2$, $m_2 = 1$, $l_2 = 2$, $m_3 = 2$, $l_3 = 3$, we get the same reliability function of the asset $A_1$, the port oil piping transportation system

$$\boldsymbol{R}_1(t,\cdot) = [1, \boldsymbol{R}_1(t,1), \boldsymbol{R}_1(\text{t}, 2)], \text{ for } t \in <0, \infty), \tag{1.128}$$

where

$$\begin{aligned}
\boldsymbol{R}_1(t,1) &= \prod_{i=1}^{3}\left[\sum_{\substack{r_1,r_2,\ldots,r_{l_i}\in\{0,1\}\\ m_i\le r_1+r_2+cldots=r_{l_i}\le l_i}}\left[\prod_{j=1}^{l_i} R_{ij}(t,1)\right]^{r_i}\left[1-\prod_{j=1}^{l_i} R_{ij}(t,1)\right]^{1-r_i}\right]\\
&= [R_{11}(t,1)[1-R_{12}(t,1)] + [1-R_{11}(t,1)]R_{12}(t,1) + R_{11}(t,1)R_{12}(t,1)]\\
&\quad\cdot [R_{21}(t,1)[1-R_{22}(t,1)] + [1-R_{21}(t,1)]R_{22}(t,1) + R_{21}(t,1)R_{22}(t,1)]\\
&\quad\cdot [[1-R_{31}(t,1)]R_{32}(t,1)R_{33}(t,1) + R_{31}(t,1)[1-R_{32}(t,1)]R_{33}(t,1)\\
&\quad+ R_{31}(t,1)R_{32}(t,1)[1-R_{33}(t,1)] + R_{31}(t,1)[R_{32}(t,1)R_{33}(t,1)]\\
&= [R_{11}(t,1) + R_{12}(t,1) - R_{11}(t,1)R_{12}(t,1)]\\
&\quad\cdot [R_{21}(t,1) + R_{22}(t,1) - R_{21}(t,1)]R_{22}(t,1)]\\
&\quad\cdot [R_{31}(t,1)R_{32}(t,1) + R_{31}(t,1)R_{33}(t,1)\\
&\quad+R_{32}(t,1)R_{33}(t,1) - 2R_{31}(t,1)R_{32}(t,1)R_{33}(t,1)]\\
&= [\exp[-0.00362t] + \exp[-0.00362t]\\
&\quad-\exp[-0.00362t]\exp[-0.00362t]]\\
&\quad\cdot[\exp[-0.01444t] + \exp[-0.01444t]\\
&\quad-\exp[-0.01444t]\exp[-0.01444t]]\\
&\quad\cdot[\exp[-0.00730t]\exp[-0.00730t]\\
&\quad+\exp[-0.00730t]\exp[-0.00874t]\\
&\quad+\exp[-0.00730t]\exp[-0.00874t] - 2\exp[-0.00730t]\\
&\quad\exp[-0.00730t]\exp[-0.00874t]] = [2\exp[-0.00362t]\\
&\quad-\exp[-0.00724t]]\cdot[2\exp[-0.01444t] - \exp[-0.02888t]]\\
&\quad\cdot[\exp[-0.01460t] + 2\exp[-0.01604t] - 2\exp[-0.02334t]]\\
&= 8\exp[-0.0341t] - 8\exp[-0.0414t] + 4\exp[-0.03266t]\\
&\quad-4\exp[-0.04854t] + 4\exp[-0.05584t] - 4\exp[-0.03772t]\\
&\quad+4\exp[-0.04502t] - 2\exp[-0.0471t] - 2\exp[-0.03628t]\\
&\quad+2\exp[-0.05216t] - 2\exp[-0.05946t]\\
&\quad+\exp[-0.05072t] \text{ for } t \in <0, \infty),
\end{aligned}\tag{1.129}$$

$$\begin{aligned}
\boldsymbol{R}_1(t,2) &= \prod_{i=1}^{3}\left[\sum_{\substack{r_1,r_2,\ldots,r_{l_i}\in\{0,1\}\\ m_i\le r_1+r_2+\cdots=r_{l_i}\le l_i}}\left[\prod_{j=1}^{l_i} R_{ij}(t,2)\right]^{r_i}\left[1-\prod_{j=1}^{l_i} R_{ij}(t,2)\right]^{1-r_i}\right]\\
&= [R_{11}(t,2)[1-R_{12}(t,2)] + [1-R_{11}(t,2)]R_{12}(t,2) + R_{11}(t,2)R_{12}(t,2)]\\
&\quad\cdot [R_{21}(t,2)[1-R_{22}(t,2)] + [1-R_{21}(t,2)]R_{22}(t,2) + R_{21}(t,2)R_{22}(t,2)]\\
&\quad\cdot [[1 \quad R_{31}(t,2)]R_{32}(t,2)R_{33}(t,2) + R_{31}(t,2)[1-R_{32}(t,2)]R_{33}(t,2)\\
&\quad+ R_{31}(t,2)R_{32}(t,2)[1-R_{33}(t,2)] + R_{31}(t,2)[R_{32}(t,2)R_{33}(t,2)]\\
&= [R_{11}(t,2) + R_{12}(t,2) - R_{11}(t,2)R_{12}(t,2)]\\
&\quad\cdot [R_{21}(t,2) + R_{22}(t,2) - R_{21}(t,2)R_{22}(t,2)]\\
&\quad\cdot [R_{31}(t,2)]R_{32}(t,2) + R_{31}(t,2)R_{32}(t,2) + R_{32}(t,2)R_{33}(t,2)
\end{aligned}$$

$$
\begin{aligned}
&- 2R_{31}(t,2)[R_{32}(t,2)R_{33}(t,2)] = [\exp[-0.00540t] \\
&+ \exp[-0.00540t - \exp[-0.00540t + \exp[-0.00540t]] \\
&\cdot [\exp[-0.02163t] + \exp[-0.02163t] \\
&- \exp[-0.02163t]\exp[-0.02163t] \\
&\cdot [\exp[-0.00912t]\exp[-0.00912t] \\
&+ \exp[-0.00912t + \exp[-0.009847] \\
&+ \exp[-0.00912t] + \exp[-0.009847] \\
&- 2\exp[-0.00912t + \exp[-0.00912, t + \exp[-0.009840] \\
= &[2\exp[-0.00540t - \exp[-0.01080t]] \\
&\cdot [2\exp[-0.02163t] - \exp[-0.04326t]] \\
&\cdot [\exp[-0.01824t] + 2\exp[-0.01896t] - 2\exp[-0.02808t]] \\
= &8\exp[-0.04598t - 8\exp[-0.05511t] + 4\exp[-0.04527t \\
&- 4\exp[-0.06762t] + 4\exp[-0.07674t] \\
&- 4\exp[-0.051390 + \exp[-0.060510 - 2\exp[-0.0669t \\
&- 2\exp[-0.082140 + 2\exp[-0.07302t] \\
&- 2\exp[-0.0506t] + \exp[-0.0723t] \text{ for } t \in <0, \infty).
\end{aligned} \tag{1.130}
$$

The graph of the reliability function of the port oil piping transportation system is given in Fig. 1.21.

Hence, applying (1.36)–(1.38), the mean values and standard deviations of the lifetimes of the port oil piping transportation system, the asset $A_1$, are:

– in the reliability state subset {1, 2}

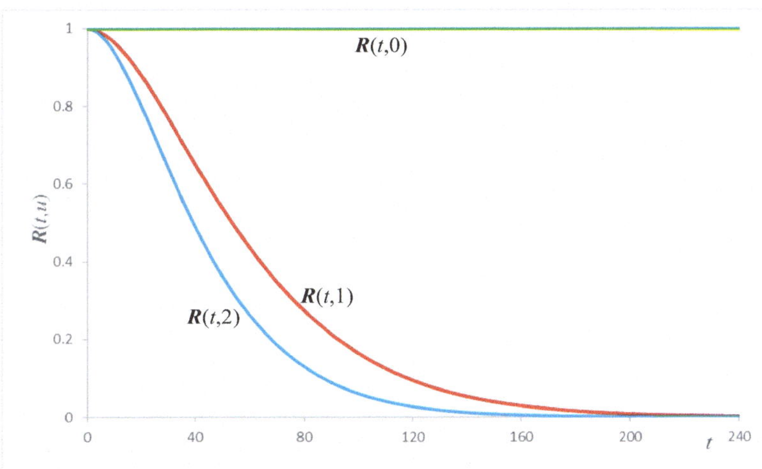


**Fig. 1.21** The graphs of the port oil piping transportation system reliability function coordinates

$$
\begin{aligned}
\boldsymbol{\mu}_1(1) &= \int_0^\infty \boldsymbol{R}_1(t, 1)dt \\
&= \int_0^\infty [8\exp[-0.0341t] - 8\exp[-0.0414t] + 4\exp[-0.03266t] \\
&\quad - 4\exp[-0.04854t] + 4\exp[-0.05584t] - 4\exp[-0.03772t] \\
&\quad + 4\exp[-0.04502t] - 2\exp[-0.0471t] - 2\exp[-0.03628t] \\
&\quad + 2\exp[-0.05216t] - 2\exp[-0.05946t] + \exp[-0.05072t]]dt \\
&\cong 62.71 \text{years}.
\end{aligned} \tag{1.131}
$$

$$
\boldsymbol{\sigma}_1(1) = \sqrt{\boldsymbol{n}_1(1) - [\boldsymbol{\mu}_1(1)]^2} \cong \sqrt{5686.4199 - (62.71)^2} \cong 41.8793 \text{ years}, \tag{1.132}
$$

where

$$
\begin{aligned}
\boldsymbol{n}_1(1) &= 2\int_0^\infty t\boldsymbol{R}_1(t, 1)dt \\
&= 2\int_0^\infty [8t\exp[-0.0341t] - 8t\exp[-0.0414t] + 4t\exp[-0.03266t] \\
&\quad - 4t\exp[-0.04854t] + 4t\exp[-0.05584t] \\
&\quad - 4t\exp[-0.03772t] + 4tt\exp[-0.04502t] - 2t\exp[-0.0471t] \\
&\quad - 2t\exp[-0.03628t] + 2t\exp[-0.05216t] \\
&\quad - 2t\exp[-0.05946t] + t\exp[-0.05072t]]dt \\
&\cong 5686.4199,
\end{aligned}
$$

- in the reliability state subset {2}

$$
\begin{aligned}
\boldsymbol{\mu}_1(2) &= \int_0^\infty \boldsymbol{R}_1(t, 2)dt \\
&= \int_0^\infty [8t\exp[-0.04598t] - 8t\exp[-0.05511t] + 4\exp[-0.04527t] \\
&\quad - 4\exp[-0.06762t] + 4\exp[-0.07674t] \\
&\quad - 4\exp[-0.05139t] + 4\exp[-0.06051t] - 2\exp[-0.0669t] \\
&\quad - 2\exp[-0.08214t] + 2\exp[-0.07302t]
\end{aligned}
$$

$$- 2\exp[-0.0506t] + \exp[-0.0723t]]dt \cong 45.84 \text{ years} \quad (1.133)$$

$$\boldsymbol{\sigma}_1(2) = \sqrt{\boldsymbol{n}_1(2) - [\boldsymbol{\mu}_1(2)]^2} \cong \sqrt{3045.9212 - (45.84)^2} \cong 30.7346 \text{ years}, \quad (1.134)$$

where

$$\boldsymbol{n}_1(2) = 2\int_0^\infty t\boldsymbol{R}_1(t,2)dt$$
$$= 2\int_0^\infty [8t\exp[-0.04598t] - 8t\exp[-0.05511t] + 4t\exp[-0.04527t]$$
$$- 4t\exp[-0.06762t] + 4t\exp[-0.07674t]$$
$$- 4t\exp[-0.05139t] + 4t\exp[-0.06051t]$$
$$- 2t\exp[-0.0669t] - 2t\exp[-0.08214t]$$
$$+ 2t\exp[-0.07302t] - 2t\exp[-0.0506t] + t\exp[-0.0723t]]dt \cong 3045.9212.$$

From (1.131) and (1.133), applying (1.43), the mean lifetimes $\bar{\boldsymbol{\mu}}(u)$, $u = 1, 2$, of the port oil piping transportation system in the particular reliability states are:

$$\bar{\mu}_1(1) = \mu_1(1) - \mu_1(2) = 16.873 \text{ years}, \bar{\mu}_1(2) = \mu_1(2) = 45.84 \text{ years}. \quad (1.135)$$

As the critical reliability state is $r = 1$, then by (1.35), considering (1.129), the port oil piping transportation system risk function is

$$\boldsymbol{r}_1(t) = 1 - \boldsymbol{R}_1(t,1) = 1 - [8\exp[-0.0341t] - 8\exp[-0.0414t]$$
$$+ 4\exp[-0.03266t] - 4\exp[-0.04854t]$$
$$+ 4\exp[-0.05584t] - 4\exp[-0.03772t]$$
$$+ 4\exp[-0.04502t] - 2\exp[-0.0471t]$$
$$- 2\exp[-0.03628t] + 2\exp[-0.05216t]$$
$$- 2\exp[-0.05946t] + \exp[-0.05072t]], \text{ for } t \in <0, \infty), \quad (1.136)$$

and by (1.37), considering the values of the risk function given in Table 1.3, the moment $\boldsymbol{\tau}_1$ of exceeding acceptable value of critical infrastructure risk function level $\delta = 0.05$ is

$$\boldsymbol{\tau}_1 = (\boldsymbol{r}_1)^{-1}(0.05) \cong 12.16 \text{ years}. \quad (1.137)$$

The graph of the port oil piping transportation system risk function is presented in Fig. 1.22.

The intensities of degradation (ageing) of the port oil piping transportation system/the intensities the port oil piping transportation system departure from the

**Table 1.3** The values of the risk function $r_1(t)$ of the port oil piping transportation system

| $t$ | $r_1(t) = 1 - R_1(t)$ |
|---|---|
| 12.00 | 0.048805 |
| 12.01 | 0.048879 |
| 12.02 | 0.048953 |
| 12.03 | 0.049027 |
| 12.04 | 0.049101 |
| 12.05 | 0.049175 |
| 12.06 | 0.049249 |
| 12.07 | 0.049323 |
| 12.08 | 0.049398 |
| 12.09 | 0.049472 |
| 12.10 | 0.049547 |
| 12.11 | 0.049621 |
| 12.12 | 0.049695 |
| 12.13 | 0.04977 |
| 12.14 | 0.049845 |
| 12.15 | 0.049919 |
| **12.16** | **0.049994** |
| 12.17 | 0.050069 |
| 12.18 | 0.050143 |
| 12.19 | 0.050218 |
| 12.20 | 0.050293 |

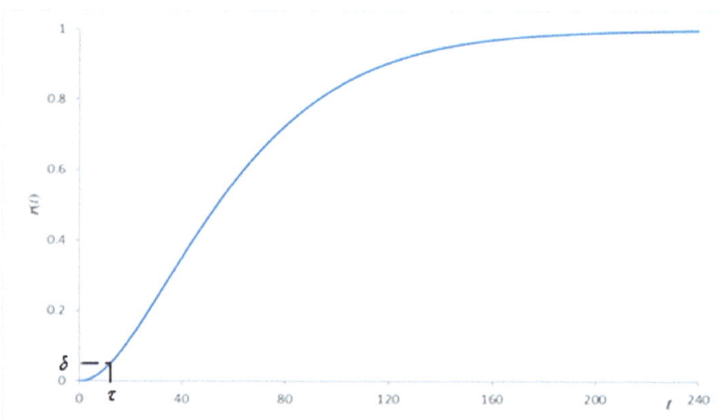


**Fig. 1.22** The graph of the port oil piping transportation system risk function

reliability state subset {1, 2}, {2}, i.e. the coordinates of the vector

$$\boldsymbol{\lambda}_1(t,\cdot) = [0, \lambda_1(t,1), \lambda_1(t,2)], t \in< 0, +\infty), \tag{1.138}$$

according to (1.45), are given by

$$\boldsymbol{\lambda}_1(t,1) = \frac{-\frac{d\boldsymbol{R}_1(t,1)}{dt}}{\boldsymbol{R}_1(t,1)}, \boldsymbol{\lambda}_1(t,2) = \frac{-\frac{d\boldsymbol{R}_1(t,2)}{dt}}{\boldsymbol{R}_1(t,2)}, t \in< 0, +\infty), \tag{1.139}$$

where $\boldsymbol{R}_1(t, 1)$ is given by the equivalent formulae (1.129) and (1.126) and $\boldsymbol{R}_1(t, 2)$ is given by the equivalent formulae (1.130) and (1.127).

The values of the intensities of degradation given by (1.139) stabilize for large time and their limit values defined by (1.46) amount:

$$\boldsymbol{\lambda}_1(1) = \lim_{t\to\infty} \boldsymbol{\lambda}_1(t,1) = \lim_{t\to\infty} \frac{4\cdot 0.03266(1+0(t))}{4(1+0(t))} = 0.03266, \tag{1.140}$$

$$\boldsymbol{\lambda}_1(2) = \lim_{t\to+\infty} \boldsymbol{\lambda}_1(t,2) = \lim_{t\to\infty} \frac{4\cdot 0.04527(1+0(t))}{4(1+0(t))} = 0.04527. \tag{1.141}$$

The graphs of the intensities of degradation of the port oil piping transportation system are given in Fig. 1.23.

The approximate mean of the intensities of degradation of the port oil piping transportation system according to (1.47) and (1.131) and (1.133) respectively are

$$\boldsymbol{\lambda}_1(t) = \frac{1}{62.71} \cong 0.01595, \tag{1.142}$$

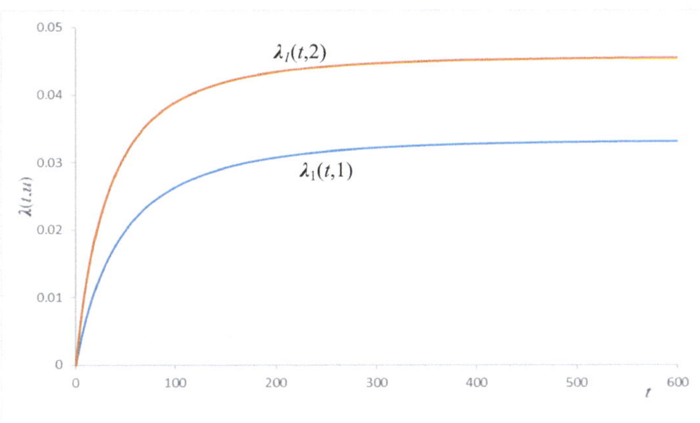


**Fig. 1.23** The graphs of the intensities of ageing of the port oil piping transportation system

$$\lambda_1(2) = \frac{1}{45.84} \cong 0.02182. \tag{1.143}$$

### 1.4.5 Port Oil Terminal Critical Infrastructure Reliability Parameters

The approximate mean values of the asset $A_1$, the port oil terminal critical infrastructure lifetimes in the reliability state subsets calculated in Sect. 1.4.4 on the basis of reliability data of its components coming from experts are:

- in reliability state subset {1, 2}

$$\mu_1(1) \approx 63\,\text{years}, \tag{1.144}$$

- in reliability state subset {2}

$$\mu_1(2) \approx 46\,\text{years}. \tag{1.145}$$

The approximate mean values of the remaining assets $A_2 - A_9$ lifetimes in the reliability state subsets evaluated approximately by experts, are as follows:

- in reliability state subset {1, 2}

$$\mu_i(1) = 80\,\text{years}, i = 2, 3, \ldots 9. \tag{1.146}$$

- in reliability state subset {2}

$$\mu_i(2) = 50\,\text{years}, i = 2, 3, \ldots 9. \tag{1.147}$$

From (1.144)–(1.147), it follows that defined by (1.51) the approximate intensities of assets departure from the reliability state subsets {1.2}, {2} respectively are:

- for asset $A_1$

$$\lambda_1(1) = \frac{1}{\mu_1(1)} = \frac{1}{63} = 0.015873, \tag{1.148}$$

$$\lambda_1(2) = \frac{1}{\mu_1(2)} = \frac{1}{46} = 0.021739; \tag{1.149}$$

- for assets $A_2 - A_9$

$$\lambda_i(1) = \frac{1}{\mu_i(1)} = \frac{1}{80} = 0.0125, i = 2, 3, \ldots, 9, \tag{1.150}$$

$$\lambda_i(2) = \frac{1}{\mu_i(2)} = \frac{1}{50} = 0.02, i = 2, 3, \ldots, 9. \tag{1.151}$$

### 1.4.6 Port Oil Terminal Critical Infrastructure Reliability Indicators

We assume that the port oil terminal critical infrastructure is a three-state (z = 2) series system and its assets have piecewise exponential reliability functions of the form,

$$R_i(t, \cdot) = [1, R_i(t, 1) R_i(t, 2)], t \geq 0, i = 1, 2, \ldots, 9,$$

where

$$R_i(t, u) = \exp[-\lambda_i(u)t], t \geq 0, u = 1, 2, i = 1, 2, \ldots, 9.$$

Thus, considering (1.143)–(1.146), we have

$$R_1(t, 1) = \exp[-0.015873t], R_1(t, 2) = \exp[-0.021739t], t \geq 0, \tag{1.152}$$

$$R_i(t, 1) = \exp[-0.0125t], R_i(t, 1) = \exp[-0.02t], t \geq 0, i = 1, 2, \ldots, 9. \tag{1.153}$$

Hence, the port oil terminal critical infrastructure reliability function (RelI1) is given by

$$\boldsymbol{R}(t, \cdot) = [1, \boldsymbol{R}(t, 1) \boldsymbol{R}(t, 2)], t \geq 0, \tag{1.154}$$

where according to the formula (1.55) for the coordinates of the series critical infrastructure reliability function and considering (1.152), (1.153), we have

$$\boldsymbol{R}(t,1)=\prod_{i=1}^{9}R_i(t,1)$$
$$=\exp[-0.015873t]\exp[-0.0125t]\exp[-0.0125t]$$
$$\exp[-0.0125t]\exp[-0.0125t]\exp[-0.0125t]$$
$$\exp[-0.0125t]\exp[-0.0125t]\exp[-0.0125t]$$
$$=\exp[-0.115873t],\ t\geq 0. \tag{1.155}$$

$$\boldsymbol{R}(t,2)=\prod_{i=1}^{9}R_i(t,2)$$
$$=\exp[-0.021739t]\exp[-0.02t]\exp[-0.02t]$$
$$\exp[-0.02t]\exp[-0.02t]\exp[-0.02t]\exp[-0.02t]$$
$$\exp[-0.02t]\exp[-0.02t]=\exp[-0.181739t],\ t\geq 0. \tag{1.156}$$

The graph of the port oil terminal critical infrastructure reliability function is shown in Fig. 1.24.

As the critical reliability state is $r=1$, then by (1.155) and (1.35), the port oil terminal critical infrastructure risk function (RelI2), is given by

$$\boldsymbol{r}(t)=1-\boldsymbol{R}(t,1)=1-\exp[-0.115873t]\ \text{for}\ t\geq 0. \tag{1.157}$$

The graph of the risk function $\boldsymbol{r}(t)$ of the oil terminal critical infrastructure, the fragility curve (RelI3), is shown in Fig. 1.25.

From (1.157) and (1.39), the moment when the port oil terminal critical infrastructure risk function exceeds a permitted level $\delta=0.05$ (RelI6), is

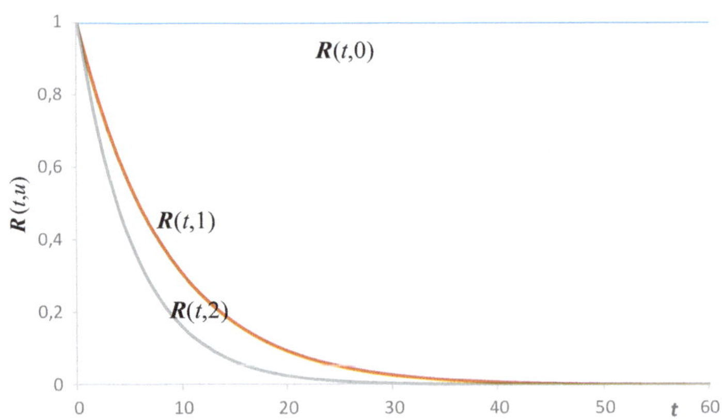


**Fig. 1.24** The graph of the port oil terminal critical infrastructure reliability function coordinates

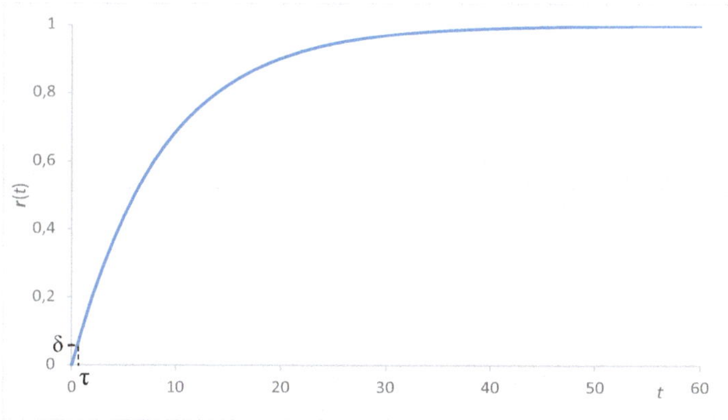


**Fig. 1.25** The graph of the risk function $r(t)$ (the fragility curve) of the port oil terminal critical infrastructure

$$\tau = -\frac{1}{0.115873} \ln(1 - 0.05) \cong 0.44 \text{ years.} \tag{1.158}$$

From (1.155), (1.156), after applying (1.40)–(1.42), the mean values (RelI4, RelI7) and standard deviations (RelI5, RelI8) of the port oil terminal critical infrastructure lifetimes in the reliability state subsets {1, 2}, {2}, respectively are:

$$\boldsymbol{\mu}(1) = \int_0^\infty \boldsymbol{R}(t, 1)dt = \int_0^\infty \exp[-0.115873t]dt = \frac{1}{0.115873} \cong 8.63\text{years,} \tag{1.159}$$

$$\boldsymbol{\mu}(2) = \int_0^\infty \boldsymbol{R}(t, 2)dt = \int_0^\infty \exp[-0.181739t]dt = \frac{1}{0.181739} \cong 5.50 \text{ years,} \tag{1.160}$$

$$\boldsymbol{\sigma}(1) = \sqrt{\boldsymbol{n}(1) - [\boldsymbol{\mu}(1)]^2} = \sqrt{148.9585696 - (8.63)^2} \cong 8.63 \text{ years,} \tag{1.161}$$

$$\boldsymbol{\sigma}(2) = \sqrt{\boldsymbol{n}(2) - [\boldsymbol{\mu}(2)]^2} = \sqrt{60.55272994 - (5.50)^2} \cong 5.50 \text{ years} \tag{1.162}$$

where

$$\boldsymbol{n}(1) = 2\int_0^\infty t\boldsymbol{R}(t, 1)dt = 2\int_0^\infty t \exp[-0.115873t]dt = \frac{2}{(0.115873)^2}$$
$$= \frac{2}{0.013426552} = 14.9585696,$$

$$\boldsymbol{n}(2) = 2\int_0^\infty t\boldsymbol{R}(t,2)dt = 2\int_0^\infty t\exp[-0.181739t]dt = \frac{2}{(0.181739)^2}$$
$$= \frac{2}{0.033029064} = 60.55272994.$$

Further from (1.43), after considering (1.159), (1.160), it follows that the mean values of the port oil terminal critical infrastructure lifetimes in the particular reliability states (RelI9), are:

$$\bar{\boldsymbol{\mu}}(1) = \boldsymbol{\mu}(1) - \boldsymbol{\mu}(2) = 8.63 - 5.50 \cong 3.13\,\text{years}, \tag{1.163}$$

$$\bar{\boldsymbol{\mu}}(2) = \boldsymbol{\mu}(2) \cong 5.50\text{years}. \tag{1.164}$$

Considering (1.155) and (1.156), the port oil terminal critical infrastructure intensities of ageing (RelI10), according to (1.45), are:

$$\boldsymbol{\lambda}(t,1) = \frac{-\frac{d\boldsymbol{R}(t,1)}{dt}}{\boldsymbol{R}(t,1)} = \frac{0.115873\exp[-0.115873t]}{\exp[-0.115873t]} = 0.115873, \tag{1.165}$$

$$\boldsymbol{\lambda}(t,2) = \frac{-\frac{d\boldsymbol{R}(t,2)}{dt}}{\boldsymbol{R}(t,2)} = \frac{0.181739\exp[-0.181739t]}{\exp[-0.181739t]} = 0.181739, \tag{1.159}$$

and by application the formulae (1.46) and (1.47) their values are the same.

The graphs of the intensities of ageing of the oil terminal critical infrastructure are presented in Fig. 1.26.

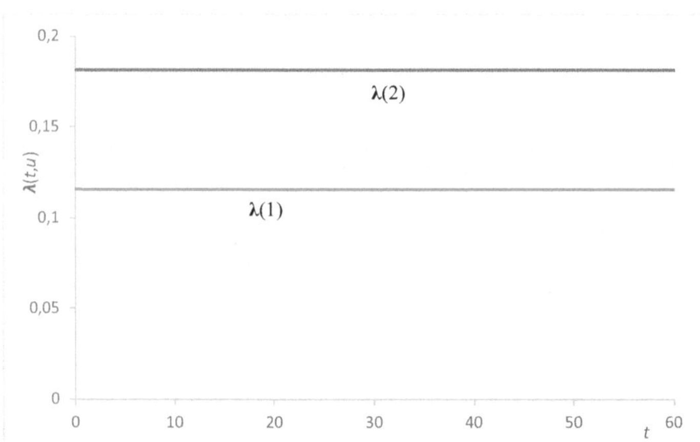


**Fig. 1.26** The graphs of the intensities of ageing of the port oil terminal critical infrastructure

## References


1. Amari SV, Misra RB (1997) Comment on: dynamic reliability analysis of coherent multistate systems. IEEE Trans Reliab 46:460–461
2. Aven T (1985) Reliability evaluation of multistate systems with multistate components. IEEE Trans Reliab 34:473–479
3. Aven T, Jensen U (1999) Stochastic models in reliability. Springer, New York
4. Aven T (1993) On performance measures for multistate monotone systems. Reliab Eng Sys Saf 41:259–266
5. Barlow RE, Wu AS (1978) Coherent systems with multi-state components. Math Oper Res 4:275–281
6. Ben P, Gouldby BP, Schultz MT, Simm JD, Wibowo JL (2010) Beyond the factor of safety: developing fragility curves to characterize system reliability, report in water resources infrastructure program ERDC SR-10-1, prepared for headquarters. Washington, U.S, Army Corps of Engineers
7. Hudson JC, Kapur KC (1982) Reliability theory for multistate systems with multistate components. Microelectr Reliab 22:1–7
8. Hudson JC, Kapur KC (1983) Reliability analysis of multistate systems with multistate components. Trans Inst Ind Eng 15:127–135
9. Hudson J, Kapur K (1985) Reliability bounds for multistate systems with multistate components. Oper Res 33:735–744
10. Kołowrocki K (1998) On applications of asymptotic reliability functions to the reliability and risk evaluation of pipelines. Int J Press Vessel Pip 75:545–558
11. Kołowrocki K (2003) An asymptotic approach to reliability evaluation of large multi-state systems with applications to piping transportation systems. Int J Press Vessel Pip 80:59–73
12. Kołowrocki K (2008) Reliability of large systems. In: Encyclopedia of quantitative risk analysis and assessment, vol. 4, Wiley
13. Kołowrocki K (2014) Reliability of large systems and complex. Elsevier, Amsterdam—Boston—Heidelberg—London—New York—Oxford—Paris—San Diego—San Francisco—Singapore—Sydney—Tokyo
14. Kołowrocki K, Soszyńska-Budny J (2011) Reliability and safety of complex technical systems and processes. Modelling—identification—prediction—optimization. Springer, London, Dordrecht, Heildeberg, New York
15. Lisnianski A, Levitin G (2003) Multi-state system reliability. Assessment, optimisation and applications. World Scientific Publishing Co. Pte. Ltd., New Jersey, London, Singapore, Hong Kong
16. Meng F (1993) Component-relevancy and characterisation in multi-state systems. IEEE Trans Reliab 42:478–483
17. Natvig B (1984) Multi-state coherent systems. In: Encyclopaedia of statistical sciences. Wiley, New York
18. Ohio F, Nishida T (1984) On multistate coherent systems. IEEE Trans Reliab 33:284–287
19. Soszyńska J (2006) Reliability and risk evaluation of a port oil pipeline transportation system in variable operation conditions. Int J Press Vessel Pip 87(2–3):81–87
20. Xue J (1985) On multi-state system analysis. IEEE Trans Reliab 34:329–337
21. Xue J, Yang K (1995) Dynamic reliability analysis of coherent multi-state systems. IEEE Trans Reliab 4(44):683–688
22. Xue J, Yang K (1995) Symmetric relations in multi-state systems. IEEE Trans Reliab 4(44):689–693
23. Yu K, Koren I, Guo Y (1994) Generalised multistate monotone coherent systems. IEEE Trans Reliab 43:242–250

# Chapter 2
# General Approach to Critical Infrastructure Safety Modelling

The methodology and general approach to critical infrastructure safety and resilience analysis is proposed. The principles of multistate approach to critical infrastructure safety analysis are introduced. There are introduced the notions of critical infrastructure basic safety indicators like, the critical infrastructure safety function, the critical infrastructure risk function and the critical infrastructure fragility curve. Basic critical infrastructure safety structures are defined and the general formulae for their safety functions and their particular forms when the critical infrastructure assets have piecewise exponential safety functions are given. The critical infrastructure safety and resilience indicators are proposed to be obtained by using of probabilistic approach to modelling of operation threats and extreme weather hazard impacts on its assets safety. The general scheme of this approach is presented together with the description of its boxes' contains. There are proposed safety and resilience indicators, crucial for operators and users of the critical infrastructure, defined as a complex system in its operating environment.

## 2.1 Multistate Approach to Critical Infrastructure Safety

Many of the terms and notions needed for the critical infrastructure safety analysis are used in different and sometimes conflicting ways across various disciplines and approaches. Some of them are incorrect. Thus, a standard set of definitions should be fixed, to support shared understanding of the foci of this book and, for use by all its readers. Therefore, definitions concerned with the methodology including notions and meanings of the critical infrastructure and its safety, the climate and weather change and the resilience should be convergent with those used in available literature. The spectrum of the terms concerned with those three main notions should be sufficiently wide and exhaustive in depth. The main fault in defining some of the terms is mixing the meaning of the defined notion with values of its parameters it

 
J. Soszyńska-Budny, *Safety Analysis of Critical Infrastructure*,
Lecture Notes in Intelligent Transportation and Infrastructure,
https://doi.org/10.1007/978-3-030-65339-2_2

is characterized by. Having in mind this state of terminology and considering its imperfection and faults, the main principles in preparation of this book are:

- to differ between the notion and values of the parameters it is defined by;
- to illustrate shortly the notion and its parameters together with short illustrations/interpretations of their meanings and their expected practical usage in order to provide a better understanding.

The first and most important term for the book is the notion of the critical infrastructure. To follow the European Commission approach, the critical infrastructure is an asset or system which is essential for the maintenance of vital societal functions. The damage of a critical infrastructure, its destruction or disruption by natural disasters, terrorism, criminal activity or malicious behavior, may have a significant negative impact for the security of the European Union and the well-being of its citizens.

The critical infrastructure is a term used by governments to describe assets that are essential for the functioning of a society and economy. Most commonly associated with the term of critical infrastructure are facilities for:

- electricity generation, transmission and distribution;
- gas production, transport and distribution;
- oil and oil products production, transport and distribution;
- telecommunication;
- water supply;
- agriculture, food production and distribution;
- heating;
- public health;
- transportation systems;
- financial services;
- security services.

Critical infrastructures are usually interconnected and mutually dependent in various and complex ways, creating critical infrastructure networks. They are interacting directly and indirectly at various levels of their complexity and operating activity [32, 56, 58, 60]. Identifying and modeling dependencies depend on the level of analysis. The selected level of analysis can vary from micro to macro level [16, 28, 29]. A holistic approach, as in [46], can be considered or, a reductionist approach in which elementary components are identified and their behavior is described. For example, Svedsen and Wolthunsen [62] focus on the components of critical infrastructure networks and they demonstrate several types of multi-dependency structures.

Considering that this book is devoted to critical infrastructure safety mathematical modelling and prediction, the critical infrastructure is defined as a complex multistate ageing system in its operating environment. Significant features of critical infrastructure are its inside dependencies and its outside impacts, that in the case of its degradation have a significant destructive influence on the health, safety and security, economics and social conditions of large human communities and territory areas.

The multistate system was introduced in [5, 6, 20, 52, 54, 61, 71]. El-Neweihi et al. [20] developed the concept of a multistate coherent structure as a generalization of binary coherent structure. Different approaches to describe multistate systems and to estimate their reliability can be found in [19, 21, 24, 27, 30, 31, 34, 37, 44, 52, 55, 57, 70]. Liu et al. [51] use fuzzy theory to assess reliability and performance for a system consisting of a single multistate element. The book of the authors Lisnianski and Levitin [50] contains a comprehensive overview and presentation of methods of multistate systems reliability analysis. Further extension of the book's results on reliability assessment and optimization with applications is described in the book [49] that presents stochastic process methods for reliability analysis of multistate systems with multistate components, including Markov, Markov reward and semi-Markov models, as well as universal generating function method. Fuzzy multi-state systems and their reliability evaluation are introduced in [51], but in that case, performance and failure rates of system components cannot be exactly determined. Ding and Lisnianski [17] propose fuzzy universal generating functions for multistate system reliability analysis, extended from the universal generating function method described in [50]. Multistate systems are also widely discussed in the literature [1–4, 13–15, 27, 33, 53, 67]. Another, practically important, approach to multistate ageing system reliability analysis considers assumption about component degradation (departures from the reliability state subsets) instead of component failures [35, 36, 40, 59, 64–70]. In real technical systems, components often degrade with time by going to states corresponding to different performance levels. Degradation of components and subsystems in case of complex systems, causes the decreasing of system reliability and its operation safety.

Considering the above performed analysis, similarly to the case of multistate approach to critical infrastructure reliability, in the multistate safety analysis in order to define the critical infrastructure with degrading/ageing components/assets, we assume that [37–46]:

- $n$ is the number of the critical infrastructure assets;
- $A_i$, $i = 1, 2, \ldots, n$, are the critical infrastructure assets;
- all assets and the critical infrastructure have the safety state set $\{0, 1, \ldots, z\}, z \geq 1$;
- the safety states are ordered, the safety state 0 is the worst and the safety state $z$ is the best;
- $r$, $r \in \{1, 2, \ldots, z\}$, is the critical safety state;
- $T_i(u)$, $u = 1, 2, \ldots, z$, $i = 1, 2, \ldots, n$, are independent random variables representing the lifetimes of assets $A_i$ in the safety state subset $\{u, u + 1, \ldots, z\}$, $u = 1, 2, \ldots, z$, while they were in the safety state $z$ at the moment $t = 0$;
- $T(u)$, $u = 1, 2, \ldots, z$, is a random variable representing the lifetime of the critical infrastructure in the safety state subset $\{u, u + 1, \ldots, z\}$, $u = 1, 2, \ldots, z$, while it was in the safety state $z$ at the moment $t = 0$;
- the safety states degrade with time $t$;
- the assets and the critical infrastructure degrade with time $t$;

- $s_i(t)$, $t \in< 0, \infty)$, $i = 1, 2, \ldots, n$, is the asset $A_i$, safety state at the moment $t$, $t \in< 0, \infty)$, given that it was in the safety state $z$ at the moment $t = 0$;
- $s(t)$, $t \in< 0, \infty)$, is the critical infrastructure safety state at the moment $t$, $t \in< 0, \infty)$, given that it was in the safety state $z$ at the moment $t = 0$.

The critical safety state $r$ means, that staying of the critical infrastructure and its assets in the safety states lower than this safety state $r$, is highly dangerous for them and for their operating environment. The above assumptions mean that the safety states of the critical infrastructure with degrading assets may be changed in time only from better to worse [40–42]. The way in which the assets and the critical infrastructure safety states change is illustrated in Figs. 2.1 and 2.2.

We denote by $T(u)$, $u = 1, 2, \ldots, z$, the critical infrastructure lifetime in the safety state subset $\{u, u + 1, \ldots, z\}$, $u = 1, 2, \ldots, z$, and we define the critical infrastructure safety function by the vector [40, 42]

$$\boldsymbol{S}(t, \cdot) = [1, \boldsymbol{S}(t, 1), \boldsymbol{S}(t, 2), \ldots, \boldsymbol{S}(t, z)], t \in< 0, \infty), \tag{2.1}$$

where

**Fig. 2.1** Illustration of a critical infrastructure safety states changing [42]

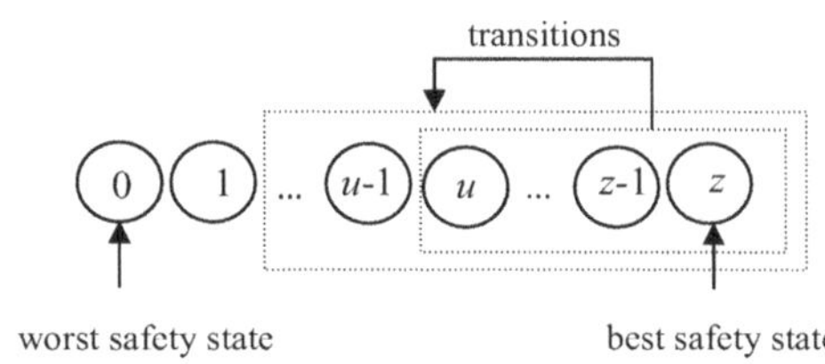


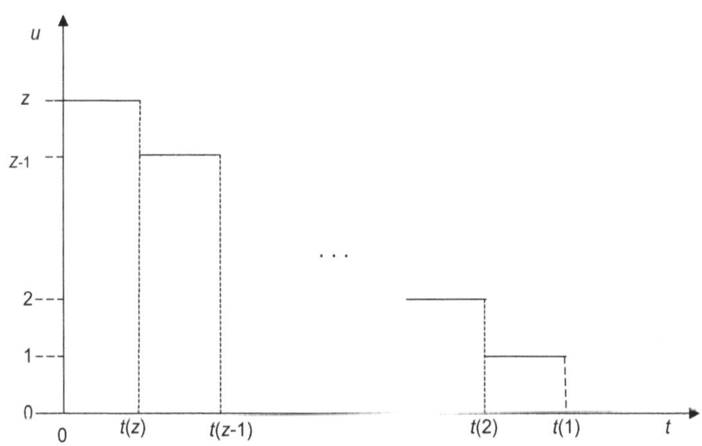


**Fig. 2.2** The relationship between the realizations $t(u)$, $u = 1, 2, \ldots, z$, of the critical infrastructure lifetime $T(u)$, $u = 1, 2, \ldots, z$, in the safety state subsets $\{u, u + 1, \ldots, z\}$, $u = 1, 2, \ldots, z$

$$S(t,u) = P(s(t) \geq u | s(0) = z) = P(T(u) > t)$$
$$\text{for } t \in < 0, \infty), u = 1, 2, \ldots, z \qquad (2.2)$$

is the probability that the critical infrastructure is in the safety state subset $\{u, u+1, \ldots, z\}$, $u = 1, 2, \ldots, z$, at the moment $t$, $t \in < 0, \infty)$, while it was in the safety state $z$ at the moment $t = 0$.

The safety functions $\boldsymbol{S}(t, u)$, $t \in < 0, \infty)$, $u = 1, 2, \ldots, z$, defined by (2.2) are called the coordinates of the critical infrastructure safety function $\boldsymbol{S}(t, \cdot)$, $t \in < 0, \infty)$ given by (2.1). Thus, the relationship between the distribution function $\boldsymbol{F}(t, u)$, $t \in < 0, \infty)$, $u = 1, 2, \ldots, z$, of the critical infrastructure lifetime $T(u)$, $u = 1, 2, \ldots, z$, in the safety state subset $\{u, u+1, \ldots, z\}$, $u = 1, 2, \ldots, z$, and the coordinate $\boldsymbol{S}(t, u)$, $t \in < 0, \infty)$ $u = 1, 2, \ldots, z$, of its safety function (2.1) is given by

$$F(t,u) = P(T(u) \leq t) = 1 - P(T(u) > t) = 1 - S(t,u)$$
$$t \in < 0, \infty), u = 1, 2, \ldots z$$

The exemplary graph of a five-state ($z = 4$) critical infrastructure safety function

$$\boldsymbol{S}(t, \cdot) = [1, \boldsymbol{S}(t,1), \boldsymbol{S}(t,2), \boldsymbol{S}(t,3), \boldsymbol{S}(t,4)], t \in < 0, \infty),$$

is shown in Fig. 2.3.

If $r$ is the critical safety state, then the critical infrastructure risk function

$$\boldsymbol{r}(t) = P(s(t) < r | s(0) = z) = P(T(r) \leq t), t \in < 0, \infty), \qquad (2.3)$$

is defined as a probability that the critical infrastructure is in the subset of safety states worse than the critical safety state $r$, $r \in \{1, \ldots, z\}$ while it was in the best safety state $z$ at the moment $t = 0$ and given by [37, 40, 42]

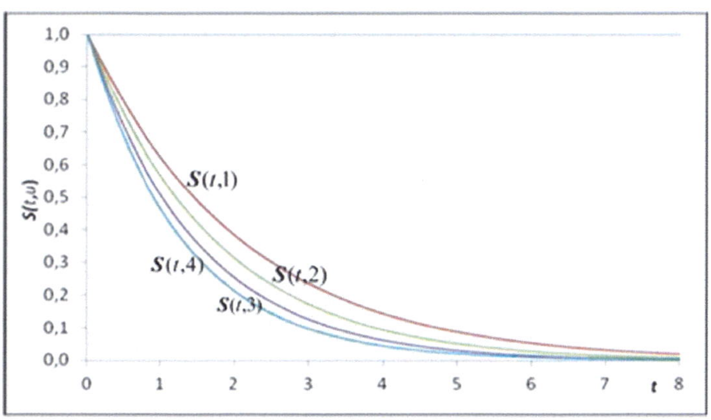


**Fig. 2.3** The graphs of a five-state critical infrastructure safety function $\boldsymbol{S}(t, \cdot)$ coordinates

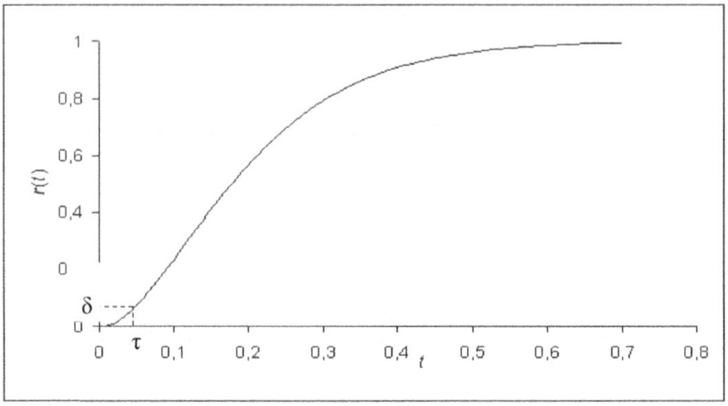


**Fig. 2.4** The graph of the exemplary critical infrastructure risk function $\boldsymbol{r}(t)$

$$\boldsymbol{r}(t) = 1 - \boldsymbol{S}(t, r), t \in < 0, \infty), \tag{2.4}$$

where $\boldsymbol{S}(t, r)$, $t \in < 0, \infty)$, is the coordinate of the critical infrastructure unconditional safety function given by (2.2) for $u = r$.

The graph of the exemplary critical infrastructure risk function is presented in Fig. 2.4.

Similarly, we define the asset $A_i$,$i = 1, 2, \ldots, n$ safety function by the vector [42]

$$S_i(t, \cdot) = [1, S_i(t, 1), S_i(t, 2), \ldots, S_i(t, z)], t \in < 0, \infty), \quad i = 1, 2, \ldots, n, \tag{2.5}$$

where

$$\begin{aligned} S_i(t, u) &= P(s_i(t) \geq u | s_i(0) = z) \\ &= P(T_i(u) > t), t \in < 0, \infty, \quad u = 1, 2, \ldots, z, \quad i = 1, 2, \ldots, n, \end{aligned} \tag{2.6}$$

is the probability that the asset $A_i$ is in the safety state subset $\{u, u + 1, \ldots, z\}$, $u = 1, 2, \ldots, z$, at the moment $t$, $t \in < 0, \infty)$, while it was in the safety state $z$ at the moment $t = 0$.

The safety functions $S_i(t, u)$, $t \in < 0, \infty)$, $u = 1, \ldots, z$, $i = 1, 2, \ldots, n$, defined by (2.6) are called the coordinates of the asset $A_i$, $i = 1, 2, \ldots, n$, safety function $\boldsymbol{S}(t, \cdot)$, $t \in < 0, \infty)$, $i = 1, 2, \ldots, n$, given by (2.5). Thus, the relationship between the distribution function $F_i(t, u)$, $t \in < 0, \infty)$, $u = 1, 2, \ldots, z$, $i = 1, 2, \ldots, n$, of the asset $A_i$, $i = 1, 2, \ldots, n$, lifetime $T_i(u)$, $u = 1, 2, \ldots, z$, $i = 1, 2, \ldots, n$, in the safety state subset $\{u, u + 1, \ldots, z\}$, $u = 1, 2, \ldots, z$, and the coordinate $S_i(t, u)$, $t \in < 0, \infty)$, $u = 1, 2, \ldots, z$, $i = 1, 2, \ldots, n$, of its safety function is given by

$$\begin{aligned} F_i(t, u) = P(T_i(u) \leq t) = 1 - P(T_i(u) > t) = 1 - S_i(t, u), t \in < 0), \infty, \\ u = 1, 2, \ldots, z, i = 1, 2, \ldots, n. \end{aligned} \tag{2.7}$$

## 2.2 Critical Infrastructure Safety Structures

On the basis of the approach to critical infrastructure safety analysis proposed in Sect. 2.1, considering definitions of reliability functions of multistate systems introduced in Chap. 1, we may similarly define basic critical infrastructure safety structures.

**Definition 2.1** A critical infrastructure composed of $n$ assets $A_i$, $i = 1, 2, \ldots, n$, is called series if its lifetime $T(u)$, $u = 1, 2, \ldots, z$, in the safety state subset $\{u, u+1, \ldots, z\}$, $u = 1, 2, \ldots, z$, is given by [42]

$$T(u) = \min_{1 \le i \le n} \{T_i(u)\},\ u = 1, 2, \ldots, z,$$

where $T_i(u)$, $u = 1, 2, \ldots, z$, $i = 1, 2, \ldots, n$, are the lifetimes of the assets $A_i$, $i = 1, 2, \ldots, n$, in the safety state subset $\{u, u+1, \ldots, z\}$, $u = 1, 2, \ldots, z$. The number $n$ is called the critical infrastructure safety structure shape parameter.

The above definition means that the series critical infrastructure is in the safety state subset $\{u, u+1, \ldots, z\}$, $u = 1, 2, \ldots, z$, if and only if all its $n$ assets are in this subset of safety states. That meaning is very close to the definition of a two-state critical infrastructure considered in a classical reliability analysis that is not failed if all its assets are not failed [39]. This fact can justify the safety structure scheme for a series critical infrastructure presented in Fig. 2.5 [42].

It is easy to work out that the safety function of the series critical infrastructure is given by the vector [37, 40, 42]

$$\boldsymbol{S}(t, \cdot) = [1, \boldsymbol{S}(t, 1), \boldsymbol{S}(t, 2), \ldots, \boldsymbol{S}(t, z)],\ t \in <0, \infty), \tag{2.8}$$

with the coordinates

$$\boldsymbol{S}(t, u) = \prod_{i=1}^{n} S_i(t, u),\ t \in <0, \infty),\ u = 1, 2, \ldots, z. \tag{2.9}$$

**Definition 2.2** A critical infrastructure composed of $n$ assets $A_i$, $i = 1, 2, \ldots, n$, is called parallel if its lifetime $T(u)$, $u = 1, 2, \ldots, z$, in the safety state subset $\{u, u+1, \ldots, z\}$, $u = 1, 2, \ldots, z$, is given by [42]

$$T(u) = \max_{1 \le i \le n} \{T_i(u)\},\ u = 1, 2, \ldots, z,$$

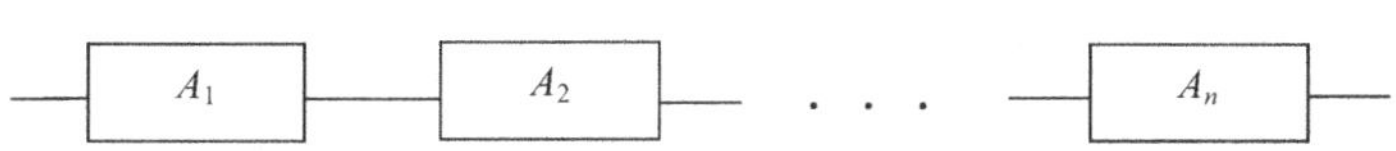


**Fig. 2.5** The scheme of a series critical infrastructure safety structure

where $T_i(u)$, $u = 1, 2, \ldots, z$, $i = 1, 2, \ldots, n$, are the lifetimes of the assets $A_i$, $i = 1, 2, \ldots, n$, in the safety state subset $\{u, u+1, \ldots, z\}$, $u = 1, 2, \ldots, z$.

The number $n$ is called the critical infrastructure structure safety shape parameter.

The above definition means that the parallel critical infrastructure is in the safety state subset $\{u, u+1, \ldots, z\}$, $u = 1, 2, \ldots, z$, if and only if at least one of its $n$ assets is in this subset of safety states. That meaning is very close to the definition of a two-state parallel system in a classical reliability analysis that is not failed if at least one of its components is not failed [39] what can justify the parallel critical infrastructure safety structure scheme presented in Fig. 2.6 [42].

The safety function of the parallel critical infrastructure is given by the vector [37, 40, 42]

$$\boldsymbol{S}(t,\cdot) = [1, \boldsymbol{S}(t,1), \boldsymbol{S}(t,2), \ldots, \boldsymbol{S}(t,z)], \boldsymbol{t} \in < 0, \infty), \tag{2.10}$$

with the coordinates

$$\boldsymbol{S}(t,u) = 1 - \prod_{i=1}^{n} [1 - S_i(t,u)], t \in < 0, \infty), u = 1, 2, \ldots, z. \tag{2.11}$$

**Definition 2.3** A critical infrastructure composed of $n$ assets $A_i$, $i = 1, 2, \ldots, n$, is called an "$m$ out of $n$" if its lifetime $T(u)$, $u = 1, 2, \ldots, z$, in the safety state subset $\{u, u+1, \ldots, z\}$, $u = 1, 2, \ldots, z$, is given by [42]

$$T(u) = T_{(n-m+1)}(u), m = 1, 2, \ldots, n, u = 1, 2, \ldots, z,$$

where $T_{(n-m+1)}(u)$, $m = 1, 2, \ldots, n$, $u = 1, 2, \ldots, z$, is the $(n\text{-}m+1)$-th order statistic in the sequence of the asset lifetimes

$$T_1(u), T_2(u), \ldots, T_n(u), u = 1, 2, \ldots, z.$$

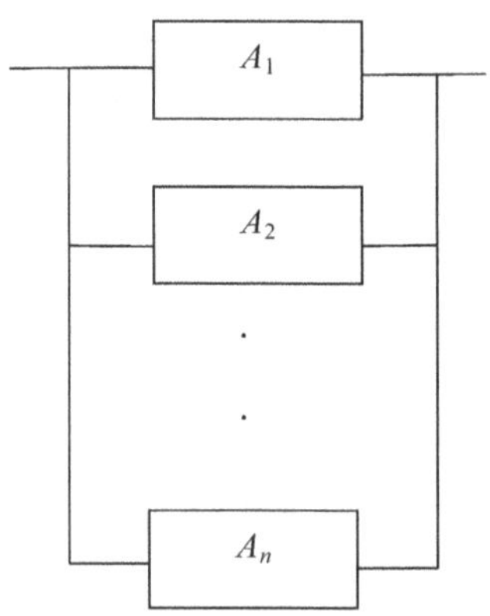


**Fig. 2.6** The scheme of a parallel critical infrastructure safety structure [42]

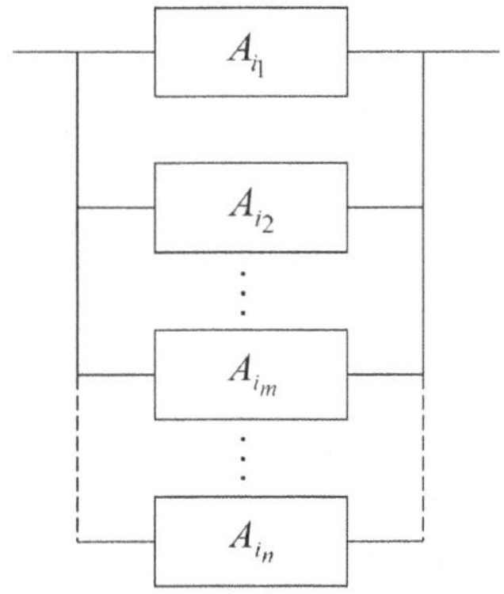


**Fig. 2.7** The scheme of an "$m$ out of $n$" critical infrastructure safety structure [42]

The above definition means that the "$m$ out of $n$" critical infrastructure is in the safety state subset $\{u, u+1, \ldots, z\}$, $u = 1, 2, \ldots, z$, if and only if at least $m$ out of its $n$ assets are in this safety state subset and it is a parallel critical infrastructure if $m = 1$ and it is a series critical infrastructure if $m = n$. The numbers $m$ and $n$ are called the critical infrastructure safety structure shape parameters. The scheme of an "$m$ out of $n$" critical infrastructure safety structure, justified in an analogous way as in the case of a series critical infrastructure and a parallel critical infrastructure, is given in Fig. 2.7, where $i_1, i_2, \ldots, i_n \in \{1, 2, \ldots, n\}$ and $i_a \neq i_b$ for $a \neq b$[42].

It can be simply shown that the safety function of the "$m$ out of $n$" critical infrastructure is given either by the vector [37, 40, 42]

$$S(t, \cdot) = [1, S(t, 1), S(t, 2), \ldots, S(t, z)], t \in <0, \infty), \tag{2.12}$$

with the coordinates

$$\mathbf{S}(t, u) = \sum_{\substack{r_1, r_2, \ldots, r_n \in \{0,1\} \\ m \le r_1 + r_2 + \ldots + r_n \le n}} [S_i(t, u)]^{r_i} [1 - S_i(t, u)]^{1-r_i}, \ t \in <0, \infty), u = 1, 2, \ldots, z, \tag{2.13}$$

or by the vector

$$S(t, \cdot) = [1, S(t, 1), S(t, 2), \ldots, S(t, z)], \tag{2.14}$$

with the coordinates

$$\mathbf{S}(t, u) = 1 - \sum_{\substack{r_1, r_2, \ldots, r_n \in \{0,1\} \\ 0 \le r_1 + r_2 + \ldots + r_n \le m-1}} [S_i(t, u)]^{r_i} [1 - S_i(t, u)]^{1-r_i},$$

$$t \in <0, \infty), u = 1, 2, \ldots, z. \tag{2.15}$$

**Definition 2.4** A critical infrastructure composed of $n$ assets $A_i$, $i = 1, 2, \ldots, n$, is called a consecutive "$m$ out of $n$: F" if it is out of the safety state subset $\{u, u+1, \ldots, z\}$, $u = 1, 2, \ldots, z$, if and only if at least its $m$ neighboring assets out of $n$ its assets arranged in a sequence of $A_1, A_2, \ldots, A_n$, are out of this safety state subset. The numbers $m$ and $n$ are called the consecutive "$m$ out of $n$: F" critical infrastructure safety structure shape parameters [42].

After denoting by

$$\boldsymbol{S}(t, u) = P(s(t) \geq u | s(0) = z) = P(T(u) > t), \ t \in< 0, \infty), u = 0, 1, \ldots, z, \tag{2.16}$$

the probability that the consecutive "$m$ out of $n$: F" critical infrastructure is in the safety state subset $\{u, u+1, \ldots, z\}$, $u = 1, 2, \ldots, z$, at the moment $t$, $t \in< 0, \infty)$, while it was in the safety state $z$ at the moment $t = 0$ and by

$$\boldsymbol{F}(t, u) = 1 - \boldsymbol{S}(t, u) = P(T(u) \leq t), \ t \in< 0, \infty), u = 1, 2, \ldots, z, \tag{2.17}$$

the distribution function of the lifetime $T(u)$, $u = 1, 2, \ldots, z$, of this critical infrastructure in the safety state subset $\{u, u+1, \ldots, z\}$, $u = 1, 2, \ldots, z$, while it was in the safety state $z$ at the moment $t = 0$, we conclude that the safety function of the consecutive "$m$ out of $n$: F" critical infrastructure is given by the vector

$$\boldsymbol{S}(t, \cdot) = [1, \boldsymbol{S}(t, 1), \boldsymbol{S}(t, 2), \ldots, \boldsymbol{S}(t, z)] \ t \in< 0, \infty), \tag{2.18}$$

with the coordinates

$$\boldsymbol{S}(t, u) = \boldsymbol{S}_n(t, u), t \in< 0, \infty), u = 1, 2, \ldots, z,$$

where $\boldsymbol{S}_n(t, u)$, $u = 1, 2, \ldots, z$, are given by the following recurrent formula [37, 40, 42]

$$\boldsymbol{S}_e(t, u) = \begin{cases} 1 - \prod_{i=1}^{e} F_i(t, u), & \text{for } m = e, \\ \boldsymbol{S}_e(t, u)\boldsymbol{S}_{e-1}(t, u) + \sum_{i=1}^{m-1} S_{e-i}(t, u)\boldsymbol{S}_{e-i-1}(t, u) & \text{for } m < e, \\ \cdot \prod_{j=e-i+1}^{n} F_j(t, u), & \end{cases} \tag{2.19}$$

for $t \in< 0, \infty)$, $u = 1, 2, \ldots, z$, $m = 1, 2, \ldots, e$, $e = 1, 2, \ldots n$.

Other basic critical infrastructure safety structures with assets degrading in time are series–parallel, parallel–series, series- "$m$ out of $k$", "$m_i$ out of $l_i$"-series, series-consecutive "$m$ out of $k$: F" and consecutive "$m_i$ out of $l_i$: F"-series.

To define them, we assume that [42]:

- $k$ is the number of the critical infrastructure subsystems;

- $l_i$, $i = 1, 2, \ldots, k$, are the numbers of assets in the subsystems;
- $A_{ij}$, $i = 1, 2, \ldots, k$, $j = 1, 2, \ldots, l_i$, $k, l_1, l_2, \ldots, l_k \in N$, are assets of a critical infrastructure;
- the critical infrastructure and all assets $A_{ij}$, $i = 1, 2, \ldots, k$, $j = 1, 2, \ldots, l_i$, have the same safety state set $\{0, 1, \ldots, z\}$, $z \geq 1$;
- $T_{ij}(u)$, $i = 1, 2, \ldots, k$, $j = 1, 2, \ldots, l_i$, $k, l_1, l_2, \ldots, l_k \in N$, are independent random variables representing the lifetimes of the assets $A_{ij}$, $i = 1, 2, \ldots, k$, $j = 1, 2, \ldots, l_i$, in the safety state subset $\{u, u+1, \ldots, z\}$ while they were in the safety state $z$ at the moment $t = 0$;
- $s_{ij}(t)$, $t \in< 0, \infty)$, $i = 1, 2, \ldots, k$, $j = 1, 2, \ldots, l_i$, is an asset $A_{ij}$, $i = 1, 2, \ldots, k$, $j = 1, 2, \ldots, l_i$, safety state at the moment $t$, $t \in< 0, \infty)$, while they were in the safety state $z$ at the moment $t = 0$.

**Definition 2.5** A vector [42]

$$S_{ij}(t, \cdot) = \left[1, S_{ij}(t, 1), S_{ij}(t, 2), \ldots, S_{ij}(t, z)\right], t \in< 0, \infty)$$
$$i = 1, 2, \ldots, k,\ j = 1, 2, \ldots, l_i, \tag{2.20}$$

where

$$S_{ij}(t, u) = P\left(s_{ij}(t) \geq u | s_{ij}(0) = z\right) = P\left(T_{ij}(u) > t\right),\ t \in< 0, \infty),$$
$$u = 1, 2, \ldots, z, \quad i = 1, 2, \ldots, k, \quad j = 1, 2, \ldots, l_i, \tag{2.21}$$

is the probability that the asset $A_{ij}$, $i = 1, 2, \ldots, k$, $j = 1, 2, \ldots, l_i$, is in the safety state subset $\{u, u+1, \ldots, z\}$, $u = 1, 2, \ldots, z$, at the moment $t$, $t \in< 0, \infty)$, while it was in the safety state $z$ at the moment $t = 0$, is called the safety function of a asset $A_{ij}$, $i = 1, 2, \ldots, k$, $j = 1, 2, \ldots, l_i$,

The safety functions $S_{ij}(t, u)$, $t \in< 0, \infty)$, $u = 1, 2, \ldots, z$, $i = 1, 2, \ldots, k$, $j = 1, 2, \ldots, l_i$, defined by (2.21) are called the coordinates of the asset $A_{ij}$, $i = 1, 2, \ldots, k$, $j = 1, 2, \ldots, l_i$, safety function $S_{ij}(t, \cdot)$, $t \in< 0, \infty)$, $i = 1, 2, \ldots, k$, $j = 1, 2, \ldots, l_i$, given by (2.20). Thus, the relationship between the distribution function $F_{ij}(t, u)$, $t \in< 0, \infty)$, $u = 1, 2, \ldots, z$, $i = 1, 2, \ldots, k$, $j = 1, 2, \ldots, l_i$, of the asset $A_{ij}$, $i = 1, 2, \ldots, k$, $j = 1, 2, \ldots, l_i$, lifetime $T_{ij}(u)$, $u = 1, 2, \ldots, z$, $i = 1, 2, \ldots, k$, $j = 1, 2, \ldots, l_i$, in the safety state subset $\{u, u+1, \ldots, z\}$, $u = 1, 2, \ldots, z$, and the coordinate $S_{ij}(t, u)$, $t \in< 0, \infty)$, $u = 1, 2, \ldots, z$, $i = 1, 2, \ldots, k$, $j = 1, 2, \ldots, l_i$, of its safety function is given by

$$F_{ij}(t, u) = P\left(T_{ij}(u) \leq t\right) = 1 - P\left(T_{ij}(u) > t\right) = 1 - S_{ij}(t, u),\ t \in< 0, \infty),$$
$$u = 1, 2, \ldots, z, i = 1, 2, \ldots, k,\ j = 1, 2, \ldots, l_i. \tag{2.22}$$

**Definition 2.6** A critical infrastructure is called series–parallel if its lifetime $T(u)$ in the state subset $\{u, u+1, \ldots, z\}$ is given by [42]

$$T(u) = \max_{1 \leq i \leq k}\left\{\min_{1 \leq j \leq l_i}\left\{T_{ij}(u)\right\}\right\}, u = 1, 2, \ldots, z.$$

The above definition means that the series–parallel critical infrastructure composed of $k$ series critical infrastructures is in the safety state subset $\{u, u+1, \ldots, z\}$, $u = 1, 2, \ldots, z$, if and only if at least one out of its $k$ series critical infrastructures is in this safety state subset. In this definition, $l_i$, $i = 1, 2, \ldots, k$, denote the numbers of assets in the series critical infrastructures. The numbers $k$ and $l_1, l_2, \ldots, l_k$ are called the critical infrastructure safety structure shape parameters. Joining the justification for the safety structure schemes of the series critical infrastructure and the parallel critical infrastructure leads to the scheme of a series–parallel critical infrastructure safety structure presented in Fig. 2.8 [42].

The safety function of the series–parallel critical infrastructure is given by the vector [37, 40, 42]

$$\boldsymbol{S}(t, \cdot) = [1, \boldsymbol{S}(t, 1), \boldsymbol{S}(t, 2), \ldots, \boldsymbol{S}(t, z)], t \in < 0, \infty), \quad (2.23)$$

with the coordinates

$$\boldsymbol{S}(t, u) = 1 - \prod_{k}^{i=1} [1 - \prod_{l_i}^{j=1} S_{ij}(t, u)], t \in < 0, \infty, u = 1, 2, \ldots, z, \quad (2.24)$$

where $k$ is the number of series critical infrastructures linked in parallel and $l_i$, $i = 1, 2, \ldots, k$, are the numbers of assets in the series critical infrastructures.

**Definition 2.7** A critical infrastructure is called parallel-series if its lifetime $T(u)$ in the safety state subset $\{u, u+1, \ldots, z\}$, $u = 1, 2, \ldots, z$, is given by [42]

$$T(u) = \min_{1 \le i \le k} \{ \max_{1 \le j \le_i} \{T_{ij}(u)\}\}, u = 1, 2, \ldots, z.$$

The above definition means that the parallel-series critical infrastructure composed of $k$ parallel critical infrastructures is in the safety state subset $\{u, u+1, \ldots, z\}$, $u = 1, 2, \ldots, z$, if and only if all its $k$ parallel critical infrastructures are in this safety state subset. In this definition $l_i$, $i = 1, 2, \ldots, k$, are the numbers of

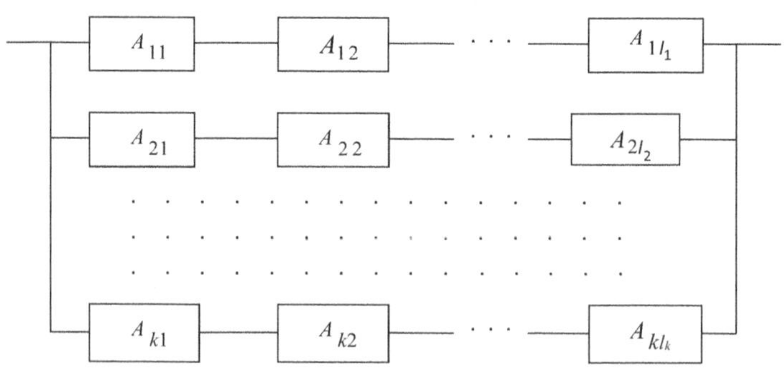


**Fig. 2.8** The scheme of a series–parallel critical infrastructure safety structure [42]

assets in the parallel critical infrastructures. The numbers $k$ and $l_1, l_2, \ldots, l_k$ are called the critical infrastructure safety structure shape parameters. Joining the justification for the safety structure schemes of the parallel critical infrastructure and the series critical infrastructure leads to the scheme of a parallel-series critical infrastructure safety structure presented in Fig. 2.9 [42].

The safety function of the parallel-series critical infrastructure is given by the vector [37, 40, 42]

$$\boldsymbol{S}(t,\cdot) = [1, \boldsymbol{S}(t,1), \boldsymbol{S}(t,2), \ldots, \boldsymbol{S}(t,z)], t \in <0, \infty), \tag{2.25}$$

with the coordinates

$$\boldsymbol{S}(t,u) = \prod_{k}^{i=1}[1 - \prod_{l_i}^{j=1} F_{ij}(t,u)], t \in <0, \infty), u = 1, 2, \ldots, z, \tag{2.26}$$

where $k$ is the number of its parallel critical infrastructures linked in series and $l_i$, $i = 1, 2, \ldots, k$, are the numbers of assets in the parallel critical infrastructures.

**Definition 2.8** A critical infrastructure is called series- "$m$ out of $k$" if its lifetime $T(u)$, $u = 1, 2, \ldots, z$, in the safety state subset $\{u, u+1, \ldots, z\}$, $u = 1, 2, \ldots, z$, is given by [42]

$$T(u) = T_{(k-m+1)}(u), u = 1, 2, \ldots, z, m = 1, 2, \ldots, k,$$

where $T_{(k-m+1)}(u)$, $u = 1, 2, \ldots, z$, $m = 1, 2, \ldots, k$, is the $(k - m + 1)$-th order statistic in the set of random variables

$$T_{ij}(u) = \min_{1 \le j \le l_i}\{T_{ij}(u)\}, u = 1, 2, \ldots, z, i = 1, 2, \ldots, k,$$

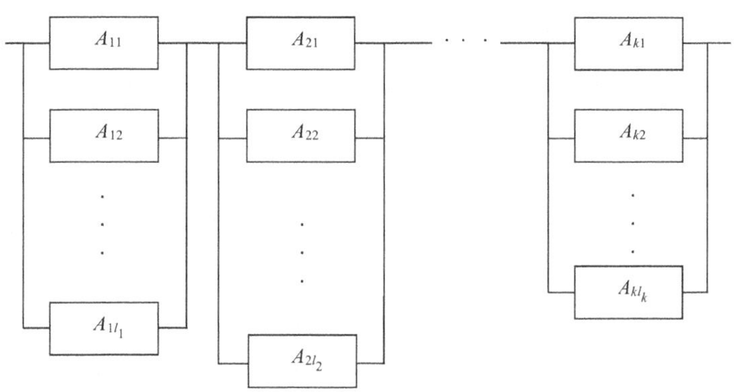


**Fig. 2.9** The scheme of a parallel-series critical infrastructure safety structure [42]

The above definition means that the series- "*m* out of *k*" critical infrastructure composed of *k* series critical infrastructures is in the safety state subset $\{u, u+1, \ldots, z\}$, $u = 1, 2, \ldots, z$, if and only if at least *m* out of its *k* series critical infrastructures are in this safety state subset. In this definition $l_i$, $i = 1, 2, \ldots, k$, denote the numbers of assets in the series critical infrastructures. The numbers *m*, *k* and $l_1, l_2, \ldots, l_k$ are called the critical infrastructure safety structure shape parameters. Joining the justification for the safety structure schemes of the series critical infrastructure and the "*m* out of *k*" critical infrastructure leads to the scheme of a series—"*m* out of *k*" critical infrastructure safety structure presented in Fig. 2.10, where $i_1, i_2, \ldots i_k \in \{1, 2, \ldots, k\} i_1$, and $i_a \neq i_b$ for $a \neq b$[42]

The safety function of the series—"*m* out of *k*" critical infrastructure is given either by the vector [37, 40, 42]

$$\boldsymbol{S}(t, \cdot) = [1, \boldsymbol{S}(t, 1), \boldsymbol{S}(t, 2), \ldots, \boldsymbol{S}(t, z)], t \in < 0, \infty), \tag{2.27}$$

with the coordinates

$$\boldsymbol{S}(t, u) = \sum_{\substack{r_1, r_2, \ldots, r_k \in \{0,1\} \\ m \le r_1 + r_2 + \ldots + r_k \le k}} \left[ \prod_{j=1}^{l_i} [S_{ij}(t, u)] \right]^{r_i} \left[ 1 - \prod_{j=1}^{l_i} S_{ij}(t, u) \right]^{1-r_i},$$

$$t \in < 0, \infty), u = 1, 2, \ldots, Z, \tag{2.28}$$

or by the vector

$$\boldsymbol{S}(t, \cdot) = [1, \boldsymbol{S}(t, 1), \boldsymbol{S}(t, 2), \ldots, \boldsymbol{S}(t, z)], \tag{2.29}$$

with the coordinates

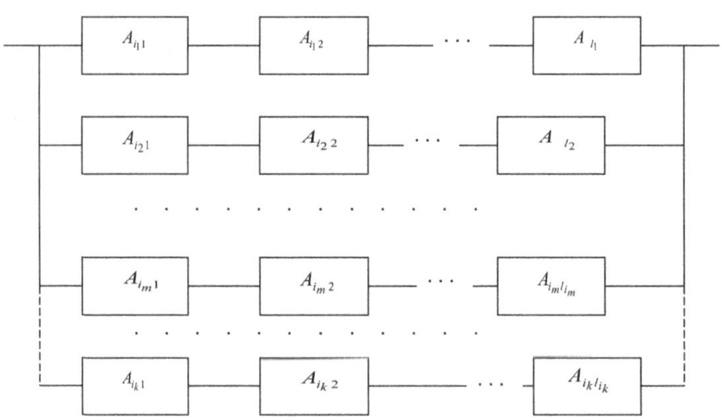


**Fig. 2.10** The scheme of a series—"*m* out of *k*" critical infrastructure safety structure [42]

$$S(t, u) = 1 - \sum_{\substack{r_1, r_2, \ldots, r_k \in \{0,1\} \\ 0 \le r_1 + r_2 + \ldots + r_k \le m-1}} \left[ \prod_{j=1}^{l_i} [S_{ij}(t, u)] \right]^{r_i} \left[ 1 - \prod_{j=1}^{l_i} S_{ij}(t, u) \right]^{1-r_k},$$

$$t \in < 0, \infty), u = 1, 2, \ldots, z, \tag{2.30}$$

**Definition 2.9** A critical infrastructure is called "$m_i$ out of $l_i$"-series if its lifetime $T(u)$ in the safety state subset $\{u, u + 1, \ldots, z\}$, $u = 1, 2, \ldots, z$, is given by [42],

$$T(u) = \min_{1 \le i \le k} T_{(l_i - m_i + 1)}(u), u = 1, 2, \ldots, z, m_i = 1, 2, \ldots, l_i, i = 1, 2, \ldots, k,$$

where $T_{(l_i - m_i + 1)}(u)$, $u = 1, 2, \ldots, z$, $m_i = 1, 2, \ldots, l_i$, $i = 1, 2, \ldots, k$, is the $(l_i - m_i + 1)$-th order statistic in the set of random variables

$$T_{i1}(u), T_{i2}(u), \ldots, T_{il_i}(u), u = 1, 2, \ldots, z, i = 1, 2, \ldots, k.$$

The above definition means that the "$m_i$ out of $l_i$"-series critical infrastructure composed of $k$ critical infrastructures that are "$m_i$ out of $l_i$" critical infrastructures is in the safety state subset $\{u, u+1, \ldots, z\}$, $u = 1, 2, \ldots, z$, if all its "$m_i$ out of $l_i$" critical infrastructures are in this safety state subset. In this definition $l_i$, $i = 1, 2, \ldots, k$, are the numbers of assets in the "$m_i$ out of $l_i$" critical infrastructures. The numbers $k$, $m_1, m_2, \ldots, m_k$ and $l_1, l_2, \ldots, l_k$ are called the critical infrastructure safety structure shape parameters. Joining the justification for the safety structure schemes of the "$m_i$ out of $l_i$" critical infrastructure and the series critical infrastructure schemes leads to the scheme of a of an "$m_i$ out of $l_i$"-series critical infrastructure safety structure presented in Fig. 2.11, where $j_1, j_2, \ldots, j_{l_i} \in \{1, 2, \ldots, l_i\}$, for $i = 1, 2, \ldots, k$ and

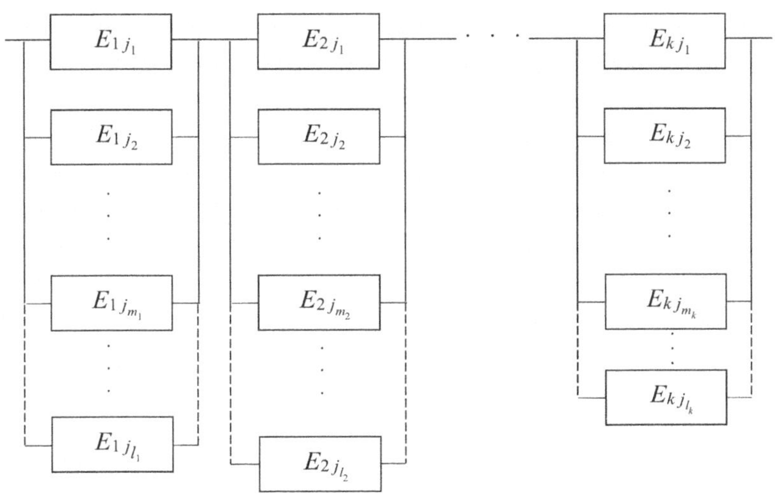


**Fig. 2.11** The scheme of an "$m_i$ out of $l_i$"-series critical infrastructure safety structure [42]

$i_a \neq i_b$ for $a \neq b$, [42].

The safety function of the "$m_i$ out of $l_i$"-series critical infrastructure is given either by the vector [37, 40, 42]

$$S(t, \cdot) = [1, S(t, 1), S(t, 2), \ldots, S(t, z)], t \in < 0, \infty), \tag{2.31}$$

where

$$S(t, u) = \prod_{i=1}^{k} \left[ \sum_{\substack{r_1, r_2, \ldots, r_{l_i} \in \{0,1\} \\ m_i \leq r_1 + r_2 + \ldots + r_{l_i} \leq l_i}} [S_{ij}(t, u)]^{r_j} [1 - S_{ij}(t, u)]^{1-r_j} \right],$$
$$t \in < 0, \infty), \; u = 1, 2, \ldots, z, \tag{2.32}$$

or by the vector

$$S(t, \cdot) = [1, S(t, 1), S(t, 2), \ldots, S(t, z)], \tag{2.33}$$

where

$$S(t, u) = \prod_{i=1}^{k} \left[ 1 - \sum_{\substack{r_1, r_2, \ldots, r_{l_i} \in \{0,1\} \\ 0 \leq r_1 + r_2 + \ldots + r_{l_i} \leq m_i - 1}} \prod_{j=1}^{l_i} [S_{ij}(t, u)]^{r_j} [1 - S_{ij}(t, u)]^{1-r_j} \right],$$
$$t \in < 0, \infty), u = 1, 2, \ldots, z, \tag{2.34}$$

To define a series-consecutive "$m$ out of $k$: F" critical infrastructure, we assume that assets $A_{i1}, A_{i2}, \ldots, A_{ili}$, $i = 1, 2, \ldots, k$, create a series critical infrastructure $S_i$, $i = 1, 2, \ldots, k$, and that these subsystems are arranged in a sequence $S_1, S_2, \ldots, S_k$ [42].

**Definition 2.10** A critical infrastructure is called series-consecutive "$m$ out of $k$: F" if it is out of the safety state subset $\{u, u + 1, \ldots, z\}$, $u = 1, 2, \ldots, z$, if and only if at least its $m$ neighboring series critical infrastructures out of $k$ its series critical infrastructures arranged in a sequence $S_1, S_2, \ldots, S_k$, are out of this safety state subset. The numbers $m$, $k$ and $l_i, l_2, \ldots, l_k$ are called the critical infrastructure safety structure shape parameters.

According to the above definition and formulae (2.5), (2.6) the coordinates of the safety function of the critical infrastructures $S_i$, $i = 1, 2, \ldots, k$, are given by [42]

$$S_{il_i}(t, u) = \prod_{j=1}^{l_i} S_{ij}(t, u), \; t \in < 0, \infty), u = 1_2 2, \ldots, z, i = 1, 2, \ldots, k, \tag{2.35}$$

and its lifetime distribution functions are given by

$$F_{il_i}(t,u) = 1 - S_{il_i}(t,u) = 1 - \prod_{j=1}^{l_i} S_{ij}(t,u), t \in < 0, \infty),$$
$$u = 1, 2, \ldots, z, i = 1, 2, \ldots, k. \tag{2.36}$$

Hence, considering (2.18), (2.19), the safety function of the series-consecutive "$m$ out of $k$: F" critical infrastructure is given by the vector [42]

$$S(t, \cdot) = [1, S(t,1), S(t,2), \ldots, S(t,z)], t \in < 0, \infty), \tag{2.37}$$

with the coordinates

$$S(t,u) = S_k(t,u), t \in < 0, \infty), u = 1, 2, \ldots, z,$$

where $S_k(t, u)$, $t \in < 0, \infty)$,, $u = 1, 2, \ldots, z$, are given by the following recurrent formula

$$S_e(t,u) = \begin{cases} 1 - \prod_{i=1}^{e} [1 - \prod_{j=1}^{l_i} S_{ij}(t,u)], & \text{for } m = e, \\ \left[\prod_{j=1}^{l_e} S_{ej}(t,u)\right] S_{e-1}(t,u) + \sum_{j=1}^{m-1} \left[\prod_{v=1}^{l_{e-j}} S_{e-j,v}(t,u)\right] S_{e-j-1}(t,u) & \\ \cdot \prod_{i=e-j+1}^{k} \left[1 - \prod_{v=1}^{l_i} S_{iv}(t,u)\right], & \text{for } m < e, \end{cases} \tag{2.38}$$

for $t \in < 0, \infty)$, $u = 1, 2, \ldots, z$, $m = 1, 2, \ldots, e$, $e = 1, 2, \ldots, k$.

To define a consecutive "$m$ out of $l$: F"-series critical infrastructure, we assume that assets $A_{i1}, A_{i2}, \ldots, A_{il_i}$, $i = 1, 2, \ldots, k$, create a consecutive "$m_i$ out of $l_i$: F" critical infrastructures $S_i$, $i = 1, 2, \ldots, k$, and that these critical infrastructures $S_1, S_2, \ldots, S_k$, create a series critical infrastructure [42].

**Definition 2.11** A critical infrastructure is called a consecutive "$m_i$ out of $l_i$: F"-series critical infrastructure if it is out of the safety state subset $\{u, u+1, \ldots, z\}$ if and only if at least one of its consecutive "$m_i$ out of $l_i$: F" critical infrastructures $S_i$, $i = 1, 2, \ldots, k$, is out of this safety state subset. The numbers $k$ and $m_i$, $l_i$, $i = 1, 2, \ldots, k$, are called the critical infrastructure structure shape parameters.

According to the above definition and formulae (2.18), (2.19), the safety function of the Z subsystem $S_i$, $i = 1, 2, \ldots, k$, is given by the vector [42]

$$S_{i,e_i}(t, \cdot) = [1, S_i(t,1) S_i t, 2), \ldots, S_i t, z)], t \in < 0, \infty), i = 1, 2, \ldots, k, \tag{2.39}$$

with the coordinates

$$S_i(t, u) = S_{il_i}(t, u), \text{ for } t \in <0, \infty), u = 1, 2, \ldots, z, i = 1, 2, \ldots, k,$$

where $S_{il_i}(t, u)$, $t \in <0, \infty)$, $u = 1, 2, \ldots, z$, $i = 1, 2, \ldots, k$, are given by the following recurrent formulae

$$S_{iei}(t, u) = \begin{cases} 1 - \prod_{j=1}^{e_i} F_{ij}(t, u), & \text{for } m_i = e_i, \\ \left[S_{ie_i}(t, u)\boldsymbol{S}_{i,e_i-1}(t, u)\right] + \sum_{j=1}^{m_i-1} S_{i,e_i-j}(t, u)\boldsymbol{S}_{i,e_i-j-1}(t, u) & \text{for } m_i < e_i, \\ \cdot \prod_{v=e_i-j+1}^{l_i} F_{iv}(t, u), & \end{cases} \tag{2.40}$$

for $t \in <0, \infty)$, $u = 1, 2, \ldots, z$, $m_i = 1, 2, \ldots, e_i$, $e_i = 1, 2, \ldots, l_i$, $i = 1, 2, \ldots, k$.

Hence, the safety function of the consecutive "$m_i$ out of $l_i$: F"-series critical infrastructure is given by the vector

$$\boldsymbol{S}(t, \cdot) = [1, \boldsymbol{S}(t, 1), \boldsymbol{S}(t, 2), \ldots, \boldsymbol{S}(t, z)], t \in <0, \infty), \tag{2.41}$$

with the coordinates

$$\boldsymbol{S}(t, u) = \prod_{k}^{i=1} \boldsymbol{S}_{i,l_i}(t, u) \text{ for } t \in <0, \infty), u = 1, 2, \ldots, z, \tag{2.42}$$

where $\boldsymbol{S}_{i,l_i}(t, u)$, $t \in <0, \infty)$, $u = 1, 2, \ldots, z$, $i = 1, 2, \ldots, k$, are given by the recurrent formula (2.40).

Further, for the critical infrastructures composed of assets having piecewise exponential safety functions, we formulate the particular cases of the above fixed formulae for the safety functions of the considered critical infrastructure structures in the form of the following proposition.

**Proposition 2.1** If assets of the critical infrastructure have the piecewise exponential safety functions.

$$S_i(t, \cdot) = [1, S_i(t, 1), \ldots, S_i(t, z)], t \in <0, \infty), i = 1, 2, \ldots, n, \tag{2.43}$$

where

$$S_i(t, u) = \exp[-\lambda_i(u)t], \text{ for } t \in <0, \infty), \lambda_i(u) > 0,$$
$$u = 1, 2, \ldots, z, i = 1, 2, \ldots, n, \tag{2.44}$$

in the case of the series, parallel, "$m$ out of $n$", consecutive "$m$ out of $n$" critical infrastructures and respectively

$$S_{ij}(t,\cdot) = [1, S_{ij}(t,1), S_{ij}(t,2), \ldots, S_{ij}(t,z)],$$
$$t \in < 0, \infty), i = 1, 2, \ldots, n, \; j = 1, 2, \ldots, l_i, \tag{2.45}$$

where

$$S_{ij}(t,u) = \exp[-\lambda_{ij}(u)t], \text{ for } t \in < 0, \infty), \lambda_{ij}(u) > 0,$$
$$u = 1, 2, \ldots, z, i = 1, 2, \ldots, n, \; j = 1, 2, \ldots, l_i, \tag{2.46}$$

in the case of the series–parallel, parallel–series, series- "$m$ out of $k$", "$m_i$ out of $l_i$"-series, series-consecutive "$m$ out of $k$" and consecutive "$m_i$ out of $l_i$"-series critical infrastructures, then the critical infrastructure safety function is given by the vector [42]:

(i) for a series critical infrastructure

$$\boldsymbol{S}(t,\cdot) = [1, \boldsymbol{S}(t,1), \boldsymbol{S}(t,2), \ldots, \boldsymbol{S}(t,z)], t \in < 0, \infty), \tag{2.47}$$

where

$$\boldsymbol{S}(t,u) = \exp\left[-\sum_{n}^{i=1} \lambda_i(u)t\right] \text{ for } t \in < 0, \infty), u = 1, 2, \ldots, z; \tag{2.48}$$

(ii) for a parallel critical infrastructure

$$\boldsymbol{S}(t,\cdot) = [1, \boldsymbol{S}(t,1), \boldsymbol{S}(t,2), \ldots, \boldsymbol{S}(t,z)], t \in < 0, \infty), \tag{2.49}$$

where

$$\boldsymbol{S}(t,u) = 1 - \prod_{n}^{i=1} [1 - \exp[-\lambda_i(u)t]] \text{ for } t \in < 0, \infty), u = 1, 2, \ldots, z; \tag{2.50}$$

(iii) for a "$m$ out of $n$" critical infrastructure

$$S(t,\cdot) = [1, S(t,1), S(t,2), \ldots, S(t,z)], t \in < 0, \infty), \tag{2.51}$$

where

$$\boldsymbol{S}(t,u) = \sum_{\substack{r_1, r_2, \ldots, r_n \in \{0,1\} \\ m \le r_1 + r_2 + \ldots + r_n \le n}}^{1} [\exp[-\lambda_i(u)t]]^{r_i} [1 - \exp[-\lambda_i(u)t]]^{1-r_i}$$
$$\text{for } t \in < 0, \infty), u = 1, 2, \ldots, z, \tag{2.52}$$

or

$$\boldsymbol{S}(t,\cdot) = [1, \boldsymbol{S}(t,1), \boldsymbol{S}(t,2), \ldots, \boldsymbol{S}(t,z)], t \in <0, \infty), \tag{2.53}$$

where

$$\boldsymbol{S}(t,u) = 1 - \sum_{\substack{r_1, r_2, \ldots, r_n \in \{0,1\} \\ 0 \le r_1 + r_2 + \ldots + r_n \le m-1}}^{1} [\exp[-\lambda_i(u)t]]^{r_i} [1 - \exp[-\lambda_i(u)t]]^{1-r_i}$$

$$\text{for } t \in <0, \infty), u = 1, 2, \ldots, z; \tag{2.54}$$

(iv) for a consecutive "*m* out of *n*: F" critical infrastructure

$$S(t,\cdot) = [1, S(t,1), S(t,2), \ldots, S(t,z)], \ t \in <0, \infty), \tag{2.55}$$

where

$$\boldsymbol{S}(t,u) = \boldsymbol{S}_s(t,u)t \in <0, \infty), u = 1, 2, \ldots, z,$$

are given by the following recurrent formula

$$S_e(t,u) = \begin{cases} 1 - \prod_{i=1}^{e} [1 - \exp[-\lambda_i(u)t]] & \text{for } m = e, \\ \exp[-\lambda_e(u)t] S_{e-1}(t,u) + \sum_{i=1}^{m-1} \exp[-\lambda_{e-i}(u)t] S_{e-i-1}(t,u) & \text{for } m < e, \\ \cdot \prod_{j=e-i+1}^{e} [1 - \exp[-\lambda_j(u)t]], & \end{cases} \tag{2.56}$$

for $t \in <0,\infty)$, $u = 1, 2, \ldots, z$, $m = 1, 2, \ldots e$, $e = 1, 2, \ldots n$;

(v) for a series–parallel critical infrastructure

$$\boldsymbol{S}(t,\cdot) = [1, \boldsymbol{S}(t,1), \boldsymbol{S}(t,2), \ldots, \boldsymbol{S}(t,z)], t \in <0, \infty), \tag{2.57}$$

where

$$\boldsymbol{S}(t,u) = 1 - \prod_{k}^{i=1} [1 - \exp[-\sum_{l_i}^{j=1} \lambda_{ij}(u)t]] \text{ for } t \in <0, \infty), u = 1, 2, \ldots, Z, \tag{2.58}$$

(vi) for a parallel-series critical infrastructure

$$\boldsymbol{S}(t,\cdot) = [1, \boldsymbol{S}(t,1), \boldsymbol{S}(t,2), \ldots, \boldsymbol{S}(t,z)], t \in <0, \infty), \tag{2.59}$$

where

$$S(t,u) = \prod_{k}^{i=1}\left[1 - \prod_{l_i}^{j=1}[1-\exp[-\lambda_{ij}(u)t]]\right] \text{ for } t \in <0,\infty),\ u = 1,2,\ldots,z; \tag{2.60}$$

(vii) for a series-"$m$ out of $k$" critical infrastructure

$$S(t,\cdot) = [1, S(t,1), S(t,2),\ldots,S(t,z)],\ t \in <0,\infty), \tag{2.61}$$

where

$$S(t,u) = \sum_{\substack{r_1,r_2,\ldots,r_k \in \{0,1\} \\ m \le r_1+r_2+\ldots+r_k \le k}} [\exp[-\sum_{j=1}^{l_i}\lambda_{ij}(u)t]]^{r_i}[1-\exp[-\sum_{j=1}^{l_i}\lambda_{ij}(u)t]]^{1-r_i}$$

$$\text{for } t \in <0,\infty),\ u = 1,2,\ldots,Z, \tag{2.62}$$

or

$$S(t,\cdot) = [1, S(t,1), S(t,2),\ldots,S(t,z)],\ t \in <0,\infty), \tag{2.63}$$

where

$$S(t,u) = 1 - \sum_{\substack{r_1,r_2,\ldots,r_k \in \{0,1\} \\ 0 \le r_1+r_2+\ldots+r_k \le m-1}} [\exp\left[-\sum_{j=1}^{l_i}\lambda_{ij}(u)t\right]^{r_i}[1-\exp\left[-\sum_{j=1}^{l_i}\lambda_{ij}(u)t\right]^{1-r_i}$$

$$\text{for } t \in <0,\infty),\ u = 1,2,\ldots,Z; \tag{2.64}$$

(viii) for a "$m_i$ out of $l_i$"-series critical infrastructure

$$S(t,\cdot) = [1, S(t,1), S(t,2),\ldots,S(t,z)],\ t \in <0,\infty), \tag{2.65}$$

where

$$S(t,u) = \prod_{i=1}^{k}\left[\sum_{\substack{r_1,r_2,\ldots,r_{l_i} \in \{0,1\} \\ m_i \le r_1+r_2+\ldots+r_{l_i} \le l_i}} [\exp[-\lambda_{ij}(u)t]\right]^{r_j}\left[1-\exp\left[-\lambda_{ij}(u)t\right]\right]^{1-r_j}$$

$$\text{for } t \in <0,\infty),\ u = 1,2,\ldots,z, \tag{2.66}$$

or

$$S(t,\cdot) = [1, S(t,1), S(t,2),\ldots,S(t,z)],\ t \in <0,\infty), \tag{2.67}$$

where

$$S(t,u)=\prod_{i=1}^{k}[1-\sum_{\substack{r_1,r_2,\ldots,r_{l_i}\in\{0,1\}\\0\le r_1+r_2+\ldots+r_{l_i}\le m_i-1}}[\exp[-\lambda_{ij}(u)t]^{r_j}[1-\exp[-\lambda_{ij}(u)t]]^{1-r_j}]$$

$$\text{for } t\in<0,\infty),\ u=1,2,\ldots,z; \tag{2.68}$$

(ix) for a series-consecutive "$m$ out of $k$: F" critical infrastructure

$$\boldsymbol{S}(t,\cdot)=[1,\boldsymbol{S}(t,1),\boldsymbol{S}(t,2),\ldots,\boldsymbol{S}(t,z)],\ t\in<0,\infty), \tag{2.69}$$

where

$$\boldsymbol{S}(t,u)=\boldsymbol{S}_e(t,u)$$

and $\boldsymbol{S}_e(t,u)$ are given by the following recurrent formula

$$S_e(t,u)=\begin{cases}1-\prod_{i=1}^{e}\left[1-\exp\left[-\sum_{j=1}^{l_i}\lambda_{ij}(u)t\right]\right] & \text{for } m=e,\\ \exp[-\sum_{j=1}^{l_e}\lambda_{ej}(u)t]S_{e-1}(t,u)+\sum_{j=1}^{m-1}\left[\exp\left[-\sum_{v=1}^{l_{e-j}}\lambda_{e-j,v}(u)t\right]\right]S_{e-j-1}(t,u) & \text{for } m<e,\\ \cdot\prod_{i=e-j+1}^{e}\left[1-\exp\left[-\sum_{v=1}^{l_i}\lambda_{iv}(u)t\right]\right], & \end{cases} \tag{2.70}$$

for $t\in\ <0,\infty)$, $u=1,2,\ldots,z$, $m=1,2,\ldots,e$, $e=1,2,\ldots k$;

(x) nnnnnfor a consecutive "$m_i$ out of $l_i$: F"-series critical infrastructure

$$\boldsymbol{S}(t,\cdot)=[1,\boldsymbol{S}(t,1),\boldsymbol{S}(t,2),\ldots,\boldsymbol{S}(t,z)],\ t\in<0,\infty), \tag{2.71}$$

where

$$\boldsymbol{S}(t,u)=\prod_{i=1}^{k}\boldsymbol{S}_{il_i}(t,u),\ \text{ for } t\in<0,\infty),\ u=1,2,\ldots,z, \tag{2.72}$$

and $\boldsymbol{S}_{il_i}(t,u)$, $u=1,2,\ldots,z$, $i=1,2,\ldots,k$, are given by the following recurrent formula

$$\boldsymbol{S}_{i,e_i}(t,u)=\begin{cases}1-\prod_{j=1}^{e_i}[1-\exp[-\lambda_{ij}(u)t)], & \text{for } m_i=e_i,\\ \exp[-\lambda_{ie_i}(u)t]\boldsymbol{S}_{i,e_i-1}(t,u)+\sum_{j=1}^{m_i-1}\exp[-\lambda_{ie_i-j}(u)t]\boldsymbol{S}_{e_i-j-1}(t,u) & \text{for } m_i<e_i,\\ \prod_{v=e_i-j+1}^{e_i}[1-\exp[-\lambda_{iv}(u)t]]], & \end{cases} \tag{2.73}$$

for $t \in <0, \infty)$, $u = 1, 2, \ldots, Z$, $m_i = 1, 2, \ldots, e_i$, $e_i = 1, 2, \ldots, l_i$.

## 2.3 General Scheme of Operation and Climate-Weather Influence on Critical Infrastructure Safety Modelling

Most real complex technical systems are strongly influenced by their, changing in time, operation conditions and climate-weather conditions at their operating areas. Time dependent interactions between the operation process, related to varying states of the climate-weather change process at the system operating area, and the system safety structure and its changing components/assets safety states, are evident features of most real technical systems, including critical infrastructures [38–46, 63]. The critical infrastructure is defined as a complex system in its operating environment that significant features are inside-system dependencies [7–11] and outside-system impacts [38, 40–42, 63]. In case of system degradation these dependencies have a significant destructive influence on health, safety, security, economics and social conditions of large human communities and territories [47].

The critical infrastructure may be affected by outside processes which have negative influence on its safety. By outside processes, we understand critical infrastructure operation process and the process of weather change in the area of operation. Therefore, critical infrastructure safety analysis, related to its operation process and climate-weather change process, has a great value in the industrial practice due to, often negative, impacts of these processes on the critical infrastructure safety and resilience [7–11, 18, 38, 41–46, 63].

The safety analysis of the critical infrastructure impacted by those processes is very complex. This problem can be solved by the multistate critical infrastructures safety modelling performed similarly to reliability modelling of multistate complex systems [40, 65–69], commonly used with the semi-Markov modelling [22, 23, 25, 26, 49] of the joint operation process and climate-weather change process [38, 63]. This approach leads to the construction of the joint general safety model of the critical infrastructure, impacted by the operation process and changing weather at its operating area. Such a model has been partly considered in [63] and it is significantly developed in this book.

This original and innovative general approach consists in combining of the critical infrastructure operation process model and the climate-weather change process model and constructing of one general joint safety model of the critical infrastructure impacted by its operation process and climate-weather process at its operating area. This approach can be basis for the formulation and development of the new solutions, which consider improvement and optimization of the safety of the critical infrastructure related to their operation processes and outside climate-weather change processes. A novel concept, which would induce further complexity to the approach, is an introduction of impacts of climate pressures in the interconnection between critical infrastructures [43–46]. New and innovative part of this approach lies in inclusion of linkages and dependencies, both internal and external, to critical infrastructures that are impacted by climate-weather hazards.

In order to organize efforts during seeking of solutions of formulated problems, the scheme of a general approach to safety and resilience analysis of critical infrastructure [43–46], including subsequent steps in the research activity, is presented in Fig. 2.12.

The content of the particular steps from the above scheme is concerned in:

- Critical Infrastructure Operation Process (CIOP) and Climate-Weather Change Process (C-WCOP), with modelling of critical infrastructure operation process and climate-weather change process in its operation area through defining parameters of these processes and giving the ways of their characteristics determination;
- Integration of CIOP and C-WCP, with creating of a joint model of critical infrastructure operation process and climate-weather change process at its operating area through defining the critical infrastructure operation process impacted by climate-weather change at its operating area, defining its parameters and giving the procedures of its characteristics determinations;
- Critical Infrastructure (free of outside impact) Safety (CIS) Modelling, with constructing of critical infrastructure multistate safety model independent of outside operation and climate-weather impacts;
- Critical Infrastructure Safety Indicators (SafIs), with introducing of practically useful critical infrastructure safety characteristics, called safety indicators;
- Integration of CIOP, C-WCP and CIOP&C-WCP Models 1–3 with CIS Model 0, with constructing critical infrastructure safety models, separately and jointly dependent on outside operation and climate-weather impacts;
- Critical Infrastructure Safety and Resilience Indicators (SafIs, ResIs), with introducing of practically useful safety characteristics of critical infrastructure impacted separately and jointly by outside operation and climate-weather conditions, called safety indicators and resilience indicators;
- Inventory of Critical Infrastructure Safety and Resilience Indicators, with creating of detailed list of safety indicators of critical infrastructure free of outside impacts and safety indicators and resilience indicators of critical infrastructure outside impacts;
- Models Application and Validation, with all proposed models and creating safety and resilience indicators application and validation in practice to port oil terminal critical infrastructure safety examination;
- Suggestion for Further Research on Critical Infrastructure Safety Examination, with possible future research and development of proposed models in creating practically important tools for critical infrastructure safety strengthening and optimization, as well as for examination and mitigation of critical infrastructure accident consequences and critical infrastructure business continuity analysis.

Thus, starting from the simplest, pure safety CIS Model 0, defined as a multistate ageing system without considering outside impacts, several functions and indicators are defined. Namely, the critical infrastructure and its assets' safety functions, the critical infrastructure mean values and variances of lifetimes in the safety state subsets

**Fig. 2.12** The scheme of general approach to modelling operation and climate-weather change influence on critical infrastructure safety and resilience

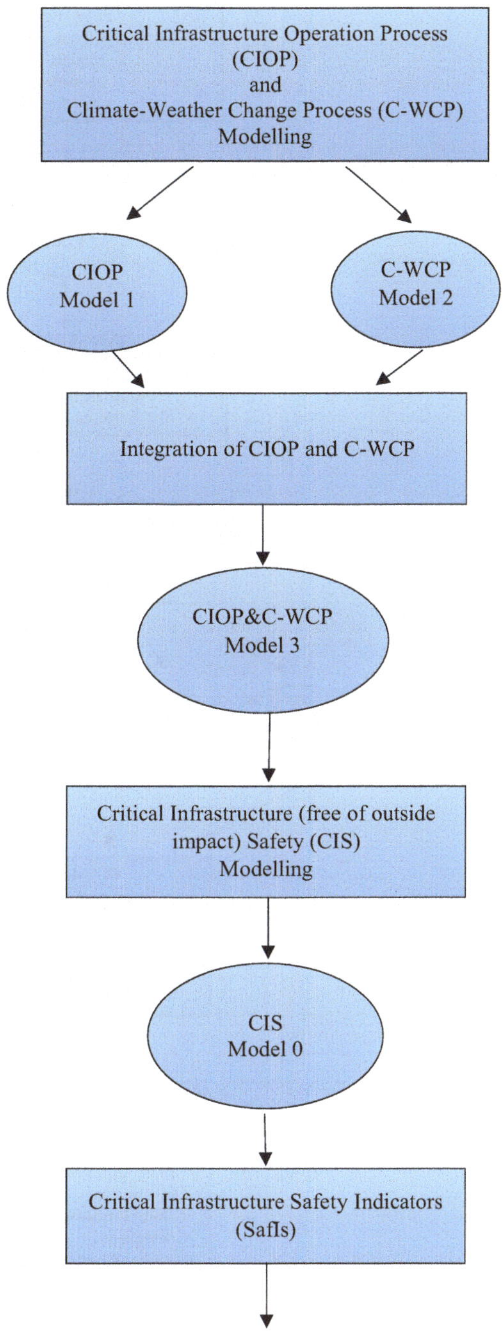

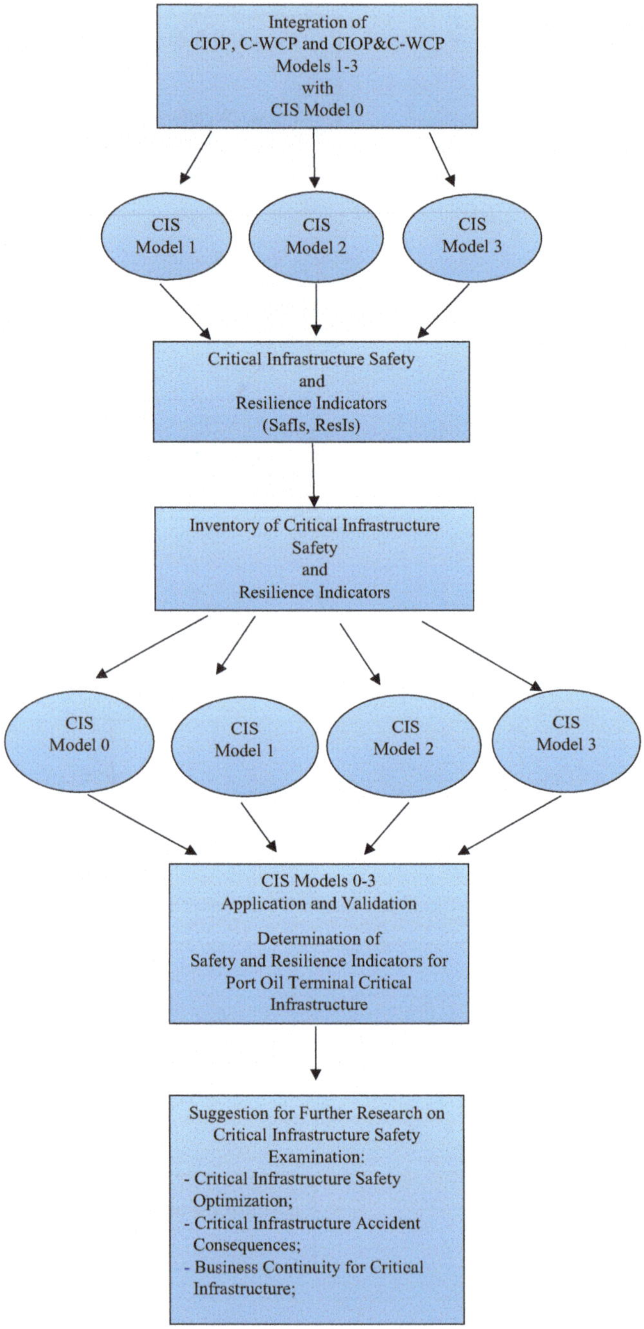


**Fig. 2.12** (continued)

and in the particular safety states, the critical infrastructure risk function, its fragility curve, the moment of exceeding by the critical infrastructure the critical safety state and its intensities of ageing/degrading, are introduced.

Next, the CIS Model 0 is combined with the critical infrastructure operation process CIOP Model 1 to create the integrated CIS Model 1, which task is safety modelling and prediction of critical infrastructure impacted by its operation process. In CIS Model 1 we define the critical infrastructure as a complex system in its operating environment that significant features are its operation impacts [40, 41, 63]. That safety model of a critical infrastructure, related to its operation process, links its multistate safety model and its operation process model in order to create the critical infrastructure operation impact safety model. Moreover, CIS Model 1 considers also variable safety structure and its components' safety parameters at different operation states [40, 41]. In this model, we introduce additional safety indicators, which are typical for the critical infrastructure and are related to its varying in time safety structures and its components' safety parameters. CIS Model 1 extends the set of safety indicators of CIS Model 0 by the components and critical infrastructure conditional intensities of ageing at particular operation states. CIS Model 1 also encompass conditional and unconditional coefficients of the operation process impact on the critical infrastructure intensities of ageing and the critical infrastructure coefficient of resilience to its operation process.

Further, an integrated safety CIS Model 2 of critical infrastructure safety is proposed. This critical infrastructure safety model is related to influence of the climate-weather change process in the critical infrastructure operating area on its safety. It is the integrated model of critical infrastructure safety, linking its multistate safety CIS Model 0 and the CIOP Model 1 of the climate-weather change process at its operating area, to create the critical infrastructure climate-weather impact safety model. The CIS Model 2 considers variable system's components safety parameters impacted by different climate-weather states. The conditional safety functions at the particular climate-weather states, the unconditional safety function and the risk function of the critical infrastructure at changing in time climate-weather conditions are defined. Other, practically significant, critical infrastructure safety indicators introduced in the CIS Model 2 are, its mean lifetime to the moment of exceeding a critical safety state, the moment when its risk function value exceeds the acceptable safety level, the intensities of ageing of the critical infrastructure related to the climate-weather change process at its operating area, coefficients of the impact of the climate-weather change process on the critical infrastructure and its components intensities of ageing and the critical infrastructure coefficient of resilience to climate-weather change process at its operating area.

Finally, the general critical infrastructure safety CIS Model 3 is proposed that simultaneously considers the operation process and the climate-weather change process influence on the safety of a critical infrastructure. It is a safety model of a critical infrastructure under the influence of the operation process related to climate-weather change at its operating area. It is an integrated model of a critical infrastructure safety, linking its multistate safety CIS Model 0 and the joint CIOP&C-WCP Model 3 of its operation process related to climate-weather change process at its operating area, what allows to create the critical infrastructure joint operation and climate-weather impact safety model. Thus, CIS Model 3 considers variable system safety structures and its components safety parameters, impacted by climate-weather states, at different operation states. The conditional safety functions at the operation and climate-weather states of the operation process related the climate-weather change, the unconditional safety function and the risk function of a critical infrastructure at changing in time operation and climate-weather conditions are defined. Other useful critical infrastructure safety indicators introduced in CIS Model 3 are, its mean lifetime up to the moment of exceeding a critical safety state and the moment when its risk function value exceeds the acceptable safety level, the intensities of ageing of the critical infrastructure and its components impacted by the operation process related to the climate-weather change process, coefficients of the operation process related to climate-weather change impact on the critical infrastructure and its components intensities of ageing and the critical infrastructure coefficient of resilience to operation process related to climate-weather change process at its operating area.

These all, above mentioned, safety indicators, proposed in CIS Models 0–3, can be used for any critical infrastructures with varying in time their safety structures and components/assets safety parameters, which are influenced by, changing in time, operation and climate-weather conditions at their operating areas.

The next step, that can be done to perform the tasks formulated in scheme items, is these models application and validation, what is practically realized in the book through the port oil terminal critical infrastructure examination in all chapters successively.

The path we should follow in our future research activity is to investigate and solve problems of improve safety and strengthening of resilience of critical infrastructure, impacted by operation and climate-weather change. This activity will lead to establishing of elaborate models of business continuity for critical infrastructure under operation and climate pressures, as well as to solving the critical infrastructure safety optimization and its degradation and accident consequences identification and mitigation [12].

All presented models are the basis for preparation of procedures, which are easy to use by the practitioners and operators of the critical infrastructures in their operation and safety analysis. The use of these procedures for real critical infrastructure is presented in details in all chapters of this book.

All created models, and procedures based on them, can be modified and developed for other problems of safety features of critical infrastructure analysis. In this context, modelling and prediction of critical infrastructure safety presented in this book, developed by considering inside dependences between the critical infrastructure assets [11], would be very important broadening to real practice in critical infrastructure safety examination to build the model considering commonly the critical infrastructure ageing, its inside dependences and outside impacts as an innovative general approach significant and breakthrough applications.

## 2.4 Critical Infrastructure Safety and Resilience Indicators

In the first step of the approach proposed in Sect. 2.3, in Chap. 3 we start with the simplest pure safety model CIS Model 0, without considering outside impacts. For the critical infrastructure (and its assets) following useful safety indicators are defined:

- the critical infrastructure safety function (SafI1);
- the critical infrastructure risk function (SafI2);
- the critical infrastructure fragility curve (SafI3);
- the mean value of the critical infrastructure lifetime up to the exceeding the critical safety state (SafI4);
- the standard deviation of the critical infrastructure lifetime up to the exceeding the critical safety state (SafI5);
- the moment of exceeding acceptable value of critical infrastructure risk function level (SafI6);
- the mean values of the critical infrastructure lifetimes in the safety state subsets (SafI7);
- the standard deviations of the critical infrastructure lifetimes in the safety state subsets (SafI8);
- the mean value of the critical infrastructure lifetimes in particular safety states (SafI9);
- the intensities of degradation (ageing) of the critical infrastructure / the intensities of critical infrastructure departure from the safety state subsets (SafI10).

In Chap. 4, in the second step of the proposed approach, the simplest safety model CIS Model 0 is combined with the critical infrastructure operation process model CIOP Model 1, in order to create a safety model CIS Model 1 of critical infrastructure related to its operating environment. Next, in Chap. 5, an impact model on critical infrastructure safety CIS Model 2 related to the climate-weather change process in its operating area is proposed. The most general safety impact model CIS Model 3 which consider jointly the operation process and climate-weather change process influence on the safety of a critical infrastructure is presented in Chapter 6. It is the integrated model of a critical infrastructure safety, linking its multistate safety model CIS Model 0 and the joint model CIOP&C-WCP of its operation process and the climate-weather change process in its operating area. These model consider variable safety structures of the critical infrastructure at different operation and climate-weather states, as well as safety parameters of critical infrastructure assets. For those models, the following safety indicators are respectively defined:

- the critical infrastructure safety function (SafI1);
- the critical infrastructure risk function (SafI2);
- the critical infrastructure fragility curve (SafI3);
- the mean value of the critical infrastructure lifetime up to the exceeding the critical safety state (SafI4);

- the standard deviation of the critical infrastructure lifetime up to the exceeding the critical safety state (SafI5);
- the moment of exceeding acceptable value of critical infrastructure risk function level (SafI6);
- the mean values of the critical infrastructure lifetimes in the safety state subsets (SafI7);
- the standard deviations of the critical infrastructure lifetimes in the safety state subsets (SafI8);
- the mean value of the critical infrastructure lifetimes in particular safety states (SafI9);
- the intensities of degradation (ageing) of the critical infrastructure / the intensities of critical infrastructure
- departure from the safety state subsets (SafI10).

These all safety indicators are defined, in general, for any critical infrastructures with varying in time their safety structures and their assets safety parameters influenced by changing in time operation conditions and climate-weather conditions at their operating areas.

We can make a next step in order to terminate methodological framework, for critical infrastructures with outside impacts to define the following critical infrastructure resilience indicators:

- the coefficients of operation process impact on the critical infrastructure intensities of degradation / the coefficients of operation process impact on critical infrastructure intensities of departure from the safety state subset (ResI1);
- the indicator of critical infrastructure resilience to operation process impact (ResI2);
- the coefficients of climate-weather change process impact on the critical infrastructure intensities of degradation / the coefficients of climate-weather change process impact on critical infrastructure intensities of departure from the safety state subset (ResI1);
- the indicator of critical infrastructure resilience to climate-weather change process impact (ResI2);
- the coefficients of operation process and climate-weather change process impact on the critical infrastructure intensities of degradation / the coefficients of operation process and climate-weather change process impact on critical infrastructure intensities of departure from the safety state subset (ResI1);
- the indicator of critical infrastructure resilience to operation process and climate-weather change process impact (ResI2).

All the proposed indicators and other safety and resilience tools can be validated through their practical application to the real critical infrastructures.

Further research activities could concentrate on investigating and solving of optimization problems for critical infrastructure safety. These research should include finding of optimal values of safety and resilience indictors, as well as analysis of resilience and strengthening of critical infrastructure to climate-weather change. This activity will result in elaboration of business continuity models for critical infrastructure under the operation and climate-weather pressures, cost-effectiveness analysis and modelling and critical infrastructure degradation and accident consequences analysis and mitigation.

## References

1. Abouammoh A, Al-Kadi M (1991) Component relevancy in multi-state reliability models. IEEE Trans Reliab 40:370–375
2. Amari SV, Misra RB (1997) Comment on: dynamic reliability analysis of coherent multistate systems. IEEE Trans Reliab 46:460–461
3. Aven T (1985) Reliability evaluation of multistate systems with multistate components. IEEE Trans Reliab 34:473–479
4. Aven T (1993) On performance measures for multistate monotone systems. Reliab Eng Sys Saf 41:259–266
5. Barlow RE, Proschan F (1975) Statistical theory of reliability and life testing. Probability Models, Holt Rinehart and Winston Inc, New York
6. Barlow SE, Wu AS (1978) Coherent systems with multi-state components. Math Oper Res 3(11):275–281
7. Blokus A (2018) Quantitative assessment of economic, social and environmental impacts of critical infrastructure disruptions. In: International conference on industrial engineering and engineering management IEEM 2018
8. Blokus-Roszkowska A, Bogalecka M, Kołowrocki K (2016) General methodology on the Baltic Sea critical infrastructure safety aspects—dictionary. J Pol Saf Reliab Assoc, Summer Safety Reliab Semin 7(2):105–128
9. Blokus-Roszkowska A, Dziula P, Kołowrocki K (2018) Safety and resilience of global baltic network of critical infrastructure networks related to cascading effects. J Pol Saf Reliab Assoc, Summer Safety Reliab Semin 9(2):79–86
10. Blokus-Roszkowska A, Guze S, Kołowrocki K, Soszyńska-Budny J (2016) Port critical infrastructure network. J Pol Saf Reliab Assoc, Summer Safety Reliab Semin 7(2):15–28
11. Blokus-Roszkowska A, Kolowrocki K (2017) Modeling dependencies in critical infrastructure networks. Safety and reliability—theory and applications. Proceedings and Monographs in Engineering, European Safety and Reliability Conference ESREL 2017:655–663
12. Bogalecka M (2019) Consequences of maritime critical infrastructure accidents—environmental impacts. Elsevier
13. Brausch A (1987) Calculation of critical importance for multi-state components. IEEE Trans Reliab 36:247–249
14. Kapur BRD, KC, (1999) Review and classification of reliability measures for multistate and continuum models. IEEE Trans 31:1117–1180
15. Butler D (1982) Bounding the reliability of multi-state systems. Oper Res 30:530–544
16. De Porcellinis S, Oliva G, Panzieri S, Setola R (2009) A holistic-reductionistic approach for modeling interdependencies. In: Palmer C, Shenoi S (eds) Critical infrastructure protection III, 3rd IFIP international conference on critical infrastructure protection (CIP-2009). Springer, USA, pp 215–227
17. Ding Y, Lisnianski A (2008) Fuzzy universal generating functions for multi-state system reliability assessment. Fuzzy Sets Syst. 159:307–324
18. Drzazga M, Kołowrocki K, Soszyńska-Budny J (2016) Methodology for oil pipeline critical infrastructures safety and resilience to climate change analysis. J Pol Saf Reliab Assoc, Summer Safety Reliab Semin 7(2):173–178
19. Ebrahimi N (1984) Multistate reliability models. Naval Res Logistics 31:671–680

20. El-Neweihi E, Proschan F, Sethuraman J (1978) Multistate coherent systems. J Appl Prob 15(12):675–688
21. Fardis MN, Cornell CA (1981) Analysis of coherent multistate systems. IEEE Trans Reliab 30:117–122
22. Ferreira F, Pacheco A (2007) Comparison of level-crossing times for Markov and semi-Markov processes. Stat Prob Lett 77(2):151–157
23. Glynn PW, Haas PJ (2006) Laws of large numbers and functional central limit theorems for generalized semi-Markov processes. Stoch Model 22(2):201–231
24. Gouldby BP, Schultz MT, Simm JD, Wibowo JL (2010) Beyond the factor of safety: developing fragility curves to characterize system reliability, report in water resources infrastructure program ERDC SR-10-1, prepared for headquarters. Washington, U.S, Army Corps of Engineers
25. Grabski F (2002) Semi-Markov models of systems reliability and operations analysis. System Research Institute, Polish Academy of Science (in Polish)
26. Grabski F (2014) Semi-Markov processes: application in system reliability and maintenance. Elsevier
27. Griffith W (1980) Multistate reliability models. J Appl Prob 17:735–744
28. Han F, Zio E (2019) A multi-perspective framework of analysis of critical infrastructures with respect to supply service, controllability and topology. Int J Critical Infrastruct Protect 24:1–13
29. Holden R, Val DV, Burkhard R, Nodwell S (2013) A network flow model for interdependent infrastructures at the local scale. Saf Sci 53(3):51–60
30. Hudson JC, Kapur KC (1982) Reliability theory for multistate systems with multistate components. Microelectron Reliab 22:1–7
31. Hudson JC, Kapur KC (1983) Reliability analysis of multistate systems with multistate components. Trans Inst Ind Eng 15:127–135
32. Kjølle GH, Utne IB, Gjerde O (2012) Risk analysis of critical infrastructures emphasizing electricity supply and interdependencies. Reliab Eng Sys Saf 105:80–89
33. Kołowrocki K (2000) On asymptotic approach to multi-state systems reliability evaluation. Chapter 11. In: Limnios N, Nikulin M (eds) Recent Advances in reliability theory: methodology, practice and inference. Birkhauser, Boston, pp 163–180
34. Kołowrocki K (2003) Asymptotic approach to reliability analysis of large systems with degrading components. Int J Reliab Qual Saf Eng 10(3):249–288
35. Kołowrocki K (2008) Reliability and risk analysis of multi-state systems with degrading components. Int J Mater Struct Relib 6(2):213–228
36. Kołowrocki K (2005) Reliability of large systems. Elsevier, Amsterdam—Boston—Heidelberg—London—New York—Oxford—Paris—San Diego—San Francisco—Singapore—Sydney—Tokyo
37. Kołowrocki K (2014) Reliability of large and complex systems (2nd edn) Elsevier, Amsterdam—Boston—Heidelberg—London—New York—Oxford—Paris—San Diego—San Francisco—Singapore—Sydney—Tokyo
38. Kołowrocki K, Kuligowska E, Soszyńska-Budny J (2018) Optimization of critical infrastructure operation and safety with considering climate-weather change influence—maximizing safety lifetime. J Pol Saf Reliab Assoc, Summer Safety Reliab Semin 9(1):1–10
39. Kołowrocki K, Soszyńska J (2006) Reliability and availability of complex systems. Qual Reliab Eng Int 22(1):79–99
40. Kołowrocki K, Soszyńska-Budny J. (2011/2014) Reliability and safety of complex technical systems and processes: modeling—identification—prediction—optimization (1st edn). Springer English/Chinese Edition
41. Kołowrocki K, Soszyńska-Budny J (2017) An overall approach to modeling operation threats and extreme weather hazards impact on critical infrastructure safety. In: Proceedings of 27th ESREL conference, Portoroz
42. Kołowrocki K, Soszyńska-Budny J (2017) Modelling safety of multistate ageing systems. J Pol Saf Reliab Assoc, Summer Safety Reliab Semin 8(3):9–22

43. Kołowrocki K, Soszyńska-Budny J (2018) Critical infrastructure safety indicators, international conference on industrial engineering and engineering management IEEM 2018
44. Kołowrocki K Soszyńska-Budny J (2018) Critical infrastructure impacted by operation safety and resilience indicators. international conference on industrial engineering and engineering management IEEM 2018
45. Kołowrocki K, Soszyńska-Budny J (2019) Safety indicators of critical infrastructure application to port oil terminal examination. In: Proceedings of 29th ISOPE conference, Honolulu
46. Kołowrocki K, Soszyńska-Budny J (2019) Safety and resilience indicators of critical infrastructure impacted by operation application to port oil terminal examination. In: Proceedings of 13th transnav conference, Gdynia
47. Lauge A, Hernantes J, Sarriegi JM (2015) Critical infrastructure dependencies: a holistic, dynamic and quantitative approach. Int J Crit Infr Protection 8:16–23
48. Li W, Pham H (2005) Reliability modeling of multi-state degraded systems with multi-competing failures and random shocks. IEEE Trans Reliab 54(2):297–303
49. Limnios N, Oprisan G (2005) Semi-Markov processes and reliability. Birkhauser, Boston
50. Lisnianski A, Frenkel I, Ding Y (2010) Multi-state systems reliability analysis and optimization for engineers and industrial managers. Springer, London
51. Lisnianski A, Levitin G (2003) Multi-state system reliability. assessment, optimisation and applications. World Scientific Publishing Co. Pte. Ltd., New Jersey, London, Singapore, Hong Kong
52. Misra KB (1992) Reliability analysis and prediction, Elsevier
53. Natvig B (1984) Multi-state coherent systems. In: Encyclopedia of statistical sciences. Wiley, New York
54. Natvig B (2007) Multi-state reliability theory. In: Ruggeri F, Kenett R, Faltin FW (eds) Encyclopedia of statistics in quality and reliability. Wiley, New York, pp 1160–1164
55. Natvig B, Morch H (2003) An application of multistate reliability theory to an offshore gas pipeline network. Int J Reliab Qual Saf Eng 10(4):361–381
56. Natvig B, Streller A (1984) The steady-state behaviour of multistate monotone systems. J Appl Prob 21:826–835
57. Nieuwenhuijs A, Luiijf E, Klaver M (2008) Modeling dependencies in critical infrastructures. In: Papa M, Shenoi S (eds) Critical infrastructure protection II, IFIP Series 253. Springer, Boston, Massachusetts, pp 205–213
58. Ohio F, Nishida T (1984) On multistate coherent systems. IEEE Trans Reliab 33:284–287
59. Ouyang M (2014) Review on modelling and simulation of interdependent critical infrastructure systems. Reliab Eng Sys Saf 121:43–60
60. Ramirez-Marqueza JE, Coit DW (2007) Multi-state component criticality analysis for reliability improvement in multi-state systems. Reliab Eng Sys Saf 92:1608–1619
61. Rinaldi S, Peerenboom J, Kelly T (2001) Identifying, understanding and analyzing critical infrastructure interdependencies. IEEE Control Syst Mag 21(6):11–25
62. Samaniego FJ (2007) System signatures and their applications in engineering reliability (Internat Ser Operat Res Manag Sci 110). Springer, New York
63. Torbicki M (2019) Safety of a critical network infrastructure exposed to operation and weather condition changes. PhD Thesis, Polish Academy of Science, System Research Institute, Warsaw
64. Wang Z, Huang HZ, Li Y, Xiao NC (2011) An approach to reliability assessment under degradation and shock process. Reliab IEEE Trans 60(4):852–863
65. Xue J (1985) On multi-state system analysis. IEEE Trans Reliab 34:329–337
66. Xue J, Yang K (1995) Dynamic reliability analysis of coherent multi-state systems. IEEE Trans Reliab 4(44):683–688
67. Xue J, Yang K (1995) Symmetric relations in multi-state systems. IEEE Trans Reliab 4(44):689–693
68. Yingkui G, Jing L (2012) Multi-state system reliability: a new and systematic review. Proc Eng 29:531–536
69. Yu K, Koren I, Guo Y (1994) Generalised multistate monotone coherent systems. IEEE Trans Reliab 43:242–250

70. Zio E (2007). An introduction to the basics of reliability and risk analysis. In: Series in quality, reliability and engineering statistics 13. World Scientific Publishing Co. Pte. Ltd., New Jersey—London—Singapore—Beijing – Shanghai—Hong Kong—Taipei—Chennai
71. Zio E (2009) Computational methods for reliability and risk analysis. World Scientific Publishing, Co. Pte. Ltd., New Jersey—London—Singapore—Beijing – Shanghai—Hong Kong—Taipei—Chennai

# Chapter 3
# Modelling Safety of Critical Infrastructure Without Outside Impacts

The safety of a critical infrastructure free of any outside impacts is discussed and modelled. The safety indicators of this critical infrastructure and its assets are defined. A real technical application of the proposed safety model to the port oil terminal critical infrastructure free of any outside impacts safety examination is performed and its safety indicators are evaluated.

## 3.1 Safety Indicators of Critical Infrastructure Without Outside Impacts

We denote the critical infrastructure free of outside impacts lifetime in the safety state subset $\{u, u+1, \ldots, z\}$, $u = 1, 2, \ldots, z$, by $T^0(u)$ and define the first safety indicator, the critical infrastructure safety function (SafI1) by the vector [5–15, 17]

$$\mathbf{S}^0(t, \cdot) = \left[1, \mathbf{S}^0(t, 1) \ldots, \mathbf{S}^0(t, z)\right], t \in < 0, \infty) \tag{3.1}$$

with the coordinates

$$\mathbf{S}^0(t, u) = P(s(t) \geq u | s(0) = z) = P\left(T^0(u) > t\right) \text{ for } t \in < 0, \infty), u = 1, 2, \ldots, z \tag{3.2}$$

defined as the probability that the critical infrastructure is in the safety state subset $\{u, u+1, \ldots, z\}$, $u = 1,2, \ldots, z$, at the moment $t$, $t \in < 0, \infty)$, while it was in the safety state $z$ at the moment $t = 0$.

Moreover, if $r$ is the critical safety state, then the second safety indicator, the critical infrastructure risk function (SafI2)



J. Soszyńska-Budny, *Safety Analysis of Critical Infrastructure*,
Lecture Notes in Intelligent Transportation and Infrastructure,
https://doi.org/10.1007/978-3-030-65339-2_3

$$r^0(t) = P\big(s(t) < r | s(0) = z\big) = P\big(T^0(r) \le t\big), t \in < 0, \infty), \tag{3.3}$$

is defined as a probability that the critical infrastructure is in the subset of safety states worse than the critical safety state $r$, $r \in \{1, \ldots, z\}$ while it was in the best safety state $z$ at the moment $t = 0$ and given by [10, 12, 16]

$$r^0(t) = 1 - S^0(t, r), \quad t \in < 0, \infty \tag{3.4}$$

where $\boldsymbol{S}^0(t, r)$ is the coordinate of the critical infrastructure safety function given by (3.2) for $u = r$.

The graph of the critical infrastructure risk function (see: Fig. 2.4) is the third safety indicator called the critical infrastructure fragility curve [4] (SafI3).

The critical infrastructure safety function (SafI1), the critical infrastructure risk function (SafI2) and the critical infrastructure fragility curve (SafI3) are proposed as main basic critical infrastructure safety indicators.

Other practically useful critical infrastructure safety indicators are:

- the mean value of the critical infrastructure lifetime $T^0(r)$ up to exceeding critical safety state $r$ (SafI4) given by

$$\boldsymbol{\mu}^0(r) = \int_0^\infty \boldsymbol{S}^0(t, r)dt, \tag{3.5}$$

where $\boldsymbol{S}^0(t, r)$ is defined by (3.2) for $u = r$;

- the standard deviation of the critical infrastructure lifetime $T^0(r)$ up to the exceeding the critical safety state $r$(SafI5) given by

$$\boldsymbol{\sigma}^0(r) = \sqrt{\boldsymbol{n}^0(r) - [\boldsymbol{\mu}^0(r)]^2}, \tag{3.6}$$

where

$$\mathbf{n}^0(r) = 2\int_0^\infty t\mathbf{S}^0(t, r)dt, \tag{3.7}$$

and $\boldsymbol{S}^0(t, r)$ is given by (3.2) for $u = r$ and $\boldsymbol{\mu}^0(r)$ is given by (3.5);

- the moment $\boldsymbol{\tau}$ of exceeding acceptable value of critical infrastructure risk function level $\delta$ (SafI6) given by

$$\tau^0 = r^{-1}(\delta), \tag{3.8}$$

and illustrated in Fig. 2.4, where $\boldsymbol{r}^{0\ -1}(t)$ is the inverse function of the risk function $\boldsymbol{r}^{0}(t)$ given by (3.4);

- the mean values of the critical infrastructure lifetimes in the safety state subsets $\{u, u+1, \ldots, z\}$, $u = 1, 2, \ldots, z$, (SafI7) given by

$$\boldsymbol{\mu}^{0}(u) = \int_{0}^{\infty} [\boldsymbol{S}^{0}(t, u)]dt, u = 1, 2, \ldots, z, \quad i = 1, 2, \ldots, n; \tag{3.9}$$

- the standard deviations of the critical infrastructure lifetimes in the safety state subsets $\{u, u+1, \ldots, z\}$, $u = 1, 2, \ldots, z$, (SafI8) given by

$$\boldsymbol{\sigma}^{0}(u) = \sqrt{\boldsymbol{n}^{0}(u) - [\boldsymbol{\mu}^{0}(u)]^2}, u = 1, 2, \ldots, z, \tag{3.10}$$

where

$$\boldsymbol{n}^{0}(u) = 2\int_{0}^{\infty} t\boldsymbol{S}^{0}(t, u)dt, u = 1, 2, \ldots, z, \quad i = 1, 2, \ldots, n; \tag{3.11}$$

- the mean lifetimes $\overline{\boldsymbol{\mu}}^{0}(u)$, $u = 1, 2, \ldots, z$, of the critical infrastructure in the particular safety states (SafI9)

$$\overline{\boldsymbol{\mu}}^{0}(u) = \boldsymbol{\mu}^{0}(u) - \boldsymbol{\mu}^{0}(u+1), u = 0, 1, \ldots, z-1, \overline{\boldsymbol{\mu}}^{0}(z) = \boldsymbol{\mu}^{0}(z); \tag{3.12}$$

- the intensities of degradation (ageing) of the critical infrastructure / the intensities of critical infrastructure departure from the safety state subset $\{u, u+1, \ldots, z\}$, $u = 1, 2, \ldots, z$, (SafI10), i.e. the coordinates of the vector

$$\lambda^{0}(t, \cdot) = \left[0, \lambda^{0}(t, 1), \ldots, \lambda^{0}(t, z)\right], t \in <0, +\infty), \tag{3.13}$$

where

$$\boldsymbol{\lambda}^{0}(t, u) = \frac{-\frac{d\boldsymbol{S}^{0}(t,u)}{dt}}{\boldsymbol{S}^{0}(t, u)}, t \in <0, +\infty), u = 1, 2, \ldots, z. \tag{3.14}$$

In the particular case, when the critical infrastructure has the piecewise exponential safety function (SafI1), i.e.

$$\boldsymbol{S}^{0}(t, \cdot) = \left[1, \boldsymbol{S}^{0}(t, 1), \ldots, \boldsymbol{S}^{0}(t, z)\right], t \in <0, +\infty), \tag{3.15}$$

where

$$S^0(t,u) = \exp[-\lambda^0(u)t], t \in < 0, +\infty), \lambda^0(u) \geq 0, u = 1, 2, \ldots, z, \quad (3.16)$$

the intensities of degradation of the critical infrastructure / the intensities of critical infrastructure departure from the safety state subset $\{u, u+1, \ldots, z\}$, $u = 1, 2, \ldots, z$, (SafI7), i.e. the coordinates of the vector

$$\lambda^0(\cdot) = \left[0, \lambda^0(1), \ldots, \lambda^0(z)\right], \quad (3.17)$$

are constant and

$$\boldsymbol{\lambda}^0(u) = \frac{1}{\boldsymbol{\mu}^0(u)}, u = 1, 2, \ldots, z, \quad (3.18)$$

where $\boldsymbol{\mu}^{(0)}(u)$ is the mean value of the critical infrastructure lifetime $T^0(u)$ in the safety state subset $\{u, u+1, \ldots, z\}$,$u = 1, 2, \ldots z$, defined by

$$\boldsymbol{\mu}^{(0)}(u) = \int_0^\infty [\boldsymbol{S}^0(t,u)]dt, \quad (3.19)$$

and $\boldsymbol{S}^0(t,u)$ is defined by (3.2) for $u = 1, 2, \ldots z$, and given in this case by (3.16).

## 3.2 Critical Infrastructure Assets Safety Parameters

We mark by $T_i^0(u)$, $u = 1, 2, \ldots, z$, the asset $A_i$ free of outside impacts lifetime $T_i(u)$ in the safety state subset $\{u, u+1, \ldots, z\}$, $u = 1, 2, \ldots, z$, and define the asset $A_i$ safety function (SafI1) by the vector [1, 2, 9–12]

$$S_i^0(t,) = \left[1, S_i^0(t,1) \ldots, S_i^0(t,-z)\right], t \in < 0, \infty), i = 1, 2, \ldots, n, \quad (3.20)$$

where

$$S_i^0(t,u) = P(s_i(t) \geq u | s_i(0) = z)\, P\left(T_i^0(u) > t\right), t \in < 0, \infty), u = 1, 2, \ldots, z, \quad (3.21)$$

is the probability that the asset $A_i$ is in the safety state subset $\{u, u+1, \ldots, z\}$, $u = 1, 2, \ldots, z$, at the moment $t$, $t \in < 0, \infty)$, while it was in the safety state $z$ at the moment $t = 0$.

The safety functions $S_i^0(t,u)$, $t \in < 0, \infty)$, $u = 1, 2, \ldots, z$, defined by (3.21) are called the coordinates of the asset $A_i$, $i = 1, 2, \ldots, n$, safety function $S_i^0(t,\cdot)$,

$t \in < 0, \infty)$, given by (3.20). Thus, the relationship between the distribution function $F_i^0(t, u)$, of the asset $A_i$, $i = 1, 2, \ldots, n$, lifetime $T_i^0(u)$, $u = 1, 2, \ldots, z$, in the safety state subset $\{u, u+1, \ldots, z\}$, $u = 1, 2, \ldots, z$, and the coordinate $S_i^0(t, u)$, $t \in < 0, \infty)$, $u = 1, 2, \ldots, z$, of its safety function is given by

$$F_i^0(t, u) = P\big(T_i^0(u) \le t\big) = 1 - P\big(T_i^0(u) > t\big) = 1 - S_i^0(t, u),$$
$$t \in < 0, \infty, u = 1, 2, \ldots, z. \tag{3.22}$$

Thus, the function

$$r_i^0(t) = 1 - S_i^0(t, r), t \in < 0, \infty), i = 1, 2, \ldots, n, \tag{3.23}$$

is the asset $A_i$, $i = 1, 2, \ldots, n$, risk function (SafI2) and its graph is the asset $A_i$, $i = 1, 2, \ldots, n$, fragility curve [3] (SafI3).

Moreover, the mean lifetime of the asset $A_i$ in the safety state subset $\{u, u + 1, \ldots, z\}$, $u = 1, 2, \ldots, z$, (SafI7) is given by

$$\mu_i^0(u) = \int_0^\infty S_i^0(t, u)dt, u = 1, 2, \ldots, z, \quad i = 1, 2, \ldots, n, \tag{3.24}$$

and the standard deviation of the asset $A_i$ lifetime in the safety state subset $\{u, u + 1, \ldots, z\}$, $u = 1, 2, \ldots, z$, (SafI8) is given by

$$\sigma_i^0(u) = \sqrt{n_i^0(u) - [\mu_i^0(u)]^2}, u = 1, 2, \ldots, z, i = 1, 2, \ldots, n, \tag{3.25}$$

where

$$n_i^0(u) = 2\int_0^\infty t S_i^0(t, u)dt, u = 1, 2, \ldots, z, i = 1, 2, \ldots, n. \tag{3.26}$$

Next, it can be justified that the mean lifetimes $\overline{\mu}_i^0(u)$, $u = 1, 2, \ldots, z$, of the asset $A_i$ in the particular safety states (SafI9) are

$$\overline{\mu}_i^0(u) = \mu_i^0(u) - \mu_i^0(u + 1), u = 1, \ldots, z - 1, \overline{\mu}_i^0(z) = \mu_i^0(z), i = 1, 2, \ldots, n. \tag{3.27}$$

In the case when the critical infrastructure asset $A_i$, $i = 1, 2, \ldots, n$, have the piecewise exponential safety functions (SafI1), i.e.

$$S_i^0(t, \cdot) = \big[1, S_i^0(t, 1), \ldots, S_i^0(t, z)\big], t \in < 0, \infty\big), i = 1, 2, \ldots, n, \tag{3.28}$$

where

$$S_i^0(t,u) = \exp[-\lambda_i^0(u)t],\ t \in <0,+\infty),\ \lambda_i^0(u) \geq 0,\ u = 1,2,\ldots,z,\ i = 1,2,\ldots,n. \quad (3.29)$$

the intensities of ageing of the critical infrastructure asset $A_i$, $i = 1, 2, \ldots, n$, / the intensities of critical infrastructure assets $A_i$, $i = 1, 2, \ldots, n$, departure from the safety state subset $\{u, u+1, \ldots, z\}$, $u = 1, 2, \ldots, z$, (SafI10), i.e. the coordinates of the vector

$$\lambda_i^0(\cdot) = [0, \lambda_i^0(1), \ldots, \lambda_i^0(z)] i = 1,2,\ldots,n, \quad (3.30)$$

are constant and given by

$$\lambda_i^0(u) = \frac{1}{\mu_i^0(u)},\ u = 1,2,\ldots z,\ i = 1,2,\ldots,n, \quad (3.31)$$

where $\mu_i^0(u)$, $i = 1, 2, \ldots, n$, is the mean value of the critical infrastructure asset $A_i$, $i = 1, 2, \ldots, n$, lifetime $T_i^0(u)$, $i = 1,2,\ldots,n$, in the safety state subset $\{u, u+1, \ldots, z\}$, $u = 1, 2, \ldots z$, defined by

$$\mu_i^0(u) = \int_0^\infty S_i^0(t,u)dt,\ u = 1,2,\ldots z,\ i = 1,2,\ldots,n, \quad (3.32)$$

and $S_i^0(t,u)$, $i = 1, 2, \ldots, n$, are defined by (3.21) for $u = 1, 2, \ldots z$, and given in this case by (3.29).

## 3.3 Application

We consider the port oil terminal critical infrastructure free of any outside impacts. The considered port oil terminal placed at the Baltic seaside is described and its reliability indicators are determined in Chap. 1. In this section, this port oil terminal safety analysis is performed and its safety indicators are evaluated.

### 3.3.1 Port Oil Terminal Critical Infrastructure Assets

The considered port oil terminal critical infrastructure is composed of three parts *A*, *B* and *C*, linked by the piping transportation system with the pier. The port oil terminal is presented in Figs. 1.17 and 1.18.

The main technical assets of the port oil terminal critical infrastructure are:

- $A_1$—port oil piping transportation system;
- $A_2$—internal pipeline technological system;
- $A_3$—supporting pump station;
- $A_4$—internal pump system;
- $A_5$—port oil tanker shipment terminal;
- $A_6$—loading railway carriage station;
- $A_7$—loading road carriage station;
- $A_8$—unloading railway carriage station;
- $A_9$—oil storage reservoir system.

The asset $A_1$, the port oil piping transportation system is operating at one of the Baltic oil terminals that is designated for the reception from ships, the storage and sending by carriages or cars the oil products. It is also designated for receiving from carriages or cars, the storage and loading the tankers with oil products such like petrol and oil. The area in the neighborhood of the port oil piping transportation system is presented in Figs. 1.17 and 1.18. The approximate length of the port oil piping transportation system is equal to around 25 km. The scheme of the port oil piping transportation system is presented in Fig. 1.19.

The asset $A_1$, operation is the main activity of the port oil terminal critical infrastructure involving the remaining assets $A_2$–$A_9$.

The considered port oil piping transportation system, the asset $A_1$, is a series system composed of series–parallel subsystems $S_1$, containing two pipelines (assets $A_{11}, A_{12}$), a series–parallel subsystems $S_2$, containing two pipelines (assets $A_{21}, A_{22}$), and one "2 out of 3" subsystem $S_3$ containing 3 pipelines (assets $A_{31}, A_{32}, A_{33}$) [10].

The subsystems $S_1, S_2$ and $S_3$ are forming a general series port oil pipeline system safety structure presented in Fig. 3.1.

The asset $A_1$, the port oil piping transportation system operating at the Baltic oil terminal consists of three subsystems:

- the subsystem $S_1$ composed of two pipelines, each composed of 176 pipe segments and 2 valves;
- the subsystem $S_2$ composed of two pipelines, each composed of 717 pipe segments and 2 valves;
- the subsystem $S_3$ composed of three pipelines, each composed of 360 pipe segments and 2 valves.

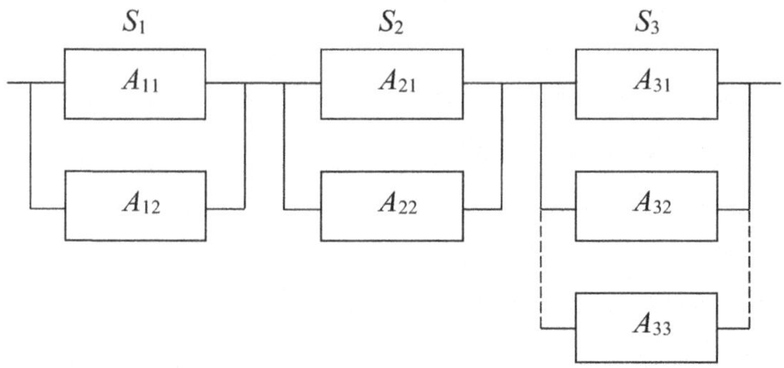


**Fig. 3.1** General scheme of the port oil piping transportation system safety structure

### *3.3.2 Port Oil Terminal Critical Infrastructure Safety States*

After considering the comments and opinions coming from experts, taking into account the effectiveness and safety aspects of the operation of the port oil terminal critical infrastructure and its assets, we fix for all of them the following parameters [10]:

- the number of safety states (excluding safety state 0) $z = 2$;
- three $(z + 1)$ safety states:
  - a safety state 2—an asset and the port oil terminal critical infrastructure operation is fully safe,
  - a safety state 1—an asset and the port oil terminal critical infrastructure is less safe and more dangerous because of the possibility of environment pollution,
  - a safety state 0—an asset and the port oil terminal critical infrastructure is destroyed;
  - and we assume that:
  - there are possible the transitions between the assets and the port oil terminal critical infrastructure safety states only from better to worse ones,
  - the critical safety state of an asset and the port oil terminal critical infrastructure is $r = 1$,
  - the port oil terminal critical infrastructure risk function permitted level $\delta = 0.05$.

### *3.3.3 Port Oil Piping Transportation System Safety Parameters*

Now, we will perform detailed safety analysis of the asset $A_1$, the port oil piping transportation system free of any outside impacts. The safety parameters of its assets are as follows:

- the mean values of the asset $A_{11}$ of the port oil piping transportation system lifetimes in the safety state subsets {1, 2}, {2}, calculated on the basis of safety data of its components coming from experts, are as follows:

  - for safety state subset {1, 2}

$$\mu_{11}^{0}(1) = 276 \text{ years}, \tag{3.33}$$

  - for safety state subset {2}

$$\mu_{11}^{0}(2) = 185 \text{ years}; \tag{3.34}$$

- the mean values of the asset $A_{12}$ of the port oil piping transportation system lifetimes in the safety state subsets {1, 2}, {2}, calculated on the basis of safety data of its components coming from experts, are as follows:

  - for safety state subset {1, 2}

$$\mu_{12}^{0}(1) = 276 \text{ years}, \tag{3.35}$$

  - for safety state subset {2}

$$\mu_{12}^{0}(2) = 185 \text{ years}; \tag{3.36}$$

- the mean values of the asset $A_{21}$ of the port oil piping transportation system lifetimes in the safety state subsets {1, 2}, {2}, calculated on the basis of safety data of its components coming from experts, are as follows:

  - for safety state subset {1, 2}

$$\mu_{21}^{0}(1) = 69 \text{ years}, \tag{3.37}$$

  - for safety state subset {2}

$$\mu_{21}^{0}(2) = 46 \text{ years}; \tag{3.38}$$

- the mean values of the asset $A_{22}$ of the port oil piping transportation system lifetimes in the safety state subsets {1, 2}, {2}, calculated on the basis of safety data of its components coming from experts, are as follows:

  - for safety state subset {1, 2}

$$\mu_{22}^{0}(1) = 69 \text{ years}, \tag{3.39}$$

– for safety state subset {2}

$$\mu_{22}^{0}(2) = 46 \text{ years;} \tag{3.40}$$

- the mean values of the asset $A_{31}$ of the port oil piping transportation system lifetimes in the safety state subsets {1, 2}, {2}, calculated on the basis of safety data of its components coming from experts, are as follows:

– for safety state subset {1, 2}

$$\mu_{31}^{0}(1) = 137 \text{ years,} \tag{3.41}$$

– for safety state subset {2}

$$\mu_{31}^{0}(2) = 110 \text{ years;} \tag{3.42}$$

- the mean values of the asset $A_{32}$ of the port oil piping transportation system lifetimes in the safety state subsets {1, 2}, {2}, calculated on the basis of safety data of its components coming from experts, are as follows:

– for safety state subset {1, 2}

$$\mu_{32}^{0}(1) = 137 \text{ years,} \tag{3.43}$$

– for safety state subset {2}

$$\mu_{32}^{0}(2) = 110 \text{ years;} \tag{3.44}$$

- the mean values of the asset $A_{33}$ of the port oil piping transportation system lifetimes in the safety state subsets {1, 2}, {2}, calculated on the basis of safety data of its components coming from experts, are as follows:

– for safety state subset {1, 2}

$$\mu_{33}^{0}(1) = 114 \text{ years,} \tag{3.45}$$

– for safety state subset {2}

$$\mu_{33}^{0}(2) = 102 \text{ years.} \tag{3.46}$$

Applying (3.33)–(3.46) and (3.31), we get the intensities of ageing of the critical infrastructure assets $A_{ij}$, $i = 1, 2, j = 1, 2, i = 3, j = 1, 2, 3,$/ the intensities of critical infrastructure assets $A_{ij}$, $i = 1, 2, j = 1, 2, i = 3, j = 1, 2, 3$, departure from the safety state subset {1, 2} and {2}, i.e. the coordinates of the vectors

$$\lambda_{ij}^{0}(\cdot) = \left[0, \lambda_{ij}^{0}(1), \lambda_{ij}^{0}(2)\right], i = 1, 2, j = 1, 2, i = 3, j = 1, 2, 3, \tag{3.47}$$

are constants and given respectively by:

- the intensities of departure of the asset $A_{11}$
- for safety state subset {1, 2}

$$\lambda_{11}^{0}(1) = \frac{1}{\mu_{11}^{0}(1)} = \frac{1}{276} = 0.00362, \tag{3.48}$$

- for safety state subset {2}

$$\lambda_{11}^{0}(2) = \frac{1}{\mu_{11}^{0}(2)} = \frac{1}{185} = 0.00540; \tag{3.49}$$

- the intensities of departure of the asset $A_{12}$
  - for safety state subset {1, 2}

$$\lambda_{12}^{0}(1) = \frac{1}{\mu_{12}^{0}(1)} = \frac{1}{276} = 0.00362, \tag{3.50}$$

  - for safety state subset {2}

$$\lambda_{12}^{0}(2) = \frac{1}{\mu_{12}^{0}(2)} = \frac{1}{185} = 0.00540; \tag{3.51}$$

- the intensities of departure of the asset $A_{21}$
  - for safety state subset {1, 2}

$$\lambda_{21}^{0}(1) = \frac{1}{\mu_{21}^{0}(1)} = \frac{1}{69} = 0.01444, \tag{3.52}$$

  - for safety state subset {2}

$$\lambda_{21}^{0}(2) = \frac{1}{\mu_{21}^{0}(2)} = \frac{1}{46} = 0.02163; \tag{3.53}$$

- the intensities of departure of the asset $A_{22}$
  - for safety state subset {1, 2}

$$\lambda_{22}^{0}(1) = \frac{1}{\mu_{22}^{0}(1)} = \frac{1}{69} = 0.01444, \tag{3.54}$$

– for safety state subset {2}

$$\lambda_{22}^{0}(2) = \frac{1}{\mu_{22}^{0}(2)} = \frac{1}{46} = 0.02163; \tag{3.55}$$

- the intensities of departure of the asset $A_{31}$

  – for safety state subset {1, 2}

$$\lambda_{31}^{0}(1) = \frac{1}{\mu_{31}^{0}(1)} = \frac{1}{137} = 0.00730, \tag{3.56}$$

  – for safety state subset {2}

$$\lambda_{31}^{0}(2) = \frac{1}{\mu_{31}^{0}(2)} = \frac{1}{110} = 0.00912; \tag{3.57}$$

- the intensities of departure of the asset $A_{32}$

  – for safety state subset {1, 2}

$$\lambda_{32}^{0}(1) = \frac{1}{\mu_{32}^{0}(1)} = \frac{1}{137} = 0.00730, \tag{3.58}$$

  – for safety state subset {2}

$$\lambda_{32}^{0}(2) = \frac{1}{\mu_{32}^{0}(2)} = \frac{1}{110} = 0.00912; \tag{3.59}$$

- the intensities of departure of the asset $A_{33}$

  – for safety state subset {1, 2}

$$\lambda_{33}^{0}(1) = \frac{1}{\mu_{33}^{0}(1)} = \frac{1}{114} = 0.00874, \tag{3.60}$$

  – for safety state subset {2}

$$\lambda_{33}^{0}(2) = \frac{1}{\mu_{33}^{0}(2)} = \frac{1}{102} = 0.00984. \tag{3.61}$$

Assuming that the assets $A_{ij}, i = 1, 2, j - 1, 2, i = 3, j = 1, 2, 3$, of the asset $A_1$, the port oil piping transportation system, have piecewise exponential safety functions of the form

$$S_{ij}^{0}(t,) = [1, S_{ij}^{0}(t, 1), S_{ij}^{0}(t, 2)], t \in < 0, \infty), i = 1, 2, j = 1, 2, i = 3, j = 1, 2, 3$$

with the coordinates

$$S_{ij}^{0}(t,u) = \exp[-\lambda_{ij}^{0}(u)t] \text{ for } t \in <0,\infty),\ u = 1,2,\ i = 1,2,$$
$$j = 1,2,\ i = 3,\ j = 1,2,3$$

where the intensities of degradation

$$\lambda_{ij}^{0}(u),\ u = 1,2,\ i = 1,2,\ j = 1,2,\ i = 3,\ j = 1,2,3$$

are respectively given by (3.48)–(3.61), we get their safety functions of the following forms:

- the safety function of the asset $A_{11}$

$$S_{11}^{0}(t,\cdot) = [1, S_{11}^{0}(t,1), S_{11}^{0}(t,2)],\ t \in <0,\infty), \tag{3.62}$$

where

$$S_{11}^{0}(t,1) = \exp[-0.00362t],\ S_{11}^{0}(t,2) = \exp[-0,00540t]t \in <0,\infty); \tag{3.63}$$

- the safety function of the asset $A_{12}$

$$S_{12}^{0}(t,) = [1, S_{12}^{0}(t,1), S_{12}^{0}(t,2)],\ t \in <0,\infty), \tag{3.64}$$

where

$$S_{12}^{0}(t,1) = \exp[-0.00362t],\ S_{12}^{0}(t,2) = \exp[-0.00540t],\ t \in <0,\infty); \tag{3.65}$$

- the safety function of the asset $A_{21}$

$$S_{21}^{0}(t;) = [1, S_{21}^{0}(t,1), S_{21}^{0}(t,2)],\ t \in <0,\infty), \tag{3.66}$$

where

$$S_{21}^{0}(t,1) = \exp[-0.01444t],\ S_{21}^{0}(t,2) = \exp[-0.02163t],\ t \in <0,\infty); \tag{3.67}$$

- the safety function of the asset $A_{22}$

$$S_{22}^{0}(t,') = [1, S_{22}^{0}(t,1), S_{22}^{0}(t,2)],\ t \in <0,\infty), \tag{3.68}$$

where

$$S_{22}^{0}(t,1)=\exp[-0.01444t],\ S_{22}^{0}(t,2)=\exp[-0.02163t],\ t\in<0,\infty); \quad (3.69)$$

- the safety function of the asset $A_{31}$

$$S_{31}^{0}(t,\cdot)=\left[1,\ S_{31}^{0}(t,1),\ S_{31}^{0}(t,2)\right],\ t\in<0,\infty), \quad (3.70)$$

where

$$S_{31}^{0}(t,1)=\exp[-0.00730t],\ S_{31}^{0}(t,2)=\exp[-0.00912t],\ t\in<0,\infty); \quad (3.71)$$

- the safety function of the asset $A_{32}$

$$S_{32}^{0}(t;)=\left[1,\ S_{32}^{0}(t,1),\ S_{32}^{0}(t,2)\right],\ t\in<0,\infty), \quad (3.72)$$

where

$$S_{32}^{0}(t,1)=\exp[-0.00730t],\ S_{32}^{0}(t,2)=\exp[-0.00912t],\ t\in<0,\infty); \quad (3.73)$$

- the safety function of the asset $A_{33}$

$$S_{33}^{0}(t,\cdot)=\left[1,\ S_{33}^{0}(t,1),\ S_{33}^{0}(t,2)\right],\ t\in<0,\infty), \quad (3.74)$$

where

$$S_{33}^{0}(t,1)=\exp[-0.00874t],\ S_{33}^{0}(t,2)=\exp[-0.00984t],\ t\in<0,\infty). \quad (3.75)$$

## 3.4 Port Oil Piping Transportation System Safety Characteristics

After applying formulae (2.31)–(2.32) for the safety function of the "$m_i$ out of $l_i$"-series critical infrastructure with the shape parameters $k=2$, $m_1=1$, $l_1=2$, $m_2=1$, $l_2=2$, $m_3=2$, $l_3=3$, and considering (3.62)–(3.75), we get the safety function of the port oil piping transportation system, of the asset $A_1$,

$$S_{1}^{0}(t,\cdot)=\left[1,\ S_{1}^{0}(t,1),\ S_{1}^{0}(t,2)\right],\ t\geq 0, \quad (3.76)$$

where

$$
\begin{aligned}
S_1^0(t,1) &= \prod_{i=1}^{3}[\sum_{\substack{r_1,r_2,\ldots,r_{l_i}\in\{0,1\}\\ m_i\le r_1+r_2+\ldots=r_{l_i}\le l_i}}[\prod_{j=1}^{l_i} S_{ij}^0(t,1)]^{r_i}[1-\prod_{j=1}^{l_i} S_{ij}^0(t,1)]^{1-r_i}]\\
&= [S_{11}^0(t,1)[1-S_{12}^0(t,1)]+[1-S_{11}^0(t,1)]S_{12}^0(t,1)+S_{11}^0(t,1)S_{12}^0(t,1)]\\
&\cdot[S_{21}^0(t,1)[1-S_{22}^0(t,1)]+[1-S_{21}^0(t,1)]S_{22}^0(t,1)+S_{21}^0(t,1)S_{22}^0(t,1)]\\
&\cdot[[1-S_{31}^0(t,1)]S_{32}^0(t,1)S_{33}^0(t,1)+S_{31}^0(t,1)[1-S_{32}^0(t,1)]\,S_{33}^0(t,1)\\
&+S_{31}^0(t,1)S_{32}^0(t,1)[1-S_{33}^0(t,1)]+S_{31}^0(t,1)S_{32}^0(t,1)S_{33}^0(t,1)]\\
&= [S_{11}^0(t,1)+S_{12}^0(t,1)-S_{11}^0(t,1)S_{12}^0(t,1)]\\
&\cdot[S_{21}^0(t,1)+S_{22}^0(t,1)-S_{21}^0(t,1)]\,S_{22}^0(t,1)]\cdot[S_{31}^0(t,1)S_{32}^0(t,1)+S_{31}^0(t,1)S_{33}^0(t,1)\\
&+S_{32}^0(t,1)S_{33}^0(t,1)-2S_{31}^0(t,1)[S_{32}^0(t,1)S_{33}^0(t,1)]\\
&= [\exp[-0.00362t]+\exp[-0.00362t]-\exp[-0.00362t]\exp[-0.00362t]]\\
&\cdot[\exp[-0.01444t]+\exp[-0.01444t]-\exp[-0.01444t]\exp[-0.01444t]]\\
&\cdot[\exp[-0.00730t]\exp[-0.00730t]+\exp[-0.00730t]\exp[-0.00874t]\\
&+\exp[-0.00730t]\exp[-0.00874t]-2\exp[-0.00730t]\exp[-0.00730t]\exp[-0.00874t]]\\
&= [2\exp[-0.00362t]-\exp[-0.00724t]]\cdot[2\exp[-0.01444t]-\exp[-0.02888t]]\\
&\cdot[\exp[-0.01460t]+2\exp[-0.01604t]-2\exp[-0.02334t]]\\
&= 8\exp[-0.0341t]-8\exp[-0.0414t]+4\exp[-0.03266t]-4\exp[-0.04854t]\\
&+4\exp[-0.05584t]-4\exp[-0.03772t]+4\exp[-0.04502t]-2\exp[-0.0471t]\\
&-2\exp[-0.03628t]+2\exp[-0.05216t]-2\exp[-0.05946t]\\
&+\exp[-0.05072t]\ \text{for}\ t\ge 0
\end{aligned}
\tag{3.77}
$$

$$
\begin{aligned}
S_1^0(t,2) &= \prod_{i=1}^{3}[\sum_{\substack{r_1,r_2,\ldots,r_{l_i}\in\{0,1\}\\ m_i\le r_1+r_2+\ldots=r_{l_i}\le l_i}}[\prod_{j=1}^{l_i} S_{ij}^0(t,2)]^{r_i}[1-\prod_{j=1}^{l_i} S_{ij}^0(t,2)]^{1-r_i}]\\
&= [S_{11}^0(t,2)[1-S_{12}^0(t,2)]+[1-S_{11}^0(t,2)]S_{12}^0(t,2)+S_{11}^0(t,2)S_{12}^0(t,2)]\\
&\cdot[S_{21}^0(t,2)[1-S_{22}^0(t,2)]+[1-S_{21}^0(t,2)]S_{22}^0(t,2)+S_{21}^0(t,2)S_{22}^0(t,2)]\\
&\cdot[[1-S_{31}^0(t,2)]S_{32}^0(t,2)S_{33}^0(t,2)+S_{31}^0(t,2)[1-S_{32}^0(t,2)]S_{33}^0(t,2)\\
&+S_{31}^0(t,2)S_{32}^0(t,2)[1-S_{33}^0(t,2)]+S_{31}^0(t,2)S_{32}^0(t,2)S_{33}^0(t,2)]\\
&= [S_{11}^0(t,2)+S_{12}^0(t,2)-S_{11}^0(t,2)S_{12}^0(t,2)]\\
&\cdot[S_{21}^0(t,2)+S_{22}^0(t,2)-S_{21}^0(t,2)]S_{22}^0(t,2)]\cdot[S_{31}^0(t,2)S_{32}^0(t,2)\\
&+S_{31}^0(t,2)S_{33}^0(t,2)+S_{32}^0(t,2)S_{33}^0(t,2)-2S_{31}^0(t,2)[S_{32}^0(t,2)S_{33}^0(t,2)]\\
&= [\exp[-0.00540t]+\exp[-0.00540t]\\
&-\exp[-0.00540t]\exp[-0.00540t]]\cdot[\exp[-0.02163t]+\exp[-0.02163t]\\
&-\exp[-0.02163t]\exp[-0.02163t]]\cdot[\exp[-0.00912t]\exp[-0.00912t]\\
&+\exp[-0.00912t]\exp[-0.00984t]+\exp[-0.00912t]\exp[-0.00984t]\\
&-2\exp[-0.00912t]\exp[-0.00912t]\exp[-0.00984t]]\\
&= [2\exp[-0.00540t]-\exp[-0.01080t]]\cdot[2\exp[-0.02163t]
\end{aligned}
$$

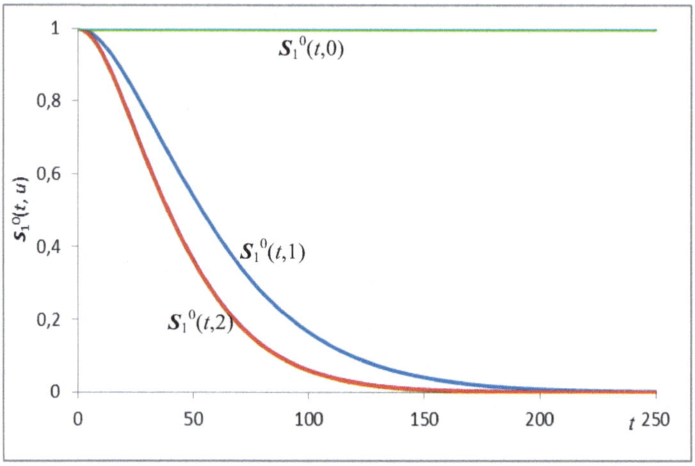


**Fig. 3.2** The graphs of the port oil piping transportation system safety function coordinates

$$
\begin{aligned}
&- \exp[-0.04326t]][\exp[-0.01824t] + 2\exp[-0.01896t] - 2\exp[-0.02808t]] \\
&= 8\exp[-0.04598t] - 8\exp[-0.05511t] + 4\exp[-0.04527t] \\
&- 4\exp[-0.06762t] + 4\exp[-0.07674t] - 4\exp[-0.05139t] + 4\exp[-0.06051t] \\
&- 2\exp[-0.0669t] - 2\exp[-0.08214t] + 2\exp[-0.07302t] \\
&- 2\exp[-0.0506t] + \exp[-0.0723t] \text{ for } t \geq 0.
\end{aligned} \tag{3.78}
$$

The graph of the safety function of the port oil piping transportation system is given in Fig. 3.2.

Considering (3.77)–(3.78) and applying (3.24)–(3.26), the mean values and standard deviations of the lifetimes of the port oil piping transportation system are:

- in the safety state subset: {1, 2}

$$
\begin{aligned}
\mu_1^0(1) &= \int_0^\infty \boldsymbol{s}_1^0(t, 1)dt \\
&\int_0^\infty 8\exp[-0.04598t] - 8\exp[-0.05511t] + 4\exp[-0.04527t] - 4\exp[-0.06762t] \\
&+ 4\exp[-0.07674t] - 4\exp[-0.05139t] + 4\exp[-0.06051t] - 2\exp[-0.0669t] \\
&- 2\exp[-0.08214t] + 2\exp[-0.07302t] - 2\exp[-0.0506t] + \exp[-0.0723t]]dt \\
&\cong 62.71 \text{ years},
\end{aligned} \tag{3.79}
$$

$$
\sigma_1^0(1) = \sqrt{\boldsymbol{n}_1^0(1) - [\boldsymbol{\mu}_1^0(1)]^2} \cong \sqrt{5686.4199 - (62.71)^2} \cong 41.8793 \text{ years}, \tag{3.80}
$$

where

$$n_1^0(1) = 2\int_0^\infty t S_1^0(t,1)dt$$

$$= 2\int_0^\infty [8t\exp[-0.0341t] - 8t\exp[-0.0414t] + 4t\exp[-0.03266t]$$
$$- 4t\exp[-0.04854t] + 4t\exp[-0.05584t] - 4t\exp[-0.03772t] + 4t\exp[-0.04502t]$$
$$- 2t\exp[-0.0471t] - 2t\exp[-0.03628t] + 2t\exp[-0.05216t]$$
$$- 2t\exp[-0.05946t] + t\exp[-0.05072t]]dt$$
$$\cong 5686.4199,$$

- in the safety state subset {2}

$$\boldsymbol{\mu}_1^0(2) = \int_0^\infty \boldsymbol{S}_1^0(t,2)dt$$
$$\int_0^\infty [8\exp[-0.04598t] - 8\exp[-0.05511t] + 4\exp[-0.04527t]$$
$$- 4\exp[-0.06762t] + 4\exp[-0.07674t] - 4\exp[-0.05139t] + 4\exp[-0.06051t]$$
$$- 2\exp[-0.0669t] - 2\exp[-0.08214t] + 2\exp[-0.07302t]$$
$$- 2\exp[-0.0506t] + \exp[-0.0723t]]dt$$
$$\cong 45.84 \text{ years,} \tag{3.81}$$

$$\boldsymbol{\sigma}_1^0(2) = \sqrt{\boldsymbol{n}_1^0(2) - [\boldsymbol{\mu}_1^0(2)]^2} \cong \sqrt{3045.9212 - (45.84)^2} \cong 30.7346 \text{ years,} \tag{3.82}$$

where

$$n_1^0(2) = 2\int_0^\infty t S_1^0(t,2)dt = 2\int_0^\infty [8t\exp[-0.04598t] - 8t\exp[-0.05511t]$$
$$+ 4t\exp[-0.04527t] - 4t\exp[-0.06762t]$$
$$+ 4t\exp[-0.07674t] - 4t\exp[-0.05139t]$$
$$+ 4t\exp[-0.06051t] - 2t\exp[-0.0669t]$$
$$- 2t\exp[-0.08214t] + 2t\exp[-0.07302t]$$
$$- 2t\exp[-0.0506t] + t\exp[-0.0723t]]dt \cong 3045.9212.$$

**Table. 3.1** The values of the risk function $\boldsymbol{r}_1(t)$ of the port oil piping transportation system

| $t$ | $\boldsymbol{r}_1{}^0(t) = 1 - \boldsymbol{S}_1{}^0(t)$ |
|---|---|
| 12.00 | 0.048805 |
| 12.01 | 0.048879 |
| 12.02 | 0.048953 |
| 12.03 | 0.049027 |
| 12.04 | 0.049101 |
| 12.05 | 0.049175 |
| 12.06 | 0.049249 |
| 12.07 | 0.049323 |
| 12.08 | 0.049398 |
| 12.09 | 0.049472 |
| 12.10 | 0.049547 |
| 12.11 | 0.049621 |
| 12.12 | 0.049695 |
| 12.13 | 0.04977 |
| 12.14 | 0.049845 |
| 12.15 | 0.049919 |
| **12.16** | **0.049994** |
| 12.17 | 0.050069 |
| 12.18 | 0.050143 |
| 12.19 | 0.050218 |
| 12.20 | 0.050293 |

From (3.79) and (3.81), applying (3.27), the mean lifetimes $\overline{\boldsymbol{\mu}}_1^0(u), u = 1, 2,$ of the port oil piping transportation system, the asset $A_1$, in the particular safety states are:

$$\overline{\boldsymbol{\mu}}_1^0(1) = \boldsymbol{\mu}_1^0(1) - \boldsymbol{\mu}_1^0(2) = 16.873 \text{ years}, \overline{\boldsymbol{\mu}}_1^0(2) = \boldsymbol{\mu}_1^0(2) = 45.840 \text{ years}. \quad (3.83)$$

As the critical safety state is $r = 1$, then by (3.4) and (3.23), the port oil piping transportation system risk function is

$$\begin{aligned} r_1^0(t) &= 1 - S_1^0(t, 1). \\ &= 1 - [8\exp[-0.0341t] - 8\exp[-0.0414t] + 4\exp[-0.03266t] \\ &- 4\exp[-0.04854t] + 4\exp[-0.05584t]. - 4\exp[-0.03772t] + 4\exp[-0.04502t] \\ &- 2\exp[-0.0471t] - 2\exp[-0.03628t] + 2\exp[-0.05216t]. - 2\exp[-0.05946t] \\ &+ \exp[-0.05072t]], t \geq 0, \end{aligned} \quad (3.84)$$

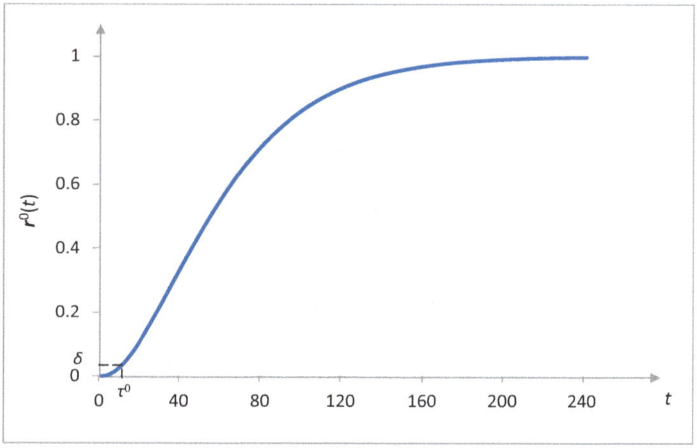


**Fig. 3.3** The graph of the port oil piping transportation system risk function

and by (3.8), considering the values of the risk function given in Table 3.1, the moment $\boldsymbol{\tau}_1^0$ of exceeding acceptable value of critical infrastructure risk function level $\delta = 0.05$ is

$$\tau_1^0 = \left(r^0\right)^{-1}(0.05) \cong 12.16 \text{ years.} \tag{3.85}$$

The graph of the port oil piping transportation system risk function is presented in Fig. 3.3.

The intensities of degradation (ageing) of the port oil piping transportation system / the intensities of the port oil piping transportation system, i.e. the asset $A_1$ departure from the safety state subsets {1, 2}, {2}, i.e. the coordinates of the vector

$$\lambda_1^0(t,\cdot) = \left[0, \lambda_1^0(t,1), \lambda_1^0(t,2)\right], t \in <0, +\infty), \tag{3.86}$$

according to (3.14) are given by

$$\boldsymbol{\lambda}_1^0(t,1) = \frac{-\frac{d\boldsymbol{S}_1^0(t,1)}{dt}}{\boldsymbol{S}_1^0(t,1)}, \boldsymbol{\lambda}_1^0(t,2) = \frac{-\frac{d\boldsymbol{S}_1^0(t,2)}{dt}}{\boldsymbol{S}_1^0(t,2)}, t \in <0, +\infty), \tag{3.87}$$

where $\boldsymbol{S}_1^0(t,1)$ is given by (3.77) and $\boldsymbol{S}_1^0(t,2)$ is given by (3.78).

The values of the intensities of degradation given by (3.87) stabilize for large time and their limit values amount:

$$\boldsymbol{\lambda}_1(1) = \lim_{t\to\infty} \boldsymbol{\lambda}_1(t,1) = \lim_{t\to\infty} \frac{4 \cdot 0.03266\,(1 + 0(t))}{4\,(1 + 0(t))} \cong 0.03266,$$

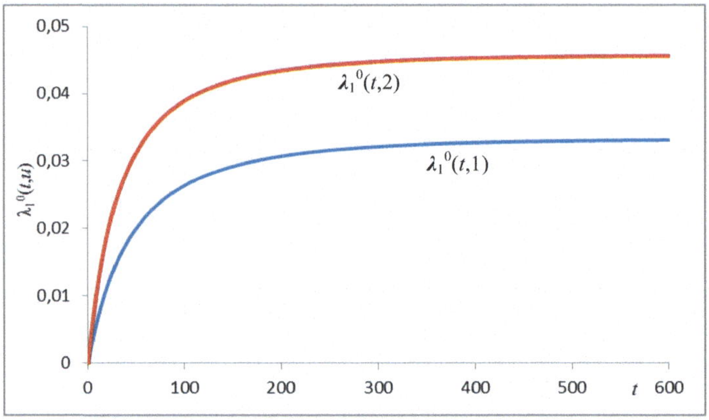


**Fig. 3.4** The graphs of the intensities of ageing of the port oil piping transportation system

$$\boldsymbol{\lambda}_1(2) = \lim_{t\to+\infty} \boldsymbol{\lambda}_1(t,2) = \lim_{t\to\infty} \frac{4 \cdot 0.04527\,(1+0(t))}{4\,(1+0(t))} \cong 0.04527. \tag{3.88}$$

$$\boldsymbol{\lambda}_1(2) = \lim_{t\to+\infty} \boldsymbol{\lambda}_1(t,2) = \lim_{t\to\infty} \frac{4 \cdot 0.04527\,(1+0(t))}{4\,(1+0(t))} \cong 0.04527. \tag{3.89}$$

The approximate mean of the intensities of degradation of the port oil piping transportation system according to (3.18) and (3.79) and (3.81) respectively are

$$\boldsymbol{\lambda}_1(t) = \frac{1}{62.71} \cong 0.01595, \tag{3.90}$$

$$\boldsymbol{\lambda}_1(2) = \frac{1}{45.84} \cong 0.02182. \tag{3.91}$$

The graphs of the intensities of degradation of the port oil piping transportation system are given in Fig. 3.4.

## 3.5 Port Oil Terminal Critical Infrastructure Safety Parameters

Thus, the approximate mean values of the asset $A_1$, the port oil terminal critical infrastructure lifetimes in the safety state subsets calculated in Sect. 3.3.4 on the basis of safety parameters of its components coming from experts are:

- in safety state subset $\{1, 2\}$

$$\mu_1^0(1) \approx 63 \text{ years}, \tag{3.92}$$

- in safety state subset {2}

$$\mu_1^0(2) \approx 46 \text{ years.} \tag{3.93}$$

The approximate mean values of the remaining assets $A_2$–$A_9$ lifetimes in the safety state subsets evaluated approximately by experts, are as follows:

- in safety state subset {1, 2}

$$\mu_i^0(1) = 80\text{years}\, i = 2, 3, \ldots, 9, \tag{3.94}$$

- in safety state subset {2}

$$\mu_i^0(2) = 50\text{years}, i = 2, 3, \ldots, 9. \tag{3.95}$$

From (3.31) and (3.92)–(3.95), it follows that the intensities of assets departure from the safety states subsets {1, 2},{2} respectively are:

- for asset $A_1$

$$\lambda_1^0(1) = \frac{1}{\mu_1^0(1)} = \frac{1}{63} \cong 0.015873, \tag{3.96}$$

$$\lambda_1^0(2) = \frac{1}{\mu_1^0(2)} = \frac{1}{46} \cong 0.021739; \tag{3.97}$$

- for assets $A_2$–$A_9$

$$\lambda_i^0(1) = \frac{1}{\mu_i^0(1)} = \frac{1}{80} = 0.0125, i = 2, 3, \ldots, 9, \tag{3.98}$$

$$\lambda_i^0(2) = \frac{1}{\mu_i^0(2)} = \frac{1}{50} 0.02, i = 2, 3, \ldots, 9. \tag{3.99}$$

## 3.6 Port Oil Terminal Critical Infrastructure Safety Indicators

Considering that the port oil terminal critical infrastructure is a three-state (z = 2) series system and assuming that the assets have picewise exponential safety functions of the form,

$$S_i^0(t, \cdot) = \left[1, S_i^0(t, 1) S_i^0(t, 2)\right], \underline{t} \geq 0,$$

where

$$S_i^0(t,u) = \exp[-\lambda_i^0(u)t], t \geq 0, u = 1,2, i = 1,2,\ldots,9,$$

and, considering (3.96)–(3.99), we have

$$S_1^0(t,1) = \exp[-0.015873t], S_1^0(t,2) = \exp[-0.021739], \underline{t} \geq 0, \quad (3.100)$$

$$S_i^0(t,1) = \exp[-0.0125t], S_i^0(t,1) = \exp[-0.02t], \underline{t} \geq 0, i = 2,\ldots,9. \quad (3.101)$$

Hence, applying the Formula (2.9) for the safety function of a series critical infrastructure, the port oil terminal critical infrastructure safety function (SafI1) is given by

$$\boldsymbol{S}^0(t,\ \cdot) = [1,\ \boldsymbol{S}^0(t,1)\boldsymbol{S}^0(t,2)], t \geq 0, \quad (3.102)$$

where

$$\begin{aligned}\boldsymbol{S}^0(t,1) &= \prod_{i=1}^{9} S_i^0(t,1) \\ &= \exp[-0.015873t]\exp[-0.0125t]\exp[-0.0125t]\exp[-0.0125t] \\ &\exp[-0.0125t]\exp[-0.0125t]\exp[-0.0125t]\exp[-0.0125t]\exp[-0.0125t] \\ &= \exp[-0.115873t], t \geq 0, \end{aligned} \quad (3.103)$$

$$\begin{aligned}\boldsymbol{S}^0(t,2) &= \prod_{i=1}^{9} S_i^0(t,2) \\ &= \exp[-0.021739t]\exp[-0.02t]\mathrm{l}\exp[-0.02t]\exp[-0.02t]\exp[-0.02t] \\ &\exp[-0.02t]\exp[-0.02t]\exp[-0.02t]\exp[-0.02t] \\ &= \exp[-0.181739t], t \geq 0. \end{aligned} \quad (3.104)$$

The graph of the port oil terminal critical infrastructure safety function is shown in Fig. 3.5.

As the critical safety state is $r = 1$, then by (3.103) and (3.4), the port oil terminal critical infrastructure risk function (SafI2), is given by

$$\begin{aligned} r^0(\mathrm{t}) &= 1 - \boldsymbol{S}^0(t,1) \\ &= 1 - \exp[-0.115873t] \text{ for } t \geq 0. \end{aligned} \quad (3.105)$$

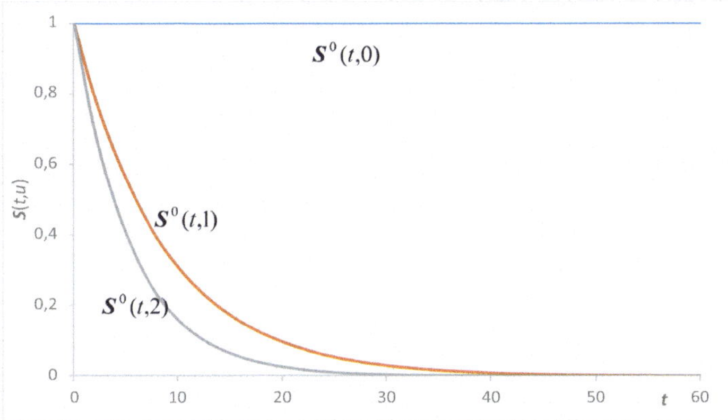


**Fig. 3.5** The graph of the port oil terminal critical infrastructure safety function $\boldsymbol{S}^0(t, \cdot)$ coordinates

The graph of the risk function $\boldsymbol{r}^0(t)$ of the port oil terminal critical infrastructure, the fragility curve (SafI3), is shown in Fig. 3.6.

From (3.105) and (3.8), the moment when the port oil terminal critical infrastructure risk function exceeds a permitted level $\delta^0 = 0.05$ (SafI6), is

$$\tau^0 = -\frac{1}{0.115873}\ln(1 - 0.05) \cong 0.44\text{years}. \tag{3.106}$$

From (3.103)–(3.104), after applying (3.5)–(3.7) and (3.9)–(3.11), the mean values (SafI4, SafI7) and standard deviations (SafI5, SafI8) of the port oil terminal critical infrastructure lifetimes in the safety state subsets {1, 2}, {2}, respectively are:

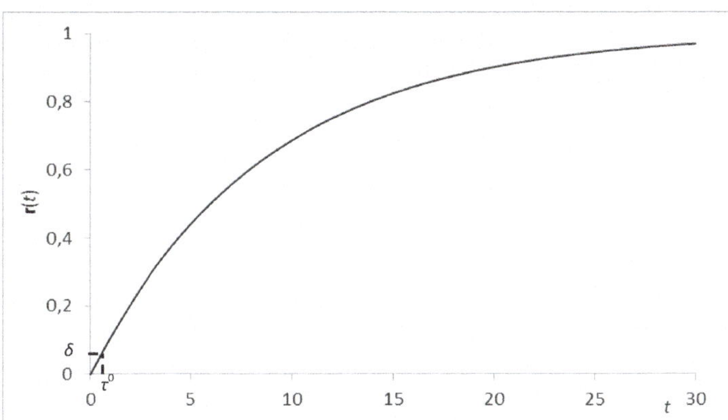


**Fig. 3.6** The graph of the risk function $\boldsymbol{r}^0(t)$ (the fragility curve) of the port oil terminal critical infrastructure

$$\boldsymbol{\mu}^0(1) = \int_0^\infty \boldsymbol{S}^0(t,1)dt = \int_0^\infty \exp[-0.115873t]dt = \frac{1}{0.115873} \cong 8.63\text{years}, \tag{3.107}$$

$$\boldsymbol{\mu}^0(2) = \int_0^\infty \boldsymbol{S}^0(t,2)dt = \int_0^\infty \exp[-0.181739t]dt = \frac{1}{0.181739} \cong 5.50\text{years}, \tag{3.108}$$

$$\boldsymbol{\sigma}^0(1) \cong \sqrt{\boldsymbol{n}^0(1) - [\boldsymbol{\mu}^0(1)]^2} = \sqrt{148.9585696 - (8.63)^2} \cong 8.63\text{years}, \tag{3.109}$$

$$\boldsymbol{\sigma}^0(2) \cong \sqrt{\boldsymbol{n}^0(2) - [\boldsymbol{\mu}^0(2)]^2} = \sqrt{60.55272994 - (5.50)^2} \cong 5.50\text{years}, \tag{3.110}$$

where

$$\boldsymbol{n}^0(1)7 = 2\int_0^\infty t\boldsymbol{S}^0(t,1)dt = 2\int_0^\infty t\exp[-0.115873t]dt$$
$$= \frac{2}{(0.115873)^2} = \frac{2}{0.013426552} = 14.9585696,$$

$$\boldsymbol{n}^0(2) = 2\int_0^\infty t\boldsymbol{S}^0(t,2)dt = 2\int_0^\infty t\exp[-0.181739t]dt$$
$$= \frac{2}{(0.181739)^2} = \frac{2}{0.033029064} = 60.55272994.$$

Further from (3.12), it follows that the mean values of the port oil terminal critical infrastructure lifetimes in the particular safety states (SafI9), are:

$$\overline{\boldsymbol{\mu}}^0(1) = \boldsymbol{\mu}^0(1) - \boldsymbol{\mu}^0(2) = 8.63 - 5.50 \cong 3.13 \text{ years}, \tag{3.111}$$

$$\overline{\boldsymbol{\mu}}^0(2) = \boldsymbol{\mu}^0(2) \cong 5.50 \text{ years}. \tag{3.112}$$

Considering (3.103) and (3.104), the port oil terminal critical infrastructure intensities of ageing (SafI10), according to (3.14), are:

$$\boldsymbol{\lambda}^0(t,1) = \frac{-\frac{d\boldsymbol{S}^0(t,1)}{dt}}{\boldsymbol{S}^0(t,1)} = \frac{0.115873\exp[-0.115873t]}{\exp[-0.115873t]} = 0.115873, \tag{3.113}$$

$$\boldsymbol{\lambda}^0(t,2) = \frac{-\frac{d\boldsymbol{S}^0(t,2)}{dt}}{\boldsymbol{S}^0(t,2)} = \frac{0.181739\exp[-0.181739t]}{\exp[-0.181739t]} = 0.181739. \tag{3.114}$$

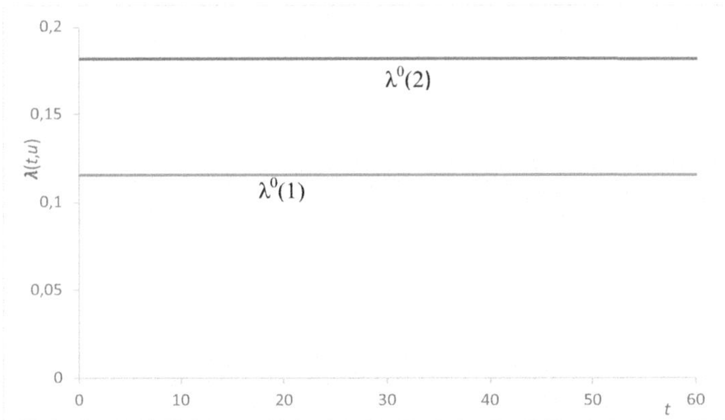


**Fig. 3.7** The graphs of the intensities of ageing of the port oil terminal critical infrastructure

The graphs of the intensities of ageing of the port oil terminal critical infrastructure are given in Fig. 3.7.

## References

1. Aven T (1985) Reliability evaluation of multistate systems with multistate components. IEEE Trans Reliab 34:473–479
2. Barlow RE, Wu AS (1978) Coherent systems with multi-state components. Math Oper Res 4:275–281
3. Brunelle RD, Kapur KC (1999) Review and classification of reliability measures for multistate and continuum models. IEEE Trans 31:1117–1180
4. Gouldby BP, Schultz MT, Simm JD, Wibowo JL (2010) Beyond the factor of safety: developing fragility curves to characterize system reliability, Report in Water Resources Infrastructure Program ERDC SR-10-1, Prepared for Headquarters. Washington, U.S, Army Corps of Engineers
5. Hudson JC, Kapur KC (1982) Reliability theory for multistate systems with multistate components. Microelectron and Reliab 22:1–7
6. Hudson JC, Kapur KC (1983) Reliability analysis of multistate systems with multistate components. Trans Inst Ind Eng 15:127–135
7. Hudson J, Kapur K (1985) Reliability bounds for multistate systems with multistate components. Oper Res 33:735–744
8. Kołowrocki K (2005) Reliability of large systems. Elsevier, Amsterdam, Boston, Heidelberg, London, New York, Oxford, Paris, San Diego, San Francisco, Singapore, Sydney, Tokyo
9. Kołowrocki K (2014) Reliability of large and complex systems, 2nd edn. Elsevier, Amsterdam, Boston, Heidelberg, London, New York, Oxford, Paris, San Diego, San Francisco, Singapore, Sydney, Tokyo
10. Kołowrocki K, Kuligowska E, Soszyńska-Budny J, Torbicki M (2018) Oil transport in port part 0 port oil piping transportation system safety without outside impacts. J Pol Saf Reliab Assoc Summer Saf Reliab Seminars 9(2):1–10
11. Kołowrocki K, Soszyńska-Budny J (2011) Reliability and safety of complex technical systems and processes. Modeling—identification—prediction—optimization. Springer, London, Dordrecht, Heildeberg, New York

12. Kołowrocki K, Soszyńska-Budny J (2018) Critical infrastructure safety indicators. In: International conference on industrial engineering and engineering management IEEM 2018
13. Kołowrocki K, Soszyńska-Budny J (2018) Critical infrastructure impacted by operation safety and resilience indicators. In: International conference on industrial engineering and engineering management IEEM 2018
14. Kossow A, Preuss W (1995) Reliability of linear consecutively-connected systems with multistate components. IEEE Trans Reliab 44:518–522
15. Lisnianski A, Levitin G (2003) Multi-state system reliability. In: Assessment, optimisation and applications. World Scientific Publishing Co. Pte. Ltd., New Jersey, London, Singapore, Hong Kong
16. Soszyńska J (2006) Reliability and risk evaluation of a port oil pipeline transportation system in variable operation conditions. Int J Press Vessel Pip 87(2–3):81–87
17. Xue J (1985) On multi-state system analysis. IEEE Trans Reliab 34:329–337

# Chapter 4
# Modelling Safety of Critical Infrastructure Impacted by Operation Process

The safety of critical infrastructure impacted by its operation process is considered. The critical infrastructure operation process and its parameters are defined and its characteristics are determined. The safety and resilience indicators of the critical infrastructure and its assets related to the operation process impact are proposed. The safety of the port oil terminal critical infrastructure impacted by its operation process is examined and its safety and resilience indicators are determined.

## 4.1 Critical Infrastructure Operation Process

We denote by $Z(t)$, $t \in< 0, \infty)$, the critical infrastructure operation process, and we assume that it is impacted in a various way at this process operation states $z_b$, $b = 1, 2, \ldots, v$, [1–3, 16]. We assume that the changes of the operation states of the critical infrastructure operation process $Z(t)$ have an influence on the critical infrastructure safety structure and on the safety of the critical infrastructure assets $A_i$, $i = 1, 2, \ldots, n$, [3–5, 11–14].

The critical infrastructure operation process is defined by the following parameters (OPP) that can be identified either statistically using the methods given in [9, 11] or evaluated approximately by experts [8, 13]:

- the number of operation states (OPP1) [13]

  $v$;

- the operation states (OPP2) [13]



J. Soszyńska-Budny, *Safety Analysis of Critical Infrastructure*,
Lecture Notes in Intelligent Transportation and Infrastructure,
https://doi.org/10.1007/978-3-030-65339-2_4

$$z_1, z_2, \ldots, z_v;$$

– the vector

$$[p_b(0)]_{1\times v} = [p_1(0), p_2(0), \ldots p_v(0)], \tag{4.1}$$

of the initial probabilities (OPP3) [13]

$$p_b(0) = P(Z(0) = z_b), b = 1, 2, \ldots, v,$$

of the critical infrastructure operation process *Z*(*t*) staying at particular operation states $z_b$ at the moment $t = 0$;

– the matrix

$$[p_{bl}]_{v\times v} = \begin{bmatrix} p_{11}\, p_{12} \cdots p_{1v} \\ p_{21}\, p_{22} \cdots p_{2v} \\ \cdots \\ p_{v1}\, p_{v2} \cdots p_{vv} \end{bmatrix} \tag{4.2}$$

of probabilities of transition (OPP4) [13]

$$p_{bl}, b, l = 1, 2, \ldots, v, \quad p_{bb} = 0,\ b = 1, 2, \ldots, v,$$

of the critical infrastructure operation process *Z*(*t*) between the operation states $z_b$ and $z_l$;

– the matrix

$$[M_{bl}]_{v\times v} = \begin{bmatrix} M_{11} & M_{12} & \ldots & M_{1v} \\ M_{21} & M_{22} & \ldots & M_{2v} \\ \ldots \\ M_{v1} & M_{v2} & \ldots & M_{vv} \end{bmatrix}, \tag{4.3}$$

of the mean values of conditional sojourn times (OPP5) [13]

$$M_{bl} = E[\theta_{bl}] = \int_0^\infty t\, dH_{bl}(t) = \int_0^\infty t\, h_{bl}(t) dt,$$

$$b, l = 1, 2, \ldots, v, b \neq l, M_{bb} = 0, b = 1, 2, \ldots, v, \tag{4.4}$$

of the critical infrastructure operation process $Z(t)$ conditional sojourn times $\theta_{bl}$ at the operation state $z_b$ when the next state is $z_l$, where

$$H_{bl}(t) = P(\theta_{bl} < t),\ t \geq 0,\ b, l = 1, 2, \ldots, v,\ b \neq l,$$

are conditional distribution functions of the critical infrastructure operation process $Z(t)$ conditional sojourn times $\theta_{bl}$ at the operation states corresponding to conditional density functions

$$h_{bl}(t) = \frac{dH_{bl}(t)}{dt}, t \geq 0, \quad b, l = 1, 2, \ldots, v, \quad b \neq l.$$

The following critical infrastructure operation process characteristics (OPC) can be either calculated analytically using the above parameters of the operation process or evaluated approximately by experts [11, 13]:

– the vector

$$[M_b]_{1xv} = [M_1, M_2, \ldots, M_v], \tag{4.5}$$

of mean values of the critical infrastructure operation process $Z(t)$ unconditional sojourn times $\theta_b$, $b = 1, 2, \ldots, v$, at the operation states (OPC1)

$$M_b = E[\theta_b] = \sum_{i=1}^{v} p_{bl} M_{bl}, \quad b = 1, 2, \ldots, v, \tag{4.6}$$

where $M_{bl}$ are defined by the formula (4.4);

– the vector

$$[p_b]_{1xv} = [p_1, p_2, \ldots, p_v], \tag{4.7}$$

of limit values of transient probabilities (OPC2)

$$p_b(t) = P(Z(t) = z_b), t \in < 0, +\infty), \quad b = 1, 2, \ldots, v,$$

of the critical infrastructure operation process $Z(t)$ at the particular operation states $z_b$, $b = 1, 2, \ldots, v$, given by [13]

$$p_b = \lim_{t \to \infty} p_b(t) = \frac{\pi_b M_b}{\sum_{i=1}^{v} \pi_l M_l}, b = 1, 2, \ldots, v, \tag{4.8}$$

where $M_b, b = 1, 2, \ldots, v$, are given by (4.6) and the steady probabilities $\pi_b$ of the vector $[\pi_b]_{1xv}$ satisfy the system of equations [3, 9]

$$\begin{cases} [\pi_b] = [\pi_b][p_{bl}] \\ \sum_{l=1}^{v} \pi_l = 1; \end{cases} \tag{4.9}$$

– the vector

$$\left[\hat{M}_b\right]_{1\times v} = \left[\hat{M}_1, \hat{M}_2, \ldots, \hat{M}_v\right], \tag{4.10}$$

of the mean values of the total sojourn times $\hat{\theta}_b$ (OPC3)

$$\hat{M}_b = E\left[\hat{\theta}_b\right] \cong p_b\theta, \quad b = 1, 2, \ldots, v, \tag{4.11}$$

of the critical infrastructure operation process $Z(t)$ at the particular operation states $z_b$, $b = 1, 2, \ldots, v$, during the fixed critical infrastructure opetation time $\theta$, where $p_b$ are given by (4.8).

## 4.2 Safety and Resilience Indicators of Critical Infrastructure Impacted by Its Operation Process

We denote by $[T^1(u)]^{(b)}$, $u = 1, 2, \ldots, z$, $b = 1, 2, \ldots, v$, the critical infrastructure conditional lifetime in the safety state subset $\{u, u+1, \ldots, z\}$, $u = 1, 2, \ldots, z$, while its operation process $Z(t)$, $t \in< 0, \infty)$, is at the operation state $z_b, b = 1, 2, \ldots, v$, and the conditional safety function of the critical infrastructure at this operation state by the vector [6, 10–15]

$$\left[\boldsymbol{S}^1(t, \cdot)\right]^{(b)} = \left[1, \left[\boldsymbol{S}^1(t, 1)\right]^{(b)}, \ldots, \left[\boldsymbol{S}^1(t, z)\right]^{(b)}\right], \tag{4.12}$$

with the coordinates

$$\left[\boldsymbol{S}^1(t, u)\right]^{(b)} = P\left(\left[T^1(u)\right]^{(b)} > t | Z(t) = z_b\right) \tag{4.13}$$

for $t \in< 0, \infty)$, $u = 1, 2, \ldots, z, b = 1, 2, \ldots, v$.

The safety function $[\boldsymbol{S}^1(t, u)]^{(b)}$, $u = 1, 2, \ldots, z$, is the conditional probability that the critical infrastructure impacted by its operation process $Z(t)$, $t \in< 0, \infty)$, lifetime $[T^1(u)]^{(b)}$, $u = 1, 2, \ldots, z$, in the safety state subset $\{u, u + 1, \ldots, z\}$,

$u = 1, 2, \ldots, z$, is greater than $t$, while the critical infrastructure operation process $Z(t)$, $t \in< 0, \infty)$, is at the operation state $z_b$, [13].

Next, we denote by $T^1(u)$,$u = 1, 2, \ldots, z$, the critical infrastructure impacted by its operation process $Z(t)$, $t \in< 0, \infty)$, unconditional lifetime in the safety state subset $\{u, u + 1, \ldots, z\}$, $u = 1, 2, \ldots, z$, and the firs safety indicator, the unconditional safety function (SafI1) of the critical infrastructure impacted by its operation process $Z(t)$,$t \in< 0, \infty)$, by the vector [13]

$$\boldsymbol{S}^1(t, \cdot) = \left[1, \boldsymbol{S}^1(t, 1) \ldots, \boldsymbol{S}^1(t, z)\right], \tag{4.14}$$

with the coordinates

$$\boldsymbol{S}^1(t, u) = P\left(T^1(u) > t\right) \tag{4.15}$$

for $t \in< 0, \infty)$, $u = 1, 2, \ldots, z$.

In the case when the system operation time $\theta$ is large enough, the coordinates of the unconditional safety function of the critical infrastructure related to the operation process $Z(t)$, $t \in< 0, \infty)$, defined by (4.15), are given by [13]

$$\boldsymbol{S}^1(t, u) \cong \sum_{b=1}^{v} p_b\left[\boldsymbol{S}^1(t, u)\right]^{(b)} \text{ for } t \geq 0, \quad u = 1, 2, \ldots, z, \tag{4.16}$$

where $[\boldsymbol{S}^1(t, u)]^{(b)}$, $u = 1, 2, \ldots, z$, $b = 1, 2, \ldots, v$, are the coordinates of the critical infrastructure impacted by its operation process $Z(t)$, $t \in< 0, \infty)$, conditional safety functions defined by (4.12), (4.13) and $p_b$, $b = 1, 2, \ldots, v$, are the critical infrastructure operation process $Z(t)$, $t \in< 0, \infty)$, limit transient probabilities at the operation states $z_b$, $b = 1, 2, \ldots, v$, given by (4.8).

If $r$ is the critical safety state, then the second safety indicator of the critical infrastructure impacted by its operation process $Z(t)$, $t \in< 0, \infty)$, the risk function (SafI2) [13]

$$\boldsymbol{r}^1(t) = P(s(t) < r|s(0) = z) = P\left(T^1(r) \leq t\right), t \in< 0, \infty), \tag{4.17}$$

is defined as a probability that the critical infrastructure impacted by its operation process $Z(t)$, $t \in< 0, \infty)$, is in the subset of safety states worse than the critical safety state $r$, $r \in \{1, \ldots, z\}$ while it was in the best safety state $z$ at the moment $t = 0$ and given by [7, 9, 13–15]

$$\boldsymbol{r}^1(t) = 1 - \boldsymbol{S}^1(t, r), t \in< 0, \infty), \tag{4.18}$$

where $\boldsymbol{S}^1(t, r)$ is the coordinate of the critical infrastructure impacted by its operation process $Z(t)$, $t \in< 0, \infty)$, unconditional safety function given by (4.16) for $u = r$.

The graph of the critical infrastructure risk function $\boldsymbol{r}^1(t)$, $t \in <0,\infty)$, defined by (4.18), is the safety indicator called the fragility curve (SafI3) of the critical infrastructure impacted by its operation process $Z(t)$, $t \in <0,\infty)$.

Other practically useful safety and resilience indicators of the critical infrastructure impacted by its operation process $Z(t)$, $t \in <0,\infty)$, are:

- the mean value of the critical infrastructure unconditional lifetime $T^1(r)$ up to exceeding critical safety state $r$ (SafI4) [13] given by

$$\boldsymbol{\mu}^1(r) = \int_0^\infty [\boldsymbol{S}^1(t,r)]dt \cong \sum_{b=1}^{v} p_b[\boldsymbol{\mu}^1(r)]^{(b)}, \tag{4.19}$$

where $[\boldsymbol{\mu}^1(r)]^{(b)}$ are the mean values of the critical infrastructure conditional lifetimes $[T^1(r)]^{(b)}$ in the safety state subset $\{r, r+1, \ldots, z\}$ at the operation state $z_b$, $b = 1, 2, \ldots, v$, given by

$$[\boldsymbol{\mu}^1(r)]^{(b)} = \int_0^\infty [\boldsymbol{S}^1(t,r)]^{(b)}dt, \quad b = 1, 2, \ldots, v, \tag{4.20}$$

and $[\boldsymbol{S}^1(t,r)]^{(b)}$, $b = 1, 2, \ldots, v$, are defined by (4.12), (4.13) and $p_b$ are given by (4.8);

- the standard deviation of the critical infrastructure lifetime $T^1(r)$ up to the exceeding the critical safety state $r$(SafI5) given by [13]

$$\boldsymbol{\sigma}^1(r) = \sqrt{\boldsymbol{n}^1(r) - [\boldsymbol{\mu}^1(r)]^2}, \tag{4.21}$$

where

$$\boldsymbol{n}^1(r) = 2\int_0^\infty t\boldsymbol{S}^1(t,r)dt, \tag{4.22}$$

and $\boldsymbol{S}^1(t,r)$ is defined by (4.13) for $u = r$ and $\boldsymbol{\mu}^1(r)$ is given by (4.19);

- the moment $\boldsymbol{\tau}^1$ of exceeding acceptable value of critical infrastructure risk function level $\delta$ (SafI6) given by [13]

$$\boldsymbol{\tau}^1 = \boldsymbol{r}^{1^{-1}}(\delta), \tag{4.23}$$

where $\boldsymbol{r}^{1\ -1}(\delta)$ is the inverse function of the risk function $\boldsymbol{r}^1(t)$ given by (4.17);

– the mean values of unconditional lifetimes of the critical infrastructure in the safety state subsets $\{u, u+1, \ldots, z\}$, $u = 1, 2, \ldots, z$, (SafI7) given by [13]

$$\boldsymbol{\mu}^1(u) = \int_0^\infty [\boldsymbol{S}^1(t, u)]dt \cong \sum_{b=1}^{v} p_b[\boldsymbol{\mu}^1(u)]^{(b)}, \; u = 1, 2, \ldots, z, \tag{4.24}$$

where $[\boldsymbol{\mu}^1(u)]^{(b)}$ are the mean values of the critical infrastructure conditional lifetimes $[T^1(u)]^{(b)}$ in the safety state subsets $\{u, u+1, \ldots, z\}$, at the operation state $z_b$, $b = 1, 2, \ldots, v$, given by

$$[\boldsymbol{\mu}^1(u)]^{(b)} = \int_0^\infty [\boldsymbol{S}^1(t, u)]^{(b)}dt, u = 1, 2, \ldots, z, \; b = 1, 2, \ldots, v, \tag{4.25}$$

and $[\boldsymbol{S}(t, u)]^{(b)}$, $u = 1, 2, \ldots, z$, $b = 1, 2, \ldots, v$, are defined by (4.12), (4.13) and $p_b$ are given by (4.8);

– the standard deviations of the critical infrastructure lifetimes in the safety state subsets $\{u, u+1, \ldots, z\}$, $u = 1, 2, \ldots, z$, (SafI8) given by [13]

$$\boldsymbol{\sigma}^1(u) = \sqrt{\boldsymbol{n}^1(u) - [\boldsymbol{\mu}^1(u)]^2}, \; u = 1, 2, \ldots z, \tag{4.26}$$

where

$$\boldsymbol{n}^1(u) = 2\int_0^\infty t\boldsymbol{S}^1(t, u)dt, u = 1, 2, \ldots z; \tag{4.27}$$

– the mean lifetimes $\bar{\boldsymbol{\mu}}^1(u)$, $u = 1, 2, \ldots, z$, of the critical infrastructure in the particular safety states (SafI9) [13]

$$\bar{\boldsymbol{\mu}}^1(u) = \boldsymbol{\mu}^1(u) - \boldsymbol{\mu}^1(u+1), \; u = 0, 1, \ldots, z-1, \; \bar{\boldsymbol{\mu}}^1(z) = \boldsymbol{\mu}^1(z); \tag{4.28}$$

– the intensities of degradation of the critical infrastructure/the intensities of critical infrastructure departure from the safety state subset $\{u, u+1, \ldots, z\}$,, $u = 1, 2, \ldots, z$, (SafI10), i.e. the coordinates of the vector [13]

$$\boldsymbol{\lambda}^1(t, \cdot) = \left[0, \boldsymbol{\lambda}^1(t, 1), \ldots, \boldsymbol{\lambda}^1(t, z)\right], t \in < 0, +\infty), \tag{4.29}$$

where

$$\boldsymbol{\lambda}^1(t,u) = \frac{-\frac{dS^1(t,u)}{dt}}{\mathbf{S}^1(t,u)},\ t \in <0,+\infty),\ u = 1,2,\ldots,z; \tag{4.30}$$

- the coefficients of operation process impact on the critical infrastructure intensities of degradation/the coefficients of operation process impact on critical infrastructure intensities of departure from the safety state subset $\{u, u+1, \ldots, z\}$ (ResI1), i.e. the coordinates of the vector [13]

$$\rho^1(t,\cdot) = \left[0, \rho^1(t,1), \ldots, \rho^1(t,z)\right], t \in <0,+\infty), \tag{4.31}$$

where

$$\boldsymbol{\lambda}^1(t,u) = \boldsymbol{\rho}^1(t,u)\cdot\boldsymbol{\lambda}^0(t,u), t \in <0,+\infty), u = 1,2,\ldots,z, \tag{4.32}$$

i.e.

$$\boldsymbol{\rho}^1(t,u) = \frac{\lambda^1(t,u)}{\lambda^0(t,u)}, t \in <0,+\infty\Big), u = 1,2,\ldots,z, \tag{4.33}$$

and $\boldsymbol{\lambda}^0(t,u)$, $t \in <0,+\infty)$, $u = 1,2,\ldots,z$, are the intensities of degradation of the critical infrastructure without of operation process impact, defined by (3.14), i.e. the coordinate of the vector [13]

$$\boldsymbol{\lambda}^0(t,\cdot) = \left[0, \boldsymbol{\lambda}^0(t,1), \ldots, \boldsymbol{\lambda}^0(t,z)\right], t \in <0,+\infty), \tag{4.34}$$

and $\boldsymbol{\lambda}^1(t,u)$, $t \in <0,+\infty)$, $u = 1,2,\ldots,z$, are the intensities of degradation of the critical infrastructure with of operation process impact, defined by (4.30), i.e. the coordinate of the vector

$$\boldsymbol{\lambda}^1(t,\cdot) = \left[0, \boldsymbol{\lambda}^1(t,1), \ldots, \boldsymbol{\lambda}^1(t,z)\right], t \in <0,+\infty); \tag{4.35}$$

- the indicator of critical infrastructure resilience to operation process impact (ResI2) defined by [13]

$$\boldsymbol{RI}^1(t,r) = \frac{1}{\boldsymbol{\rho}^1(t,r)}, \quad t \in <0,+\infty), \tag{4.36}$$

where $\boldsymbol{\rho}^1(t,r)$, $t \in <0,+\infty)$, is the coefficients of operation process impact on the critical infrastructure intensities of degradation given by (4.32), (4.33) for $u = r$.

In the case, when the critical infrastructure have the piecewise exponential safety functions, i.e.

$$\boldsymbol{S}^1(t,\cdot) = \left[0, \boldsymbol{S}^1(t,1), \ldots, \boldsymbol{S}^1(t,z)\right], t \in <0, +\infty), \tag{4.37}$$

where

$$\boldsymbol{S}^1(t,u) = \exp\left[-\boldsymbol{\lambda}^1(u)t\right], \quad t \in <0, +\infty), \quad \boldsymbol{\lambda}^1(u) \geq 0, u = 1, 2, \ldots, z, \tag{4.38}$$

the critical infrastructure safety and resilience indicators defined by (4.29)–(4.36) take forms:

- the intensities of degradation of the critical infrastructure related to the operation process impact, (SafII10), i.e. the coordinates of the vector

$$\boldsymbol{\lambda}^1(\cdot) = \left[0, \boldsymbol{\lambda}^1(1), \ldots, \boldsymbol{\lambda}^1(z)\right], \tag{4.39}$$

are constant and

$$\boldsymbol{\lambda}^1(u) = \frac{1}{\boldsymbol{\mu}^1(u)}; \tag{4.40}$$

- the coefficients of the operation process impact on the critical infrastructure intensities of degradation/the coefficients of the operation process impact on critical infrastructure intensities of departure from the safety state subset, (ResI1) i.e. the coordinates of the vector [13]

$$\boldsymbol{\rho}^1(\cdot) = \left[0, \boldsymbol{\rho}^1(1), \ldots, \boldsymbol{\rho}^1(z)\right], \tag{4.41}$$

where

$$\boldsymbol{\rho}^1(u) = \frac{\boldsymbol{\lambda}^1(u)}{\boldsymbol{\lambda}^0(u)} = \frac{\boldsymbol{\mu}^0(u)}{\boldsymbol{\mu}^1(u)}, \quad u = 1, 2, \ldots, z, \tag{4.42}$$

and $\boldsymbol{\lambda}^0(u)$, $u = 1, 2, \ldots, z$, are the intensities of degradation of the critical infrastructure without of operation process impact, given by (3.18), i.e. the coordinate of the vector [13]

$$\boldsymbol{\lambda}^0(u) = \left[0, \boldsymbol{\lambda}^0(1), \ldots, \boldsymbol{\lambda}^0(z)\right] \tag{4.43}$$

and $\boldsymbol{\lambda}^1(u)$, $u = 1, 2, \ldots, z$, are the intensities of degradation of the critical infrastructure related to the operation impact, given by (4.40), i.e. the coordinates of the vector

$$\boldsymbol{\lambda}^1(\cdot) = \left[0, \boldsymbol{\lambda}^1(1), \ldots, \boldsymbol{\lambda}^1(z)\right]; \tag{4.44}$$

- the indicator of critical infrastructure resilience to operation process impact (ResI2) defined by

$$\boldsymbol{RI}^1(r) = \frac{1}{\boldsymbol{\rho}^1(r)}, t \in < 0, +\infty), \tag{4.45}$$

where $\boldsymbol{\rho}^1(r)$ is the coefficient of operation process impact on the critical infrastructure intensities of degradation given by (4.42) for $u = r$.

## 4.3 Critical Infrastructure Assets Safety Parameters

We denote by $[T_i^1(u)]^{(b)}$, $u = 1, 2, \ldots, z$, $b = 1, 2, \ldots, v$, $i = 1, 2, \ldots, n$, the critical infrastructure asset $A_i$, $i = 1, 2, \ldots, n$, conditional lifetime in the safety state subset $\{u, u+1, \ldots, z\}$ while the critical infrastructure is at the operation state $z_b$, $b = 1, 2, \ldots, v$, and its conditional safety function (SafI1) by the vector [9, 11–14]

$$\left[S_i^1(t, \cdot)\right]^{(b)} = \left[1, \left[S_i^1(t, 1)\right]^{(b)}, \ldots, \left[S_i^1(t, z)\right]^{(b)}\right],$$
$$t \in < 0, \infty), \quad b = 1, 2, \ldots, v, \quad i = 1, 2, \ldots, n, \tag{4.46}$$

with the coordinates

$$[S_i^1(t, u)]^{(b)} = P([T_i^1(u)]^{(b)} > t | Z(t) = z_b) \tag{4.47}$$

for $t \in < 0, \infty)$, $u = 1, 2, \ldots, z$, $b = 1, 2, \ldots, v$, $i = 1, 2, \ldots, n$.

The safety function $[S_i^1(t, u)]^{(b)}$, $t \in < 0, \infty)$, $u = 1, 2, \ldots, z$, $b = 1, 2, \ldots, v$, $i = 1, 2, \ldots, n$, is the conditional probability that the asset $A_i$ lifetime $[T_i^1(u)]^{(b)}$ $t \in < 0, \infty)$, $u = 1, 2, \ldots, z$, $b = 1, 2, \ldots, v$, $i = 1, 2, \ldots, n$, in the safety state subset $\{u, u+1, \ldots, z\}$ is greater than $t$, while the critical infrastructure is at the operation state $z_b$, $b = 1, 2, \ldots, v$.

The conditional safety functions $[S_i^1(t, u)]^{(b)}$, $t \in < 0, \infty)$, $u = 1, 2, \ldots, z$, $b = 1, 2, \ldots, v$, $i = 1, 2, \ldots, n$, defined by (4.47) are called the coordinates of the asset $A_i$, $i = 1, 2, \ldots, n$, conditional safety function $[S_i^1(t, \cdot)]^{(b)}$, $t \in < 0, \infty)$, $b = 1, 2, \ldots, v$, $i = 1, 2, \ldots, n$, given by (4.46), while the critical infrastructure operation process $Z(t)$ is at the operation state $z_b$, $b = 1, 2, \ldots, v$. Thus, the relationship between the conditional distribution function $[F_i^1(t, u)]^{(b)}$, $t \in < 0, \infty)$, $u = 1, 2, \ldots, z$, $b = 1, 2, \ldots, v$, $i = 1, 2, \ldots, n$, of the asset $A_i$, $i = 1, 2, \ldots, n$, lifetime $[T_i^1(u)]^{(b)}$, $u = 1, 2, \ldots, z$, $b = 1, 2, \ldots, v$, $i = 1, 2, \ldots, n$, in the safety state subset $\{u, u+1, \ldots, z\}$, $u = 1$,

2, ...,z, and the coordinate $[S_i^1(t,u)]^{(b)}$, $t \in< 0, \infty)$, $u = 1, 2, \ldots, z$, $b = 1, 2, \ldots, v$, $i = 1, 2, \ldots, n$, of its conditional safety function (4.46) is given by

$$[F_i^1(t,u)]^{(b)} = P\left([T_i^1(u)]^{(b)} \le t\right), = 1 - P\left([T_i^1(u)]^{(b)} > t\right)$$
$$= 1 - [S_i^1(t,u)]^{(b)}, \; t \in< 0, \infty),$$
$$u = 1, 2, \ldots, z, \quad b = 1, 2, \ldots, v, \quad i = 1, 2, \ldots, n. \tag{4.48}$$

The function

$$[r_i^1(t)]^{(b)} = 1 - [S_i^1(t,r)]^{(b)}, \; t \in< 0, \infty), \; b = 1, 2, \ldots, v, \quad i = 1, 2, \ldots, n. \tag{4.49}$$

is the asset $A_i$, $i = 1, 2, \ldots, n$, conditional risk function (SafI2) and its graph is the asset $A_i$, $i = 1, 2, \ldots, n$, fragility curve (SafI3) while the critical infrastructure is at the operation state $z_b$, $b = 1, 2, \ldots, v$.

Moreover, the conditional mean lifetime of the asset $A_i$, $i = 1, 2, \ldots, n$, in the safety state subset $\{u, u + 1, \ldots, z\}$, $u = 1, 2, \ldots, z$, (SafI4, SafI7), while the critical infrastructure is at the operation state $z_b$, $b = 1, 2, \ldots, v$. is given by

$$[\mu_i^1(u)]^{(b)} = \int_0^\infty [S_i^1(t,u)]^{(b)} dt,$$
$$u = 1, 2, \ldots, z, \quad b = 1, 2, \ldots, v, \quad i = 1, 2, \ldots, n, \tag{4.50}$$

and the conditional standard deviation of the asset $A_i$ lifetime in the safety state subset $\{u, u + 1, \ldots, z\}$, $u = 1, 2, \ldots, z$, while the critical infrastructure is at the operation state $z_b$,$b = 1, 2, \ldots, v$, (SafI5, SafI8), is given by

$$[\sigma_i^1(u)]^{(b)} = \sqrt{[n_i^1(u)]^{(b)} - [[\mu_i^1(u)]^{(b)}]^2},$$
$$u = 1, 2, \ldots, z, \quad b = 1, 2, \ldots, v, \quad i = 1, 2, \ldots, n, \tag{4.51}$$

where

$$[n_i^1(u)]^{(b)} = 2\int_0^\infty t[S_i^1(t,u)]^{(b)} dt,$$
$$u = 1, 2, \ldots, z, \quad b = 1, 2, \ldots, v, \quad i = 1, 2, \ldots, n. \tag{4.52}$$

It can be justified [9] that the conditional mean lifetimes $[\bar{\mu}_i^1(u)]^{(b)}$, $u = 1, 2, \ldots, z$, $b = 1, 2, \ldots, v$, $i = 1, 2, \ldots, n$, of the asset $A_i$, 1, 2, ..., $n$, in the particular safety states, (SafI9), while the critical infrastructure is at the operation state $z_b$,

$b = 1, 2, \ldots, \upsilon$, are

$$[\bar{\mu}_i^1(u)]^{(b)} = [\mu_i^1(u)]^{(b)} - [\mu_i^1(u+1)]^{(b)},$$
$$u = 1, \ldots, z-1, \quad b = 1, 2, \ldots, \upsilon, \quad i = 1, 2, \ldots, n \tag{4.53}$$

$$[\bar{\mu}_i^1(z)]^{(b)} = [\mu_i^1(z)]^{(b)}, \; b = 1, 2, \ldots, \upsilon, \; i = 1, 2, \ldots, n. \tag{4.54}$$

In the case, when the critical infrastructure assets $A_i$, $i = 1, 2, \ldots, n$, at the critical infrastructure operation process $Z(t)$ states $z_b$, $b = 1, 2, \ldots, \upsilon$, have the piecewise exponential safety functions, the coordinates (4.47) of the vector (4.46) are given by

$$[S_i^1(t, u)]^{(b)} = P([T_i^1(u)]^{(b)} > t | Z(t) = z_b) = \exp[-[\lambda_i^1(u)]^{(b)} t], \; t \in < 0, \infty), \tag{4.55}$$

$u = 1, 2, \ldots, z$, $b = 1, 2, \ldots, \upsilon$, $i = 1, 2, \ldots, n$.

Existing in (4.55) the intensities of degradation of the critical infrastructure asset $A_i$, $i = 1, 2, \ldots, n$, with the critical infrastructure operation process impact at the critical infrastructure operation states $z_b$, $b = 1, 2, \ldots, \upsilon$, (SafI10), i.e. the coordinates of the vector

$$[\lambda_i^1(\cdot)]^{(b)} = \left[0, [\lambda_i^1(1)]^{(b)}, \ldots, [\lambda_i^1(z)]^{(b)}\right],$$
$$t \in < 0, +\infty), \quad b = 1, 2, \ldots, \upsilon, \quad i = 1, 2, \ldots, n, \tag{4.56}$$

are constant and given by

$$[\lambda_i^1(u)]^{(b)} = \frac{1}{[\mu_i^1(u)]^{(b)}}, \; u = 1, 2, \ldots, z, \quad b = 1, 2, \ldots, \upsilon, \quad i = 1, 2, \ldots, n, \tag{4.57}$$

and moreover

$$[\lambda_i^1(u)]^{(b)} = [\rho_i^1(u)]^{(b)} \cdot \lambda_i^0(u), \; u = 1, 2, \ldots, z, \; b = 1, 2, \ldots, \upsilon, \; i = 1, 2, \ldots, n, \tag{4.58}$$

where $\lambda_i^0(u)$ are the intensities of degradation of the critical infrastructure asset $A_i$, $i = 1, 2, \ldots, n$, without operation process impact, given by (3.31), i.e. the coordinate of the vector

$$\lambda_i^0(\cdot) = [0, \lambda_i^0(1), \ldots, \lambda_i^0(z)], i = 1, 2, \ldots, n, \tag{4.59}$$

and $[\rho_i^1(u)]^{(b)}$, $u = 1, 2, \ldots, z$, $b = 1, 2, \ldots, \upsilon$, $i = 1, 2, \ldots, n$, are the coefficients of operation process impact on the critical infrastructure asset $A_i$, $i = 1, 2, \ldots,$

$n$, intensities of degradation at the critical infrastructure operation states $z_b$, $b = 1, 2, \ldots, v$, (ResI1), i.e. the coordinate of the vector

$$[\rho_i^1(\cdot)]^{(b)} = \left[0, [\rho_i^1(1)]^{(b)}, \ldots, [\rho_i^1(z)]^{(b)}\right], \; b = 1, 2, \ldots, v, \quad i = 1, 2, \ldots, n. \tag{4.60}$$

where by (4.58), (4.57) and (3.31)

$$[\rho_i^1(u)]^{(b)} = \frac{[\lambda_i^1(u)]^{(b)}}{\lambda_i^0(u)} = \frac{\mu_i^0(u)}{[\mu_i^1(u)]^{(b)}},$$
$$u = 1, 2, \ldots, z, \quad b = 1, 2, \ldots, v, \quad i = 1, 2, \ldots, n. \tag{4.61}$$

The indicator of critical infrastructure asset $A_i$, $i = 1, 2, \ldots, n$, resilience to operation process impact (ResI2) at the critical infrastructure operation states $z_b$, $b = 1, 2, \ldots, v$, can be defined by

$$[RI_i^1(r)]^{(b)} = \frac{1}{[\rho_i^1(r)]^{(b)}}, \; t \in < 0, +\infty), \; b = 1, 2, \ldots, v, \quad i = 1, 2, \ldots, n, \tag{4.62}$$

where $[\rho_i^1(r)]^{(b)}$, $b = 1, 2, \ldots, v$, $i = 1, 2, \ldots, n$, are the coefficients of operation process impact on the critical infrastructure assets intensities of degradation at the critical infrastructure operation states $z_b$, $b = 1, 2, \ldots, v$, given by (4.61) for $u = r$.

The conditional safety function (4.46), (4.47), the conditional risk function (4.49), the conditional intensities' of degradation (4.57) and the coefficients of the operation process impact on the critical infrastructure assets intensities of degradation (4.61) are main critical infrastructure asset safety and resilience indicators.

## 4.4 Application

We consider the port oil terminal critical infrastructure impacted by its operation process. This port oil terminal main technical assets are presented in Chaps. 1 and 3. The considered port oil terminal placed at the Baltic seaside is described and its safety indicators without considering its operation process impact are determined in Chap. 3. In this section, the parameters and characteristics of the considered port oil terminal operation process are evaluated. Next under the assumption of this operation process influence on the port oil terminal safety, its safety analysis is performed and its safety and resilience indicators are evaluated.

### 4.4.1 Parameters and Characteristics of Port Oil Terminal Critical Infrastructure Operation Process

On the basis of the statistical data, given in [9], and expert opinions, it is possible to fix and to evaluate the unknown basic parameters (OPP) of the port oil terminal critical infrastructure operation process. All evaluated parameters are given in [9]. These parameters are the number of operation process states (OPP1) $\upsilon = 7$ and the operation process states (OPP2) defined as follows:

the operation state $z_1$—transport of one kind of medium from the terminal part $B$ to part $C$ using two out of three pipelines of the subsystem $S_3$ of the asset $A_1$ illustrated in Fig. 4.1 and assets $A_2$, $A_4$, $A_6$, $A_7$, $A_9$ [13];
the operation state $z_2$—transport of one kind of medium from the terminal part C to part B using one out of three pipelines of the subsystem $S_3$ of the asset $A_1$ illustrated in Fig. 4.2 and assets $A_2$, $A_4$, $A_8$, $A_9$ [13];
the operation state $z_3$—transport of one kind of medium from the terminal part $B$ through part $A$ to pier using one out of two pipelines of the subsystem $S_1$ and one out of two pipelines of the subsystem $S_2$ of the asset $A_1$ illustrated in Fig. 4.3 and assets $A_2$, $A_4$, $A_5$, $A_9$ [13];
the operation state $z_4$—transport of one kind of medium from the pier through parts $A$ and $B$ to part $C$ using one out of two pipelines of the subsystem $S_1$, one out

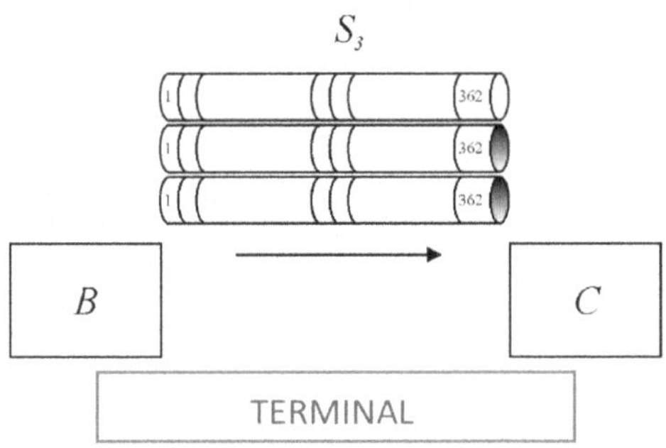


**Fig. 4.1** The scheme of the port oil piping transportation system at the operation state $z_1$

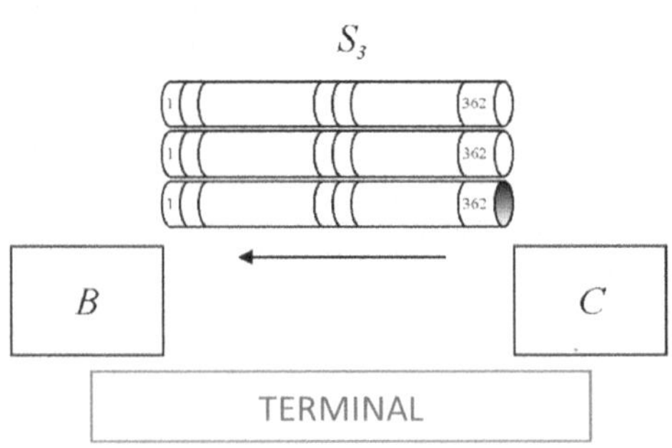


**Fig. 4.2** The scheme of the port oil piping transportation system at the operation state $z_2$

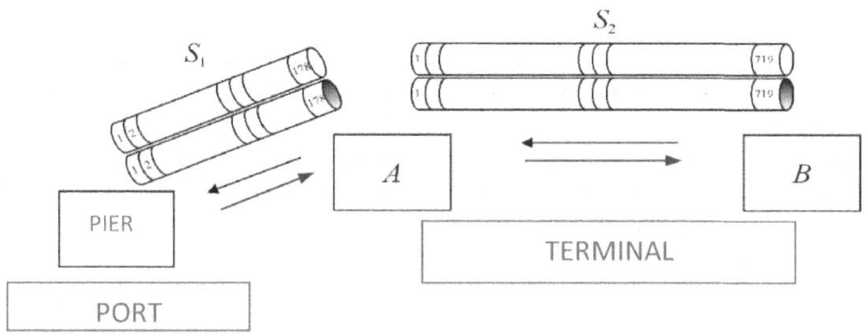


**Fig. 4.3** The scheme of port oil piping transportation system at the operation state $z_3$

of two pipelines in subsystem $S_2$ and two out of three pipelines of the subsystem $S_3$ of the asset $A_1$ illustrated in Figs. 4.4 and assets $A_2, A_3, A_4, A_5, A_6, A_7, A_9$ [13];
the operation state $z_5$—transport of one kind of medium from the pier through part $A$ to $B$ using one out of two pipelines of the subsystem $S_1$ and one out of two pipelines of the subsystem $S_2$ of the asset $A_1$ illustrated in Fig. 4.5 and assets $A_2$, $A_3, A_4, A_5, A_9$ [13];
the operation state $z_6$—transport of one kind of medium from the terminal part $B$ to $C$ using two out of three pipelines of the subsystem $S_3$, and simultaneously transport one kind of medium from the pier through part $A$ to $B$ using one out of

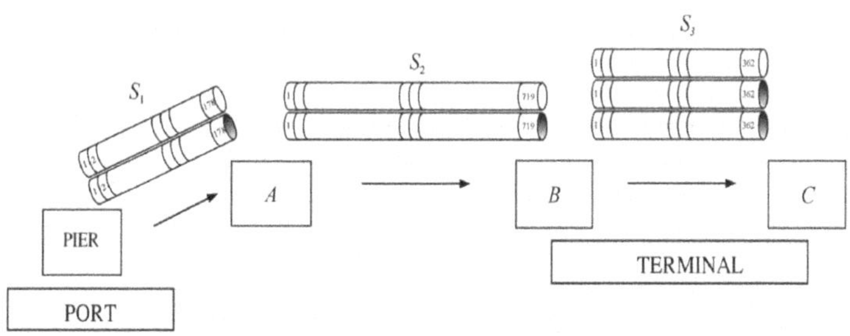


**Fig. 4.4** The scheme of the port oil piping transportation system at the operation state $z_4$

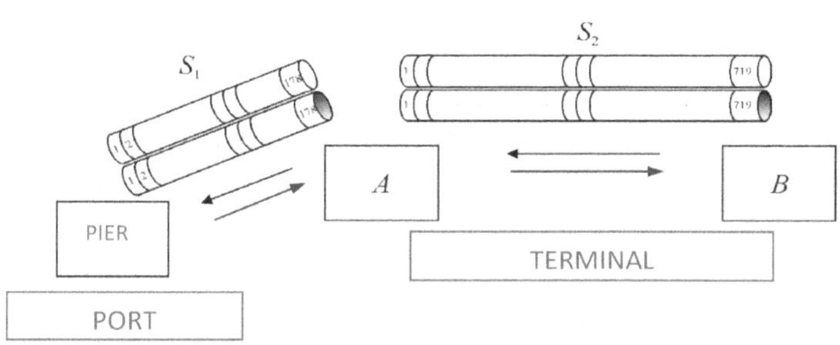


**Fig. 4.5** The scheme of port oil piping transportation system at the operation state $z_5$

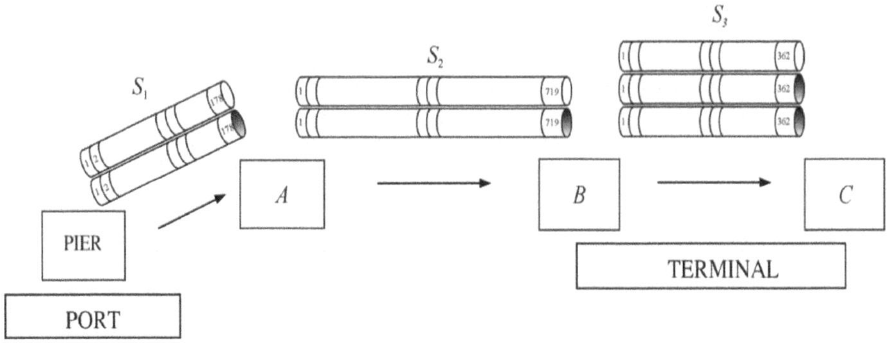


**Fig. 4.6** The scheme of the port oil piping transportation system at the operation state $z_6$

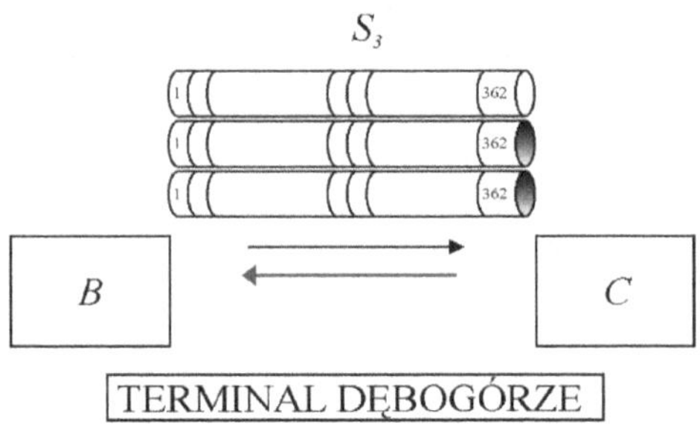


**Fig. 4.7** The scheme of the port oil piping transportation system at the operation states $z_7$

two pipelines of the subsystem $S_1$ and one out of two pipelines of the subsystem $S_2$ of the asset $A_1$ illustrated in Fig. 4.6 and assets $A_2, A_3, A_4, A_5, A_6, A_7, A_9$ [13]; the operation state $z_7$—transport of one kind of medium from the terminal part $B$ to $C$ using one out of three pipelines of the subsystem $S_3$, and simultaneously transport second kind of medium from the terminal part $C$ to $B$ using one out of three pipelines of the subsystem $S_3$ of the asset $A_1$ illustrated in Figs. 4.7 and assets $A_2, A_4, A_6, A_7, A_8, A_9$ [13].

Using the results given in [6, 9] concerned with the port oil piping transportation system operation process, obtained on the basis of the statistical data given in [9], the port oil terminal critical infrastructure operation process $Z(t)$ basic parameters (OPP) are evaluated as follows:

- the vector of the initial probabilities (OPP3) of the port oil terminal critical infrastructure operation process $Z(t)$ staying at particular operation states $z_b$, $b = 1, 2, \ldots 7$ at the moment $t = 0$

$$[p(0)]_{1x7} = [0.34,\ 0.05,\ 0,\ 0,\ 0.23,\ 0.19,\ 0.19];$$

– the matrix of probabilities (OPP4) of the port oil terminal critical infrastructure operation process of transitions from the operation state $z_b$ into the operation state $z_l$, $b, l = 1, 2, \ldots, 7$, $b \neq l$

$$[p_{bl}]_{7x7} = \begin{bmatrix} 0 & 0.022 & 0.022 & 0 & 0.534 & 0.111 & 0.311 \\ 0.2 & 0 & 0 & 0 & 0 & 0 & 0.8 \\ 1 & 0 & 0 & 0 & 0 & 0 & 0 \\ 0 & 0 & 0 & 0 & 0 & 0 & 1 \\ 0.488 & 0.023 & 0 & 0.023 & 0 & 0.233 & 0.233 \\ 0.095 & 0 & 0 & 0 & 0.667 & 0 & 0.238 \\ 0.531 & 0.062 & 0 & 0 & 0.219 & 0.188 & 0 \end{bmatrix}; \quad (4.63)$$

– the matrix of mean values $M_{bl} = E[\theta_{bl}]$ of the port oil terminal critical infrastructure operation process conditional sojourn times $\theta_{bl}$,(OPP5) at the particular operation states

$$[M_{bl}]_{7x7} = \begin{bmatrix} 0 & 1920 & 480 & 0 & 1999.4 & 1250 & 1129.6 \\ 9960 & 0 & 0 & 0 & 0 & 0 & 810 \\ 575 & 0 & 0 & 0 & 0 & 0 & 0 \\ 0 & 0 & 0 & 0 & 0 & 0 & 380 \\ 874.7 & 480 & 0 & 300 & 0 & 436.3 & 1042.5 \\ 325 & 0 & 0 & 0 & 510.7 & 0 & 438 \\ 850.9 & 510 & 0 & 0 & 2585.7 & 2380 & 0 \end{bmatrix}. \quad (4.64)$$

This way, the port oil terminal critical infrastructure operation process is approximately defined and we may predict its main characteristics (OPC). Namely, applying (4.6), (4.63) and (4.64), we have:

– the mean values of the port oil terminal critical infrastructure operation process unconditional sojourn times $\theta_b$,$b = 1, 2, \ldots, 7$, at the particular operation states (OPC1):

$$\begin{aligned} M_1 &= E[\theta_1] = p_{12}M_{12} + p_{13}M_{13} + p_{15}M_{15} + p_{16}M_{16} + p_{17}M_{17} \\ &= 0.022 - 1920 + 0.022 \cdot 480 + 0.534 \cdot 19994 + 0.111 \cdot 1250 + 0.311 \cdot 11296 \\ &\cong 161052, \end{aligned}$$

$$M_2 = E[\theta_2] = p_{21}M_{21} + p_{27}M_{27} = 0.2 \cdot 9960 + 0.8 \cdot 810 \cong 2640,$$

$$M_3 = E[\theta_3] = p_{31}M_{31} = 1 \cdot 575 = 575,$$

$$M_4 = E[\theta_4] = p_{47}M_{47} = 1 \cdot 380 = 380,$$

$$\begin{aligned} M_5 &= E[\theta_5] = p_{51}M_{51} + p_{52}M_{52} + p_{54}M_{54} + p_{56}M_{56} + p_{57}M_{57} \\ &= 0.488 \cdot 8747 + 0.023 \cdot 480 + 0.023 \cdot 300 + 0.233 \cdot 4363 + 0.233 \cdot 10425 \\ &\cong 789.35, \end{aligned}$$

$$\begin{aligned} M_6 &= E[\theta_6] = p_{61}M_{61} + p_{65}M_{65} + p_{67}M_{67} \\ &= 0.095 \cdot 325 + 0.667.5107 + 0.238 \cdot 438 + \cong 475.76, \end{aligned}$$

$$\begin{aligned} M_7 &= E[\theta_7] = p_{71}M_{71} + p_{72}M_{72} + p_{75}M_{75} + p_{76}M_{76} \\ &= 0.531.8509 + 0.062.510 + 0.219 \cdot 25857 + 0.188 \cdot 2380 \cong 1497.16; \end{aligned} \tag{4.65}$$

- the limit values of transient probabilities (OPC2) of the port oil terminal critical infrastructure operation process at the particular operation states $z_b$, $b = 1, 2, \ldots, 7$, [13]:

$$\begin{aligned} &p_1 = 0.395, \quad p_2 = 0.060, \quad p_3 = 0.003, \\ &p_4 = 0.002, \quad p_5 = 0.200, \quad p_6 = 0.058, \quad p_7 = 0.282, \end{aligned} \tag{4.66}$$

obtained after considering (4.63) in the system of Eqs. (4.9) that takes the form

$$\begin{cases} [\pi_1, \pi_2, \pi_3, \pi_4, \pi_3, \pi_6, \pi_7] = [\pi_1, \pi_2, \pi_3, \pi_4, \pi_5, \pi_6, \pi_7][p_{bl}]_{7\times 7} \\ \pi_1 + \pi_2 + \pi_3 + \pi_4 + \pi_5 + \pi_6 + \pi_7 = 1, \end{cases}$$

equivalent to the system of equations

$$\begin{cases} 0.2\pi_2 + 1\pi_3 + 0.488\pi_5 + 0.095\pi_6 + 0.531\pi_7 = \pi_1 \\ 0.022\pi_1 + 0.023\pi_5 + 0.062\pi_7 = \pi_2 \\ 0.022\pi_1 = \pi_3 \\ 0.023\pi_5 = \pi_4 \\ 0.534\pi_1 + 0.667\pi_6 + 0.219\pi_7 = \pi_5 \\ 0.111\pi_1 + 0.233\pi_5 + 0.188\pi_7 = \pi_6 \\ 0.311\pi_1 + 0.8\pi_2 + 1\pi_4 + 0.233\pi_5 + 0.238\pi_6 = \pi_7 \\ \pi_1 + \pi_2 + \pi_3 + \pi_4 + \pi_5 + \pi_6 + \pi_7 = 1, \end{cases}$$

and getting its following solution

$$\pi_1 \cong 0.291,\ \pi_2 \cong 0.027,\ \pi_3 \cong 0.006,$$
$$\pi_4 \cong 0.007,\ \pi_5 \cong 0.301,\ \pi_6 \cong 0.144,\ \pi_7 \cong 0.224. \tag{4.67}$$

and next resulting, after applying (4.8) to (4.65) and (4.67), in determining the limit values of the port oil terminal critical infrastructure operation process transient probabilities $p_b(t)$ at the operation states $z_b$, $b = 1, 2, \ldots, 7$, that are given by (4.66);

- the expected values of the total sojourn times $\hat{\theta}_b$, $b = 1, 2, \ldots, 7$, (OPC3) of the port oil terminal critical infrastructure operation process at the particular operation states $z_b$, $b = 1, 2, \ldots, 7$, during the fixed operation time $\theta = 1$ year $= 365$ days, determined according to (4.11) and (4.66):

$$\hat{M}_1 \cong E[\hat{\theta}] = 0.395\,\text{year} = 144.2\,\text{days},\ \hat{M}_2 \cong E\left[\hat{\theta}_2\right] = 0.060\,\text{year} = 21.9\,\text{days},$$
$$\hat{M}_3 \cong E\left[\hat{\theta}_3\right] = 0.003\,\text{year} = 1.1\,\text{day},\ \hat{M}_4 \cong E\left[\hat{\theta}_4\right] = 0.002\,\text{year} = 0.7\,\text{day},$$
$$\hat{M}_3 \cong E\left[\hat{\theta}_s\right] = 0.20\,\text{year} = 73\,\text{days},\ \hat{M}_6 \cong E\left[\hat{\theta}_6\right] = 0.058\,\text{year} = 21.2\,\text{days},$$
$$\hat{M}_7 \cong E[\hat{\theta}_7] = 0.282\,\text{year} = 102.93\,\text{days}. \tag{4.68}$$

### 4.4.2 Safety Parameters of Port Oil Terminal Critical Infrastructure Impacted by Its Operation Process

In this section, we accept assumptions and agreements such as in Sect. 3.3.2.

The coefficients of the operation process impact on the port oil terminal critical infrastructure intensities of ageing at the operation states $z_b$, $b = 1, 2, \ldots, 7$, evaluated approximately by experts are as follows [6, 13]:

- for asset $A_1$

$$\left[\rho_1^1(1)\right]^{(b)} = 1.10,\ \left[\rho_1^1(2)\right]^{(b)} = 1.10,\ b = 1, 2, 7,$$
$$\left[\rho_1^1(1)\right]^{(b)} = 1.20,\ \left[\rho_1^1(2)\right]^{(b)} = 1.20,\ b = 3, 5,$$
$$\left[\rho_1^1(1)\right]^{(b)} = 1.30,\ \left[\rho_1^1(2)\right]^{(b)} = 1.30,\ b = 4, 6; \tag{4.69}$$

- for asset $A_2$

$$[\rho_2^1(1)]^{(b)} = 1.10, [\rho_2^1(2)]^{(b)} = 1.10, \ b = 1, 2, 7,$$
$$[\rho_2^1(1)]^{(b)} = 1.20, [\rho_2^1(2)]^{(b)} = 1.20, b = 3, 5,$$
$$[\rho_2^1(1)]^{(b)} = 1.30, [\rho_2^1(2)]^{(b)} = 1.30, b = 4, 6; \quad (4.70)$$

- for asset $A_3$

$$[\rho_3^1(1)]^{(b)} = 1, [\rho_3^1(2)]^{(b)} = 1, b = 1, 2, 3, 7,$$
$$[\rho_3^1(1)]^{(b)} = 1.20, [\rho_3^1(2)]^{(b)} = 1.20, b = 5,$$
$$[\rho_3^1(1)]^{(b)} = 1.30, [\rho_3^1(2)]^{(b)} = 1.30, b = 4, 6; \quad (4.71)$$

- for asset $A_4$

$$[\rho_4^1(1)]^{(b)} = 1.10, [\rho_4^1(2)]^{(b)} = 1.10, b = 1, 2, 7,$$
$$[\rho_4^1(1)]^{(b)} = 1.20, [\rho_4^1(2)]^{(b)} = 1.20, b = 3, 5,$$
$$[\rho_4^1(1)]^{(b)} = 1.30, [\rho_4^1(2)]^{(b)} = 1.30, b = 4, 6; \quad (4.72)$$

- for asset $A_5$

$$[\rho_5^1(1)]^{(b)} = 1, [\rho_5^1(2)]^{(b)} = 1, b = 1, 2, 7,$$
$$[\rho_5^1(1)]^{(b)} = 1.20, [\rho_5^1(2)]^{(b)} = 1.20, b = 3, 5,$$
$$[\rho_5^1(1)]^{(b)} = 1.30, [\rho_5^1(2)]^{(b)} = 1.30, b = 4, 6; \quad (4.73)$$

- for asset $A_6$

$$[\rho_6^1(1)]^{(b)} = 1, [\rho_6^1(2)]^{(b)} = 1, b = 2, 5,$$
$$[\rho_6^1(1)]^{(b)} = 1.10, [\rho_6^1(2)]^{(b)} = 1.10, b = 1, 7,$$
$$[\rho_6^1(1)]^{(b)} = 1.20, [\rho_6^1(2)]^{(b)} = 1.20, b = 3,$$
$$[\rho_6^1(1)]^{(b)} = 1.30, [\rho_6^1(2)]^{(b)} = 1.30, b = 4, 6; \quad (4.74)$$

– for asset $A_7$

$$
\begin{aligned}
&\left[\rho_7^1(1)\right]^{(b)} = 1, \left[\rho_7^1(2)\right]^{(b)} = 1, b = 2, 3, 5,\\
&\left[\rho_7^1(1)\right]^{(b)} = 1.10, \left[\rho_7^1(2)\right]^{(b)} = 1.10, b = 1, 7,\\
&\left[\rho_7^1(1)\right]^{(b)} = 1.30, \left[\rho_7^1(2)\right]^{(b)} = 1.30, b = 4, 6;
\end{aligned} \tag{4.75}
$$

– for asset $A_8$

$$
\begin{aligned}
&\left[\rho_8^1(1)\right]^{(b)} = 1, \left[\rho_8^1(2)\right]^{(b)} = 1, b = 1, 3, 4, 5, 6,\\
&\left[\rho_8^1(1)\right]^{(b)} = 1.10, \left[\rho_8^1(2)\right]^{(b)} = 1.10, b = 2, 7;
\end{aligned} \tag{4.76}
$$

– for asset $A_9$

$$
\begin{aligned}
&\left[\rho_9^1(\mathrm{I})\right]^{(b)} = 1.10, \left[\rho_9^1(2)\right]^{(b)} = 1.10, b = 1, 2, 7\\
&\left[\rho_9^1(1)\right]^{(b)} = 1.20, \left[\rho_9^1(2)\right]^{(b)} = 1.20, b = 3, 5,\\
&\left[\rho_9^1(1)\right]^{(b)} = 1.30, \left[\rho_9^1(2)\right]^{(b)} = 1.30, b = 4, 6.
\end{aligned} \tag{4.77}
$$

Similarly to Sect. 3.3.5, we assume that the port oil terminal critical infrastructure assets $A_i$, $i = 1, 2, \ldots, 9$, at the critical infrastructure operation process $Z(t)$ states $z_b$, $b = 1, 2, \ldots, 7$, conditional safety functions (4.46), (4.47) are piecewise exponential with the coordinates $[S_i^1(t, u)]^{(b)}$, $t \in< 0, \infty)$, $u = 1, 2$, $b = 1, 2, \ldots, 7$, $i = 1, 2, \ldots, 9$, given by (4.55). Whereas, the intensities of ageing $[\lambda_i^1(u)]^{(b)}$, $u = 1, 2, b = 1, 2, \ldots, 7, i = 1, 2, \ldots, 9$, for the port oil terminal critical infrastructure assets $A_i$, $i = 1, 2, \ldots, 9$, at the system operation states $z_b$,$b = 1, 2, \ldots, 7$, existing in (4.55) are given by (4.58), where $\lambda_i^0(u)$, $u = 1, 2$, $i = 1, 2, \ldots, 9$, are the intensities of ageing of the port oil terminal critical infrastructure assets $A_i$, $i = 1, 2, \ldots, 9$, without operation process impact and $[\rho_i^1(u)]^{(b)}$, $u = 1, 2$, $b = 1, 2, \ldots, 7$, $i = 1, 2, \ldots, 9$, are the coefficients of operation process impact on the port oil terminal critical infrastructure assets $A_i$, $i = 1, 2, \ldots, 9$, intensities of ageing at the operation states $z_b$,$b = 1, 2, \ldots, 7$.

After that, from (4.58), considering (3.93)–(3.96) and (4.69)–(4.77), it follows that the intensities of assets departure from the safety states subset {1, 2}, {2}, of the port oil terminal critical infrastructure impacted by its operation process are [13]:

- for asset $A_1$

$$
\begin{aligned}
[\lambda_1^1(1)]^{(b)} &= [\rho_1^1(1)]^{(b)} \cdot \lambda_1^0(1) = 1.10 \cdot 0.015873 = 0.017460,\\
[\lambda_1^1(2)]^{(b)} &= [\rho_1^1(2)]^{(b)} \cdot \lambda_1^0(2) = 1.10 \cdot 0.021739\\
&= 0.023913,\ b = 1, 2, 7,\\
[\lambda_1^1(1)]^{(b)} &= [\rho_1^1(1)]^{(b)} \cdot \lambda_1^0(1) = 1.20 \cdot 0.015873 = 0.019048,\\
[\lambda_1^1(2)]^{(b)} &= [\rho_1^1(2)]^{(b)} \cdot \lambda_1^0(2) = 1.20 \cdot 0.021739\\
&= 1.20 \cdot 0.021739 = 0.026087,\ b = 3, 5,\\
[\lambda_1^1(1)]^{(b)} &= [\rho_1^1(1)]^{(b)} \cdot \lambda_1^0(1) = 1.30 \cdot 0.015873 = 0.019048,\\
[\lambda_1^1(2)]^{(b)} &= [\rho_1^1(2)]^{(b)} \cdot \lambda_1^0(2) = 1.30 \cdot 0.021739\\
&= 0.028261,\ b = 4, 6;
\end{aligned}
\tag{4.78}
$$

- for asset $A_2$

$$
\begin{aligned}
[\lambda_2^1(1)]^{(b)} &= [\rho_2^1(1)]^{(b)} \cdot \lambda_2^0(1) = 1.10 \cdot 0.0125 = 0.01375,\\
[\lambda_2^1(2)]^{(b)} &= [\rho_2^1(2)]^{(b)} \cdot \lambda_2^0(2) = 1.10 \cdot 0.02 = 0.022,\ b = 1, 2, 7,\\
[\lambda_2^1(1)]^{(b)} &= [\rho_2^1(1)]^{(b)} \cdot \lambda_2^0(1) = 1.20 \cdot 0.0125 = 0.015,\\
[\lambda_2^1(2)]^{(b)} &= [\rho_2^1(2)]^{(b)} \cdot \lambda_2^0(2) = 1.20 \cdot 0.02 = 0.024,\ b = 3, 5,\\
[\lambda_2^1(1)]^{(b)} &= [\rho_2^1(1)]^{(b)} \cdot \lambda_2^0(1) = 1.30 \cdot 0.0125 = 0.01625,\\
[\lambda_2^1(2)]^{(b)} &= [\rho_2^1(2)]^{(b)} \cdot \lambda_2^0(2) = 1.30 \cdot 0.02 = 0.026,\ b = 4, 6;
\end{aligned}
\tag{4.79}
$$

- for asset $A_3$

$$
\begin{aligned}
[\lambda_3^1(1)]^{(b)} &= [\rho_3^1(1)]^{(b)} \cdot \lambda_3^0(1) = 1 \cdot 0.0125 = 0.0125,\\
[\lambda_3^1(2)]^{(b)} &= [\rho_3^1(2)]^{(b)} \cdot \lambda_3^0(2) = 1 \cdot 0.02 = 0.02,\ b = 1, 2, 3, 7,\\
[\lambda_3^1(1)]^{(b)} &= [\rho_3^1(1)]^{(b)} \cdot \lambda_3^0(1) = 1.20 \cdot 0.0125 = 0.015,\\
[\lambda_3^1(2)]^{(b)} &= [\rho_3^1(2)]^{(b)} \cdot \lambda_3^0(2) = 1.20 \cdot 0.02 = 0.024,\ b = 5,\\
[\lambda_3^1(1)]^{(b)} &= [\rho_3^1(1)]^{(b)} \cdot \lambda_3^0(1) = 1.30 \cdot 0.0125 = 0.01625,\\
[\lambda_3^1(2)]^{(b)} &= [\rho_3^1(2)]^{(b)} \cdot \lambda_3^0(2) = 1.30 \cdot 0.02 = 0.026,\ b = 4, 6;
\end{aligned}
\tag{4.80}
$$

– for asset $A_4$

$$
\begin{aligned}
[\lambda_4^1(1)]^{(b)} &= [\rho_4^1(1)]^{(b)} \cdot \lambda_4^0(1) = 1.10 \cdot 0.0125 = 0.01375, \\
[\lambda_4^1(2)]^{(b)} &= [\rho_4^1(2)]^{(b)} \cdot \lambda_4^0(2) = 1.10 \cdot 0.02 = 0.022, \ b = 1, 2, 7 \\
[\lambda_4^1(1)]^{(b)} &= [\rho_4^1(1)]^{(b)} \cdot \lambda_4^0(1) = 1.20 \cdot 0.0125 = 0.015, \\
[\lambda_4^1(2)]^{(b)} &= [\rho_4^1(2)]^{(b)} \cdot \lambda_4^0(2) = 1.20 \cdot 0.02 = 0.024, \ b = 3, 5, \\
[\lambda_4^1(1)]^{(b)} &= [\rho_4^1(1)]^{(b)} \cdot \lambda_4^0(1) = 1.30 \cdot 0.0125 = 0.01625, \\
[\lambda_4^1(2)]^{(b)} &= [\rho_4^1(2)]^{(b)} \cdot \lambda_4^0(2) = 1.30 \cdot 0.02 = 0.026, \ b = 4, 6;
\end{aligned} \tag{4.81}
$$

– for asset $A_5$

$$
\begin{aligned}
[\lambda_5^1(1)]^{(b)} &= [\rho_5^1(1)]^{(b)} \cdot \lambda_5^0(1) = 1 \cdot 0.0125 = 0.0125, \\
[\lambda_5^1(2)]^{(b)} &= [\rho_5^1(2)]^{(b)} \cdot \lambda_5^0(2) = 1 \cdot 0.02 = 0.02, \ b = 1, 2, 7, \\
[\lambda_5^1(1)]^{(b)} &= [\rho_5^1(1)]^{(b)} \cdot \lambda_5^0(1) = 1.20 \cdot 0.0125 = 0.015, \\
[\lambda_5^1(2)]^{(b)} &= [\rho_5^1(2)]^{(b)} \cdot \lambda_5^0(2) = 1.20 \cdot 0.02 = 0.024, \ b = 3, 5, \\
[\lambda_5^1(1)]^{(b)} &= [\rho_5^1(1)]^{(b)} \cdot \lambda_5^0(1) = 1.30 \cdot 0.0125 = 0.01625, \\
[\lambda_5^1(2)]^{(b)} &= [\rho_5^1(2)]^{(b)} \cdot \lambda_5^0(2) = 1.30 \cdot 0.02 = 0.026, \ b = 4, 6;
\end{aligned} \tag{4.82}
$$

– for asset $A_6$

$$
\begin{aligned}
[\lambda_6^1(1)]^{(b)} &= [\rho_6^1(1)]^{(b)} \cdot \lambda_6^0(1) = 1 \cdot 0.0125 = 0.0125, \\
[\lambda_6^1(2)]^{(b)} &= [\rho_6^1(2)]^{(b)} \cdot \lambda_6^0(2) = 1 \cdot 0.02 = 0.02, \ b = 2, 5, \\
[\lambda_6^1(1)]^{(b)} &= [\rho_6^1(1)]^{(b)} \cdot \lambda_6^0(1) = 1.10 \cdot 0.0125 = 0.01375, \\
[\lambda_6^1(2)]^{(b)} &= [\rho_6^1(2)]^{(b)} \cdot \lambda_6^0(2) = 1.10 \cdot 0.02 = 0.022, \ b = 1, 7, \\
[\lambda_6^1(1)]^{(b)} &= [\rho_6^1(1)]^{(b)} \cdot \lambda_6^0(1) = 1.20 \cdot 0.0125 = 0.015, \\
[\lambda_6^1(2)]^{(b)} &= [\rho_6^1(2)]^{(b)} \cdot \lambda_6^0(2) = 1.20 \cdot 0.02 = 0.024, \ b = 3, \\
[\lambda_6^1(1)]^{(b)} &= [\rho_6^1(1)]^{(b)} \cdot \lambda_6^0(1) = 1.30 \cdot 0.0125 = 0.01625, \\
[\lambda_6^1(2)]^{(b)} &= [\rho_6^1(2)]^{(b)} \cdot \lambda_6^0(2) = 1.30 \cdot 0.02 = 0.026, \ b = 4, 6;
\end{aligned} \tag{4.83}
$$

– for asset $A_7$

$$
[\lambda_7^1(1)]^{(b)} = [\rho_7^1(1)]^{(b)} \cdot \lambda_7^0(1) = 1 \cdot 0.0125 = 0.0125,
$$

$$[\lambda_7^1(2)]^{(b)} = [\rho_7^1(2)]^{(b)} \cdot \lambda_7^0(2) = 1 \cdot 0.02 = 0.02, \; b = 2, 3, 5,$$
$$[\lambda_7^1(1)]^{(b)} = [\rho_7^1(1)]^{(b)} \cdot \lambda_7^0(1) = 1.10 \cdot 0.0125 = 0.01375,$$
$$[\lambda_7^1(2)]^{(b)} = [\rho_7^1(2)]^{(b)} \cdot \lambda_7^0(2) = 1.10 \cdot 0.02 = 0.022, \; b = 1$$
$$[\lambda_7^1(2)]^{(b)} = [\rho_7^1(2)]^{(b)} \cdot \lambda_7^0(2) = 1.30 \cdot 0.02 = 0.026, \; b = 4, 6; , 7,$$
$$[\lambda_7^1(1)]^{(b)} = [\rho_7^1(1)]^{(b)} \cdot \lambda_7^0(1) = 1.30 \cdot 0.0125 = 0.01625, \tag{4.84}$$

– for asset $A_8$

$$[\lambda_8^1(1)]^{(b)} = [\rho_8^1(1)]^{(b)} \cdot \lambda_8^0(1) = 1 \cdot 0.0125 = 0.0125,$$
$$[\lambda_8^1(2)]^{(b)} = [\rho_8^1(2)]^{(b)} \cdot \lambda_8^0(2) = 1 \cdot 0.02 = 0.02, \; b = 1, 3, 4, 5, 6,$$
$$[\lambda_8^1(1)]^{(b)} = [\rho_8^1(1)]^{(b)} \cdot \lambda_8^0(1) = 1.10 \cdot 0.0125 = 0.01375,$$
$$[\lambda_8^1(2)]^{(b)} = [\rho_8^1(2)]^{(b)} \cdot \lambda_8^0(2) = 1.10 \cdot 0.02 = 0.022, \; b = 2, 7; \tag{4.85}$$

– for asset $A_9$

$$[\lambda_9^1(1)]^{(b)} = [\rho_9^1(1)]^{(b)} \cdot \lambda_9^0(1) = 1.10 \cdot 0.0125 = 0.01375,$$
$$[\lambda_9^1(2)]^{(b)} = [\rho_9^1(2)]^{(b)} \cdot \lambda_9^0(2) = 1.10 \cdot 0.02 = 0.022, \; b = 1, 2, 7$$
$$[\lambda_9^1(1)]^{(b)} = [\rho_9^1(1)]^{(b)} \cdot \lambda_9^0(1) = 1.20 \cdot 0.0125 = 0.015,$$
$$[\lambda_9^1(2)]^{(b)} = [\rho_9^1(2)]^{(b)} \cdot \lambda_9^0(2) = 1.20 \cdot 0.02 = 0.024, \; b = 3, 5,$$
$$[\lambda_9^1(1)]^{(b)} = [\rho_9^1(1)]^{(b)} \cdot \lambda_9^0(1) = 1.30 \cdot 0.0125 = 0.01625,$$
$$[\lambda_9^1(2)]^{(b)} = [\rho_9^1(2)]^{(b)} \cdot \lambda_9^0(2) = 1.30 \cdot 0.02 = 0.026, \; b = 4, 6. \tag{4.86}$$

### *4.4.3 Safety and Resilience Indicators of Port Oil Terminal Critical Infrastructure Impacted by Its Operation Process*

Considering that the coordinates $[S_i^1(t, u)]^{(b)}$, $t \geq 0$, $u = 1, 2$, $i = 1, 2, \ldots 9$, $b = 1, 2, \ldots 7$, of the conditional safety functions (4.46), (4.47) for the port oil terminal critical infrastructure assets $A_i$, $i = 1, 2, \ldots 9$, are of the form (4.55), i.e. they are given by the vectors

$$[S_i^1(t, \cdot)]^{(b)} = [1, \cdot [S_i^1(t, 1)]^{(b)}, [S_i^1(t, 2)]^{(b)}],$$
$$t \in <0, \infty), \quad b = 1, 2, \ldots, 7, \quad i = 1, 2, \ldots, 9,$$

with the coordinates

$$[S_i^1(t,u)]^{(b)} = \exp[-[\lambda_i^1(u)]^{(b)}t],\ t \in <0,\infty),$$
$$u = 1,2,\quad b = 1,2,\ldots,7,\quad i = 1,2,\ldots,9,$$

and the intensities of ageing $[\lambda_i^1(u)]^{(b)}$, $t \in <0,\infty)$, $u = 1,2$, $i = 1,2,\ldots,9$, at the operation states $z_b$, $b = 1,2,\ldots,7$, are given respectively by (4.78)–(4.86) we get their coordinates following final forms:

– for asset $A_1$

$$\begin{aligned}
&[S_1^1(t,\ 1)]^{(b)} = \exp[-0.017460t],\ [S_1^1(t,\ 2)]^{(b)} = \exp[-0.023913t],\ b = 1,2,7,\\
&[S_1^1(t,\ 1)]^{(b)} = \exp[-0.019048t],\ [S_1^1(t,\ 2)]^{(b)} = \exp[-0.026087t],\ b = 3,5,\\
&[S_1^1(t,\ 1)]^{(b)} = \exp[-0.020635t],\ [S_1^1(t,\ 2)]^{(b)} = \exp[-0.028261t],\ b = 4,6;
\end{aligned} \tag{4.87}$$

– for asset $A_2$

$$\begin{aligned}
&[S_2^1(t,\ 1)]^{(b)} = \exp[-0.01375t],\ [S_2^1(t,\ 2)]^{(b)} = \exp[-0.022t],\ b = 1,2,7,\\
&[S_2^1(t,\ 1)]^{(b)} = \exp[-0.015t],\ \left[S_2^1(t,2)\right]^{(b)} = \exp[-0.024t],\ b = 3,5,\\
&[S_2^1(t,\ 1)]^{(b)} = \exp[-0.01625t],\ \left[S_2^1(t,2)\right]^{(b)} = \exp[-0.026t],\ b = 4,6;
\end{aligned} \tag{4.88}$$

– for asset $A_3$

$$\begin{aligned}
&[S_3^1(t,\ 1)]^{(b)} = S_3^0(t,\ 1) = \exp[-0.0125t],\\
&\left[S_3^1(t,2)\right]^{(b)} = S_3^0(t,2) = \exp[-0.02t],\ b = 1,2,3,7,\\
&[S_3^1(t,\ 1)]^{(b)} = \exp[-0.015t],\ \left[S_3^1(t,2)\right]^{(b)} = \exp[-0.024t],\ b = 5,\\
&[S_3^1(t,\ 1)]^{(b)} = \exp[-0.01625t],\ \left[S_3^1(t,2)\right]^{(b)} = \exp[-0.026t],\quad b = 4,6;
\end{aligned} \tag{4.89}$$

– for asset $A_4$

$$\begin{aligned}
&[S_4^1(t,\ 1)]^{(b)} = \exp[-0.01375t],\ \left[S_4^1(t,2)\right]^{(b)} = \exp[-0.022t],\quad b = 1,2,7\\
&[S_4^1(t,\ 1)]^{(b)} = \exp[-0.015t],\ \left[S_4^1(t,2)\right]^{(b)} = \exp[-0.024t],\quad b = 3,5,\\
&[S_4^1(t,\ 1)]^{(b)} = \exp[-0.01625t],\ \left[S_4^1(t,2)\right]^{(b)} = \exp[-0.026t],\quad b = 4,6;
\end{aligned} \tag{4.90}$$

- for asset $A_5$

$$
\begin{aligned}
&[S_5^1(t,\ 1)]^{(b)} = S_5^0(t,1) = \exp[-0.0125t],\\
&\left[S_5^1(t,2)\right]^{(b)} = S_5^0(t,2) = \exp[-0.02t],\, b = 1,2,7,\\
&[S_5^1(t,\ 1)]^{(b)} = \exp[-0.015t],\, \left[S_5^1(t,2)\right]^{(b)} = \exp[-0.024t],\, b = 3,5,\\
&[S_5^1(t,\ 1)]^{(b)} = \exp[-0.01625t],\, \left[S_5^1(t,2)\right]^{(b)} = \exp[-0.026t],\, b = 4,6;
\end{aligned}
\tag{4.91}
$$

- for asset $A_6$

$$
\begin{aligned}
&[S_6^1(t,\ 1)]^{(b)} = S_6^0(t,1) = \exp[-0.0125],\\
&\left[S_6^1(t,2)\right]^{(b)} = S_6^0(t,1) = \exp[-0.02t],\, b = 2,5,\\
&[S_6^1(t,\ 1)]^{(b)} = \exp[-0.01375t],\, \left[S_6^1(t,2)\right]^{(b)} = \exp[-0.022t], \quad b = 1,7,\\
&[S_6^1(t,\ 1)]^{(b)} = \exp[-0.015t],\, \left[S_6^1(t,2)\right]^{(b)} = \exp[-0.024t],\, b = 3,\\
&[S_6^1(t,\ 1)]^{(b)} = \exp[-0.01625t],\, \left[S_6^1(t,2)\right]^{(b)} = \exp[-0.026t], \quad b = 4,6;
\end{aligned}
\tag{4.92}
$$

- for asset $A_7$

$$
\begin{aligned}
&[S_7^1(t,\ 1)]^{(b)} = S_7^0(t,1) = \exp[-0.0125t],\\
&\left[S_7^1(t,2)\right]^{(b)} = S_7^0(t,2) = \exp[-0.02t], \quad b = 2,3,5,\\
&[S_7^1(t,\ 1)]^{(b)} = \exp[-0.01375t],\, \left[S_7^1(t,2)\right]^{(b)} = \exp[-0.022t],\, b = 1,7,\\
&[S_7^1(t,\ 1)]^{(b)} = \exp[-0.01625t],\, \left[S_7^1(t,2)\right]^{(b)} = \exp[-0.026t], \quad b = 4,6;
\end{aligned}
\tag{4.93}
$$

- for asset $A_8$

$$
\begin{aligned}
&[S_8^1(t,\ 1)]^{(b)} = S_8^0(t,1) = \exp[-0.0125t],\\
&\left[S_8^1(t,2)\right]^{(b)} = S_8^0(t,2) = \exp[-0.02t],\, b = 1,3,4,5,6,\\
&[S_8^1(t,\ 1)]^{(b)} = \exp[-0.01375t],\, \left[S_8^1(t,2)\right]^{(b)} = \exp[-0.022t], \quad b = 2,7;
\end{aligned}
\tag{4.94}
$$

– for asset $A_9$

$$[S_9^1(t,\ 1)]^{(b)} = \exp[-0.01375t],\ \left[S_9^1(t,2)\right]^{(b)} = \exp[-0.022t],\ b = 1, 2, 7$$
$$[S_9^1(t,\ 1)]^{(b)} = \exp[-0.015t],\ \left[S_9^1(t,2)\right]^{(b)} = \exp[-0.024t],\ b = 3, 5,$$
$$[S_9^1(t,\ 1)]^{(b)} = \exp[-0.01625t],\ \left[S_9^1(t,2)\right]^{(b)} = \exp[-0.026t],\ b = 4, 6. \quad (4.95)$$

Since the oil terminal critical infrastructure is a three-state (z = 2) series system, then considering (4.87)–(4.95) and applying (2.8)–(2.9) its conditional safety functions at the particular operation states $z_b$, $b = 1, 2, \ldots, 7$, are respectively given by:

– at the operation states $z_1$

$$[\boldsymbol{S}^1(t,\ \cdot)]^{(1)} = \left[1, \left[\boldsymbol{S}^1(t,1)\right]^{(1)}, \left[\boldsymbol{S}^1(t,2)\right]^{(1)}\right], t \geq 0, \quad (4.96)$$

where

$$\begin{aligned}\left[\boldsymbol{S}^1(t,1)\right]^{(1)} &= \prod_{i=1}^{9}\left[S_i^1(t,1)\right]^{(1)} \\ &= \left[S_1^1(t,1)\right]^{(1)} \cdot \left[S_2^1(t,1)\right]^{(1)} \cdot S_3^0(t,1) \cdot \left[S_4^1(t,1)\right]^{(1)} \cdot S_5^0(t,1) \\ &\quad \cdot \left[S_6^1(t,1)\right]^{(1)} \cdot \left[S_7^1(t,1)\right]^{(1)} \cdot S_8^0(t,1) \cdot \left[S_9^1(t,1)\right]^{(1)} \\ &= \exp[-0.017460] \exp[-0.01375] \\ &\quad \exp[-0.0125] \exp[-0.01375] \exp[-0.0125] \\ &\quad \exp[-0.01375] \exp[-0.01375] \exp[-0.0125] \exp[-0.01375] \\ &= \exp[-0.12371t], \end{aligned} \quad (4.97)$$

$$\begin{aligned}\left[\boldsymbol{S}^1(t,2)\right]^{(1)} &= \prod_{i=1}^{9}\left[S_i^1(t,2)\right]^{(1)} \\ &= \left[S_1^1(t,2)\right]^{(1)} \cdot \left[S_2^1(t,2)\right]^{(1)} \cdot S_3^0(t,2) \cdot \left[S_4^1(t,2)\right]^{(1)} \cdot S_5^0(t,2) \\ &\quad \cdot \left[S_6^1(t,2)\right]^{(1)} \cdot \left[S_7^1(t,2)\right]^{(L)} \cdot S_8^0(t,2) \cdot \left[S_9^1(t,2)\right]^{(1)} \\ &= \exp[-0.023913] \exp[-0.022] \exp[-0.02] \\ &\quad \exp[-0.022] \exp[-0.02] \exp[-0.022] \\ &\quad \exp[-0.022] \exp[-0.02] \exp[-0.022] \\ &= \exp[-0.193913t]; \end{aligned} \quad (4.98)$$

– at the operation states $z_2$

$$\left[\boldsymbol{S}^1(t,\cdot)\right]^{(2)} = \left[1, \left[\boldsymbol{S}^1(t,1)\right]^{(2)}, \left[\boldsymbol{S}^1(t,2)\right]^{(2)}\right], t \geq 0, \tag{4.99}$$

where

$$\begin{aligned}\left[\boldsymbol{S}^1(t,1)\right]^{(2)} &= \prod_{i=1}^{9} \left[S_i^1(t,1)\right]^{(2)} \\ &= \left[S_1^1(t,1)\right]^{(2)} \cdot \left[S_2^1(t,1)\right]^{(2)} \cdot S_3^0(t,1) \cdot \left[S_4^1(t,1)\right]^{(2)} \cdot S_5^0(t,1) \\ &\quad \cdot S_6^0(t,1) \cdot S_7^0(t,1) \cdot \left[S_8^1(t,1)\right]^{(2)} \cdot \left[S_9^1(t,1)\right]^{(2)} \\ &= \exp[-0.017460]\exp[-0.01375]\exp[-0.0125] \\ &\quad \exp[-0.01375]\exp[-0.0125]\exp[-0.0125] \\ &\quad \exp[-0.0125]\exp[-0.01375]\exp[-0.01375] \\ &= \exp[-0.12246t],\end{aligned} \tag{4.100}$$

$$\begin{aligned}\left[\boldsymbol{S}^1(t,2)\right]^{(2)} &= \prod_{i=1}^{9} \left[S_i^1(t,2)\right]^{(2)} \\ &= \left[S_1^1(t,2)\right]^{(2)} \cdot \left[S_2^1(t,2)\right]^{(2)} \cdot S_3^0(t,2) \cdot \left[S_4^1(t,2)\right]^{(2)} \cdot S_5^0(t,2) \\ &\quad \cdot S_6^0(t,2) \cdot S_7^0(t,2) \cdot \left[S_8^1(t,2)\right]^{(2)} \cdot \left[S_9^1(t,2)\right]^{(2)} \\ &= \exp[-0.023913]\exp[-0.022]\exp[-0.02] \\ &\quad \exp[-0.022]\exp[-0.02]\exp[-0.02] \\ &\quad \exp[-0.02]\exp[-0.022]\exp[-0.022] \\ &= \exp[-0.191913t];\end{aligned} \tag{4.101}$$

– at the operation states $z_3$

$$\left[\boldsymbol{S}^1(t,\cdot)\right]^{(3)} = \left[1, \left[\boldsymbol{S}^1(t,1)\right]^{(3)}, \left[\boldsymbol{S}^1(t,2)\right]^{(3)}\right], t \geq 0, \tag{4.102}$$

where

$$\begin{aligned}\left[\boldsymbol{S}^1(t,1)\right]^{(3)} &= \prod_{i=1}^{9} \left[S_i^1(t,1)\right]^{(3)} \\ &= \left[S_1^1(t,1)\right]^{(3)} \cdot \left[S_2^1(t,1)\right]^{(3)} \cdot S_3^0(t,1) \cdot \left[S_4^1(t,1)\right]^{(3)} \cdot \left[S_5^1(t,1)\right]^{(3)} \\ &\quad \cdot S_6^0(t,1) \cdot S_7^0(t,1) \cdot S_8^0(t,1) \cdot \left[S_9^1(t,1)\right]^{(3)}\end{aligned}$$

$$= \exp[-0.019048]\exp[-0.015]\exp[-0.0125]$$
$$\exp[-0.015]\exp[-0.015]\exp[-0.0125]$$
$$\exp[-0.0125]\exp[-0.0125]\exp[-0.015]$$
$$= \exp[-0.131548t], \tag{4.103}$$

$$[\boldsymbol{S}^1(t,2)]^{(3)} = \prod_{i=1}^{9}[S_i^1(t,2)]^{(3)}$$
$$= [S_1^1(t,2)]^{(3)} \cdot [S_2^1(t,2)]^{(3)} \cdot S_3^0(t,2) \cdot [S_4^1(t,2)]^{(3)} \cdot [S_5^1(t,2)]^{(3)}$$
$$\cdot S_6^0(t,2) \cdot S_7^0(t,2) \cdot S_8^0(t,2) \cdot [S_9^1(t,2)]^{(3)}$$
$$= \exp[-0.026087]\exp[-0.024]\exp[-0.02]$$
$$\exp[-0.024]\exp[-0.024]\exp[-0.024]$$
$$\exp[-0.02]\exp[-0.02]\exp[-0.024]$$
$$= \exp[-0.206087t]; \tag{4.104}$$

– at the operation states $z_4$

$$[\boldsymbol{S}^1(t,-)]^{(4)} = \left[1, [\boldsymbol{S}^1(t,1)]^{(4)}, [\boldsymbol{S}^1(t,2)]^{(4)}\right], t \geq 0, \tag{4.105}$$

where

$$[\boldsymbol{S}^1(t,1)]^{(4)} = \prod_{i=1}^{9}[S_i^1(t,1)]^{(4)}$$
$$= [S_1^1(t,1)]^{(4)} \cdot [S_2^1(t,1)]^{(4)} \cdot [S_3^1(t,1)]^{(4)} \cdot [S_4^1(t,1)]^{(4)} \cdot [S_5^1(t,1)]^{(4)}$$
$$\cdot [S_6^1(t,1)]^{(4)} \cdot [S_7^1(t,1)]^{(4)} \cdot S_8^0(t,1) \cdot [S_9^1(t,1)]^{(4)}$$
$$= \exp[-0.020635]\exp[-0.01625]\exp[-0.01625]$$
$$\exp[-0.01625]\exp[-0.01625]\exp[-0.01625]$$
$$\exp[-0.01625]\exp[-0.0125]\exp[-0.01625]$$
$$= \exp[-0.146885t], \tag{4.106}$$

$$[\boldsymbol{S}^1(t,2)]^{(4)} = \prod_{i=1}^{9}[S_i^1(t,2)]^{(4)}$$
$$= [S_1^1(t,2)]^{(4)} \cdot [S_2^1(t,2)]^{(4)} \cdot [S_3^1(t,2)]^{(4)} \cdot [S_4^1(t,2)]^{(4)} \cdot [S_5^1(t,2)]^{(4)}$$
$$\cdot [S_6^1(t,2)]^{(4)} \cdot [S_7^1(t,2)]^{(4)} \cdot S_8^0(t,2) \cdot [S_9^1(t,2)]^{(4)}$$
$$= \exp[-0.028261]\exp[-0.026]\exp[-0.026]$$

$$\exp[-0.026]\exp[-0.026]\exp[-0.026]$$
$$\exp[-0.026]\exp[-0.02]\exp[-0.026]$$
$$= \exp[-0.230261t]; \quad (4.107)$$

– at the operation states $z_5$

$$[\boldsymbol{S}^1(t,-)]^{(5)} = \left[1, [\boldsymbol{S}^1(t,1)]^{(5)}, [\boldsymbol{S}^1(t,2)]^{(5)}\right], t \geq 0, \quad (4.108)$$

where

$$[\boldsymbol{S}^1(t,1)]^{(5)} = \prod_{i=1}^{9} [S_i^1(t,1)]^{(5)}$$
$$= [S_1^1(t,1)]^{(5)} \cdot [S_2^1(t,1)]^{(5)} \cdot [S_3^1(t,1)]^{(5)} \cdot [S_4^1(t,1)]^{(5)} \cdot [S_5^1(t,1)]^{(5)}$$
$$\cdot S_6^0(t,1) \cdot S_7^0(t,1) \cdot S_8^0(t,1) \cdot [S_9^1(t,1)]^{(5)}$$
$$= \exp[-0.019048]\exp[-0.015]\exp[-0.015]$$
$$\exp[-0.015]\exp[-0.015]\exp[-0.0125]$$
$$\exp[-0.0125]\exp[-0.0125]\exp[-0.015]$$
$$= \exp[-0.131548t], \quad (4.109)$$

$$[\boldsymbol{S}^1(t,2)]^{(5)} = \prod_{i=1}^{9} [S_i^1(t,2)]^{(5)}$$
$$= [S_1^1(t,2)]^{(5)} \cdot [S_2^1(t,2)]^{(5)} \cdot [S_3^1(t,2)]^{(5)} \cdot [S_4^1(t,2)]^{(5)} \cdot [S_5^1(t,2)]^{(5)}$$
$$\cdot S_6^0(t,2) \cdot S_7^0(t,2) \cdot S_8^0(t,2) \cdot [S_9^1(t,2)]^{(5)}$$
$$= \exp[-0.026087]\exp[-0.024]\exp[-0.024]$$
$$\exp[-0.024]\exp[-0.024]\exp[-0.02]$$
$$\exp[-0.02]\exp[-0.02]\exp[-0.024]$$
$$= \exp[-0.206087t]; \quad (4.110)$$

– at the operation states $z_6$

$$[\boldsymbol{S}^1(t,-)]^{(6)} = \left[1, [\boldsymbol{S}^1(t,1)]^{(6)}, [\boldsymbol{S}^1(t,2)]^{(6)}\right], t \geq 0, \quad (4.111)$$

where

$$\left[\boldsymbol{S}^1(t,1)\right]^{(6)} = \prod_{i=1}^{9}\left[S_i^1(t,1)\right]^{(6)}$$
$$= \left[S_1^1(t,1)\right]^{(6)} \cdot \left[S_2^1(t,1)\right]^{(6)} \cdot \left[S_3^1(t,1)\right]^{(6)} \cdot \left[S_4^1(t,1)\right]^{(6)} \cdot \left[S_5^1(t,1)\right]^{(6)}$$
$$\cdot \left[S_6^1(t,1)\right]^{(6)} \cdot \left[S_7^1(t,1)\right]^{(6)} \cdot S_8^0(t,1) \cdot \left[S_9^1(t,1)\right]^{(6)}$$
$$= \exp[-0.020635]\exp[-0.01625]\exp[-0.01625]$$
$$\exp[-0.01625]\exp[-0.01625]\exp[-0.01625]$$
$$\exp[-0.01625]\exp[-0.0125]\exp[-0.01625]$$
$$= \exp[-0.146885t], \tag{4.112}$$

$$\left[\boldsymbol{S}^1(t,2)\right]^{(6)} = \prod_{i=1}^{9}\left[S_i^1(t,2)\right]^{(6)}$$
$$= \left[S_1^1(t,2)\right]^{(\theta)} \cdot \left[S_2^1(t,2)\right]^{(6)} \cdot \left[S_3^1(t,2)\right]^{(6)} \cdot \left[S_4^1(t,2)\right]^{(6)} \cdot \left[S_5^1(t,2)\right]^{(6)}$$
$$\cdot \left[S_6^1(t,2)\right]^{(6)} \cdot \left[S_7^1(t,2)\right]^{(6)} \cdot S_8^0(t,2) \cdot \left[S_9^1(t,2)\right]^{(6)}$$
$$= \exp[-0.02826]\exp[-0.026]\exp[-0.026]$$
$$\exp[-0.026]\exp[-0.026]\exp[-0.026]$$
$$\exp[-0.026]\exp[-0.02]\exp[-0.026]$$
$$= \exp[-0.23026t]; \tag{4.113}$$

– at the operation states $z_7$

$$\left[\boldsymbol{S}^1(t,-)\right]^{(7)} = \left[1, \left[\boldsymbol{S}^1(t,1)\right]^{(7)}, \left[\boldsymbol{S}^1(t,2)\right]^{(7)}\right], t \geq 0, \tag{4.114}$$

where

$$\left[\boldsymbol{S}^1(t,1)\right]^{(7)} = \prod_{i=1}^{9}\left[S_i^1(t,1)\right]^{(7)}$$
$$= \left[S_1^1(t,1)\right]^{(7)} \cdot \left[S_2^1(t,1)\right]^{(7)} \cdot S_3^0(t,1) \cdot \left[S_4^1(t,1)\right]^{(7)} \cdot S_5^0(t,1)$$
$$\cdot \left[S_6^1(t,1)\right]^{(7)} \cdot \left[S_7^1(t,1)\right]^{(7)} - \left[S_8^1(t,1)\right]^{(7)} \cdot \left[S_9^1(t,1)\right]^{(7)}$$
$$= \exp[-0.017460]\exp[-0.01375]\exp[-0.0125]$$
$$\exp[-0.01375]\exp[-0.0125]\exp[-0.01375]$$
$$\exp[-0.01375]\exp[-0.01375]\exp[-0.01375]$$
$$= \exp[-0.12496t], \tag{4.115}$$

$$\begin{aligned}
[\boldsymbol{S}^1(t,2)]^{(7)} &= \prod_{i=1}^{9} [S_i^1(t,2)]^{(7)} \\
&= [S_1^1(t,2)]^{(7)} \cdot [S_2^1(t,2)]^{(7)} \cdot S_3^0(t,2) \cdot [S_4^1(t,2)]^{(7)} \cdot S_5^0(t,2) \\
&\quad \cdot [S_6^1(t,2)]^{(7)} \cdot [S_7^1(t,2)]^{(7)} \cdot [S_8^1(t,2)]^{(7)} \cdot [S_9^1(t,2)]^{(7)} \\
&= \exp[-0.023913]\exp[-0.022]\exp[-0.02] \\
&\quad \exp[-0.022]\exp[-0.02]\exp[-0.022] \\
&\quad \exp[-0.022]\exp[-0.022]\exp[-0.022] \\
&= \exp[-0.195913\mathrm{t}].
\end{aligned} \tag{4.116}$$

Hence, applying (4.24), the mean values of the port oil terminal critical infrastructure lifetimes in the safety state subsets {1, 2},{2}, at the operation states $z_b$, $b = 1, 2, \ldots 7$, respectively are:

- at the operation states $z_1$

$$[\boldsymbol{\mu}^1(1)]^{(1)} = \int_{\delta}^{\infty} [\boldsymbol{S}^1(t,1)]^{(1)} dt = \int_0^{\infty} \exp[-0.12371t]dt \cong 8.08 \text{ years},$$

$$[\boldsymbol{\mu}^1(2)]^{(1)} = \int_0^{\infty} [\boldsymbol{S}^1(t,2)]^{(1)} dt = \int_0^{\infty} \exp[-0.193913t]dt \cong 5.16 \text{ years}; \tag{4.117}$$

- at the operation states $z_2$

$$[\boldsymbol{\mu}^1(1)]^{(2)} = \int_0^{\infty} [\boldsymbol{S}^1(t,1)]^{(2)} dt = \int_0^{\infty} \exp[-0.12246t]dt \cong 8.17 \text{ years},$$

$$[\boldsymbol{\mu}^1(2)]^{(2)} = \int_0^{\infty} [\boldsymbol{S}^1(t,2)]^{(2)} dt = \int_0^{\infty} \exp[-0.191913t]dt \cong 5.21 \text{ years}; \tag{4.118}$$

- at the operation states $z_3$

$$[\boldsymbol{\mu}^1(1)]^{(3)} = \int_0^{\infty} [\boldsymbol{S}^1(t,1)]^{(3)} dt = \int_0^{\infty} \exp[-0.131548t]dt \cong 7.60 \text{ years},$$

$$[\boldsymbol{\mu}^1(2)]^{(3)} = \int_0^\infty [\boldsymbol{S}^1(t,2)]^{(3)} dt = \int_0^\infty \exp[-0.206087t]dt \cong 4.85 \text{ years}; \quad (4.119)$$

– at the operation states $z_4$

$$[\boldsymbol{\mu}^1(1)]^{-4)} = \int_0^\infty [\boldsymbol{S}^1(t,1)]^{(4)} dt = \int_0^\infty \exp[-0.146885t]dt \cong 6.81 \text{ years},$$

$$[\boldsymbol{\mu}^1(2)]^{(4)} = \int_0^\infty [\boldsymbol{S}^1(t,2)]^{(4)} dt = \int_0^\infty \exp[-0.230261t]dt \cong 4.34 \text{ years}; \quad (4.120)$$

– at the operation states $z_5$

$$[\boldsymbol{\mu}^1(1)]^{(5)} = \int_0^\infty [\boldsymbol{S}^1(t,1)]^{(5)} dt = \int_0^\infty \exp[-0.131548t]dt \cong 7.60 \text{ years},$$

$$[\boldsymbol{\mu}^1(2)]^{(5)} = \int_0^\infty [\boldsymbol{S}^1(t,2)]^{(5)} dt = \int_0^\infty \exp[-0.206087t]dt \cong 4.85 \text{ years}; \quad (4.121)$$

– at the operation states $z_6$

$$[\boldsymbol{\mu}^1(1)]^{(0)} = \int_0^\infty [\boldsymbol{S}^1(t,1)]^{(6)} dt = \int_0^\infty \exp[-0.146885t]dt \cong 6.81 \text{ years},$$

$$[\boldsymbol{\mu}^1(2)]^{(6)} = \int_0^\infty [\boldsymbol{S}^1(t,2)]^{(6)} dt = \int_0^\infty \exp[-0.23026t]dt \cong 4.34 \text{ years}; \quad (4.122)$$

– at the operation states $z_7$

$$[\boldsymbol{\mu}^1(1)]^{(7)} = \int_0^\infty [\boldsymbol{S}^1(t,1)]^{(7)} dt = \int_0^\infty \exp[-0.12496t]dt \cong 8.00 \text{ years},$$

$$[\mu^1(2)]^{(7)} = \int_0^\infty [S^1(t,2)]^{(7)} dt = \int_0^\infty \exp[-0.195913t]dt \cong 5.10\,\text{years}. \quad (4.123)$$

From the results (4.66) and (4.96)–(4.116), after applying (4.16), the port oil terminal critical infrastructure unconditional safety function (SafI1) is given by

$$\boldsymbol{S}^1(t,-) = \left[1, \boldsymbol{S}^1(t,1)\boldsymbol{S}^1(t,2)\right], t \geq 0, \quad (4.124)$$

where

$$\begin{aligned} \boldsymbol{S}^1(t,1) &\cong \sum_{\partial=1}^{7} p_b\left[\boldsymbol{S}^1(t,1)\right]^{(b)} \\ &= 0.395\exp[-0.12371t] + 0.060\exp[-0.12246t] \\ &\quad + 0.003\exp[-0.131548t] + 0.002\exp[-0.146885t] \\ &\quad + 0.200\exp[-0.131548t] + 0.058\exp[-0.146885t] \\ &\quad + 0.282\exp[-0.12496t], \end{aligned} \quad (4.125)$$

$$\begin{aligned} \boldsymbol{S}^1(t,2) &\cong \sum_{b=1}^{7} p_b\left[\boldsymbol{S}^1(t,2)\right]^{(b)} \\ &= 0.395\exp[-0.193913t] + 0.060\exp[-0.191913t] \\ &\quad + 0.003\exp[-0.206087t] + 0.002\exp[-0.230261t] \\ &\quad + 0.200\exp[-0.206087t] + 0.058\exp[-0.23026t] \\ &\quad + 0.282\exp[-0.195913\text{t}]. \end{aligned} \quad (4.126)$$

The graph of the port oil terminal critical infrastructure safety function is shown in Fig. 4.8.

As the critical safety state is $r = 1$, then by (4.18) and (4.125), the port oil terminal critical infrastructure risk function (SafI2), is given by

$$\begin{aligned} \boldsymbol{r}^1(t) &= 1 - \boldsymbol{S}^1(t,1) \\ &= 1 - \{0.395\exp[-0.12371t] + 0.060\exp[-0.12246t] \\ &\quad + 0.003\exp[-0.131548t] + 0.002\exp[-0.146885t] \\ &\quad + 0.200\exp[-0.131548t] + 0.058\exp[-0.146885t] \\ &\quad + 0.282\exp[-0.12496t]\} \text{ for } t \geq 0. \end{aligned} \quad (4.127)$$

The values of the above risk function are given in Table 4.1.

The graph of the risk function $\boldsymbol{r}^1(t)$ of the port oil terminal critical infrastructure, the fragility curve (SafI3), is shown in Fig. 4.9.

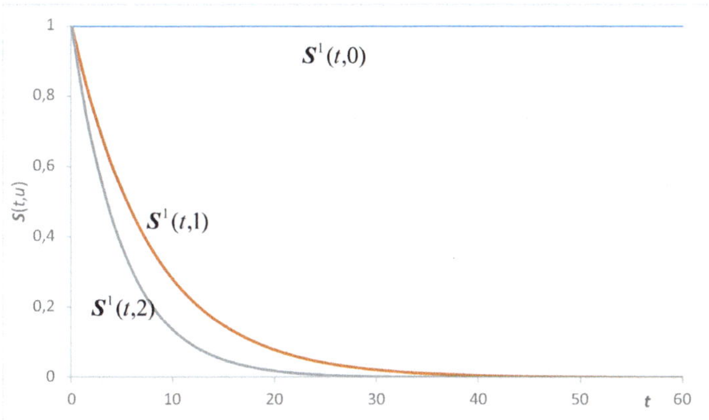


**Fig. 4.8** The graph of the port oil terminal critical infrastructure safety function $S^1(t, \cdot)$ coordinates

**Table 4.1** The values of the port oil terminal critical infrastructure risk function

| $t$ | $r^1(t) = 1 - S^1(t, 1)$ |
|---|---|
| 0.400 | 0.04952 |
| 0.401 | 0.04964 |
| 0.402 | 0.04976 |
| 0.403 | 0.04988 |
| **0.404** | **0.05000** |
| 0.405 | 0.05012 |
| 0.406 | 0.05024 |
| 0.407 | 0.05036 |
| 0.408 | 0.05048 |
| 0.409 | 0.05060 |
| 0.410 | 0.05072 |
| 0.411 | 0.05084 |
| 0.412 | 0.05096 |
| 0.413 | 0.05108 |
| 0.414 | 0.05120 |
| 0.415 | 0.05133 |
| 0.416 | 0.05145 |
| 0.417 | 0.05157 |
| 0.418 | 0.05169 |
| 0.419 | 0.05181 |
| 0.420 | 0.05193 |

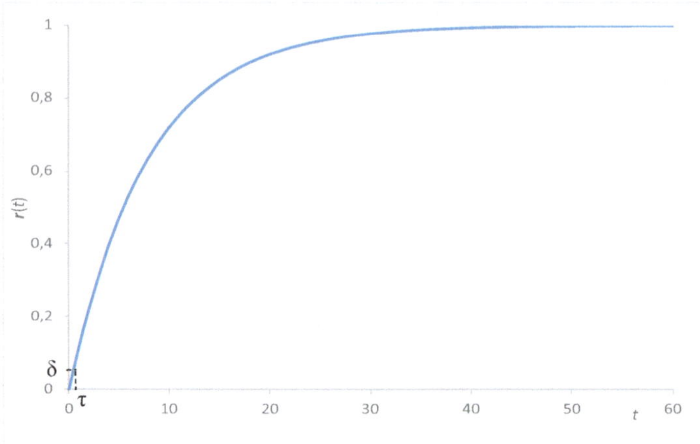


**Fig. 4.9** The graph of the risk function $\boldsymbol{r}^1(t)$ (the fragility curve) of the oil terminal critical infrastructure

From (4.23) and (4.127) and Table 4.1 the moment when the port oil terminal critical infrastructure risk function exceeds a permitted level $\delta = 0.05$ (SafI6), is

$$\boldsymbol{\tau}^1 = \boldsymbol{r}^{1-1}(\delta) \cong 0.404 \text{ years.} \tag{4.128}$$

Considering (4.66) and (4.117)–(4.123) and applying (4.19), (4.21), (4.22) and (4.24), (4.26), (4.27), the expected values (SafI4, SafI7) and standard deviations (SafI5, SafI8) of the port oil terminal critical infrastructure lifetimes in the safety state subsets $\{1, 2\}$, $\{2\}$, respectively are:

$$\begin{aligned}\boldsymbol{\mu}^1(1) &\cong \sum_{b=1}^{7} p_b\left[\mu^1(1)\right]^{(b)} \\ &\cong 0.395 \cdot 8.08 + 0.060 \cdot 8.17 + 0.003 \cdot 7.60 + 0.002 \cdot 6.81 \\ &\quad + 0.200 \cdot 7.60 + 0.058 \cdot 6.81 + 0.282 \cdot 8.00 \\ &= 7.89 \text{ years,}\end{aligned} \tag{4.129}$$

$$\begin{aligned}\boldsymbol{\mu}^1(2) &\cong \sum_{b=1}^{7} p_b\left[\mu^1(2)\right]^{(b)} \\ &\cong 0.395 \cdot 5.16 + 0.060 \cdot 5.21 + 0.003 \cdot 4.85 + 0.002 \cdot 4.34 \\ &\quad + 0.200 \cdot 4.85 + 0.058 \cdot 4.34 + 0.282 \cdot 5.1 \\ &= 5.03 \text{ years,}\end{aligned} \tag{4.130}$$

$$\boldsymbol{\sigma}^1(1) \cong 7.91, \ \boldsymbol{\sigma}^1(2) \cong 5.05 \text{ years.} \tag{4.131}$$

The detailed way of finding the above standard deviations (4.131) is illustrated below.

According to (4.22) or (4.27), using (4.125), (4.126), we have

$$\begin{aligned}
\boldsymbol{n}^1(1) &= 2\int_0^\infty t\boldsymbol{S}^1(t,1)dt,\\
&= 2\int_0^\infty t\{0.395\exp[-0.12371t] + 0.060\exp[-0.12246t]\\
&\quad + 0.003\exp[-0.131548t] + 0.002\exp[-0.146885t]\\
&\quad + 0.200\exp[-0.131548t] + 0.058\exp[-0.146885t]\\
&\quad + 0.282\exp[-0.12496t]\}dt\\
&= 124.8202,
\end{aligned} \tag{4.132}$$

and

$$\begin{aligned}
\boldsymbol{n}^1(2) &= 2\int_0^\infty t\boldsymbol{S}^1(t,2)dt,\\
&= 2\int_0^\infty t\{0.395\exp[-0.193913t]\\
&\quad + 0.060\exp[-0.191913t] + 0.003\exp[-0.206087t]\\
&\quad + 0.002\exp[-0.230261t] + 0.200\exp[-0.206087t]\\
&\quad + 0.058\exp[-0.230261t] + 0.282\exp[-0.195913\text{t}]\}dt\\
&= 50.8034.
\end{aligned} \tag{4.133}$$

Hence, after applying (4.21) or (4.26), we get

$$\boldsymbol{\sigma}^1(1) = \sqrt{n^1(1) - \left[\boldsymbol{\mu}^1(1)\right]^2} = \sqrt{124.8202 - [7.98]^2} \cong 7.91 \text{ years},$$

$$\boldsymbol{\sigma}^1(2) = \sqrt{n^1(2) - \left[\boldsymbol{\mu}^1(2)\right]^2} = \sqrt{50.8034 - [5.03]^2} \cong 5.05 \text{ years}.$$

Further, it follows that the mean values of the port oil terminal critical infrastructure lifetimes in the particular safety states (SafI9), according (4.28) and using (4.129), (4.130), are:

$$\bar{\boldsymbol{\mu}}^1(1) = \boldsymbol{\mu}^1(1) - \boldsymbol{\mu}^1(2) \cong 2.86,\ \bar{\boldsymbol{\mu}}^1(2) = \boldsymbol{\mu}^1(2) \cong 5.03 \text{ years}. \tag{4.134}$$

By (4.125), (4.126), applying (4.30), the oil terminal critical infrastructure intensities of ageing (SafI10) are:

$$\lambda^1(t,1) = \frac{-\frac{dS^1(t,1)}{dt}}{S^1(t,1)}$$
$$= \{0.12371 \cdot 0.395 \exp[-0.12371t] + 0.12246 \cdot 0.060 \exp[-0.12246t]$$
$$+ 0.131548 \cdot 0.003 \exp[-0.131548t] + 0.002 \exp[-0.146885t]$$
$$+ 0.131548 \cdot 0.200 \exp[-0.131548t] + 0.146885$$
$$\cdot 0.058 \exp[-0.146885t] + 0.12496 \cdot 0.282 \exp[-0.12496t]\}$$
$$/\{0.395 \exp[-0.12371t] + 0.060 \exp[-0.12246t]$$
$$+ 0.003 \exp[-0.131548t] + 0.002 \exp[-0.146885t]$$
$$+ 0.200 \exp[-0.131548t] + 0.058 \exp[-0.146885t]$$
$$+ 0.282 \exp[-0.12496t], \tag{4.135}$$

$$\lambda^1(t,2) = \frac{-\frac{dS^1(t,2)}{dt}}{S^1(t,2)}$$
$$= \{0.193913 \cdot 0.395 \exp[-0.193913t] + 0.191913$$
$$\cdot 0.060 \exp[-0.191913t] + 0.206087 \cdot 0.003 \exp[-0.206087t]$$
$$+ 0.230261 \cdot 0.002 \exp[-0.230261t] + 0.206087$$
$$\cdot 0.200 \exp[-0.206087t] + 0.230261 \cdot 0.058 \exp[-0.230261t]$$
$$+ 0.195913 - 0.282 \exp[-0.195913\text{t}]\}$$
$$/\{0.395 \exp[-0.193913t] + 0.060 \exp[-0.191913t]$$
$$+ 0.003 \exp[-0.206087t] + 0.002 \exp[-0.230261t]$$
$$+ 0.200 \exp[-0.206087t] + 0.058 \exp[-0.230261t]$$
$$+ 0.282 \exp[-0.195913\text{t}]\}. \tag{4.136}$$

The graphs of the intensities of ageing of the port oil terminal critical infrastructure are shown in Fig. 4.10.

The values of the intensities of degradation given by (4.135), (4.136) stabilize for large time and their limit values defined by (1.46) amount:

$$\boldsymbol{\lambda}_1(1) = \lim_{t\to\infty} \boldsymbol{\lambda}_1(t,1) = \lim_{t\to\infty} \frac{0.060 \cdot 0.12246(1 + 0(t))}{0.060(1 + 0(t))} = 0.12246, \tag{4.137}$$

$$\boldsymbol{\lambda}_1(2) = \lim_{t\to+\infty} \boldsymbol{\lambda}_1(t,2) = \lim_{t\to\infty} \frac{0.060 \cdot 0.191913(1 + 0(t))}{0.060(1 + 0(t))} = 0.191913. \tag{4.138}$$

The approximate mean of the intensities of degradation of the port oil piping transportation system according to (1.47) and (4.129) and (4.130) respectively are

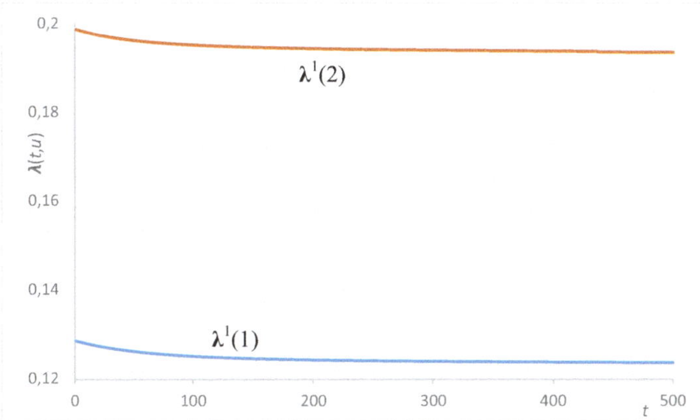


**Fig. 4.10** The graphs of the intensities of ageing of the port oil terminal critical infrastructure

$$\boldsymbol{\lambda}_1(t) = \frac{1}{7.89} \cong 0.12674, \tag{4.139}$$

$$\boldsymbol{\lambda}_1(2) = \frac{1}{5.03} \cong 0.19881. \tag{4.140}$$

Considering (4.137), (4.138), (4.139), (4.140) and the values of the port oil terminal critical infrastructure intensities of ageing without of operation impact, determined in Chap. 3:

$$\boldsymbol{\lambda}^0(t, 1) = 0.115873, \quad \boldsymbol{\lambda}^0(t, 2) = 0.181739, \tag{4.141}$$

and applying (4.33), the coefficients of the operation process impact on the oil terminal critical infrastructure intensities of ageing (ResI1) are:

$$\boldsymbol{\rho}^1(t, 1) = \frac{\boldsymbol{\lambda}^1(t, 1)}{\boldsymbol{\lambda}^0(t, 1)} \cong \frac{0.12246}{0.115873} \cong 1.057, \tag{4.142}$$

$$\boldsymbol{\rho}^1(t, 1) = \frac{\boldsymbol{\lambda}^1(t, 1)}{\boldsymbol{\lambda}^0(t, 1)} \cong \frac{0.12674}{0.115873} \cong 1.094, \tag{4.143}$$

$$\boldsymbol{\rho}^1(t, 2) = \frac{\boldsymbol{\lambda}^1(t, 2)}{\boldsymbol{\lambda}^0(t, 2)} \cong \frac{0.191913}{0.181739} \cong 1.056, \tag{4.144}$$

$$\boldsymbol{\rho}^1(t, 2) = \frac{\boldsymbol{\lambda}^1(t, 2)}{\boldsymbol{\lambda}^0(t, 2)} \cong \frac{0.19881}{0.181739} \cong 1.094. \tag{4.145}$$

Finally, by (4.45) and (4.142), the port oil terminal critical infrastructure resilence indicator (ResI2), i.e. the coefficient of the port oil terminal critical infrastructure resilience to operation process impact, is

$$\boldsymbol{RI}^1(t,1) = \frac{1}{\boldsymbol{\rho}^1(t,1)} \cong 0.946 \cong 95\%. \, t \in < 0, +\infty). \tag{4.144}$$

The comparison of safety indicators (3.102)–(3.114) and (4.124)–(4.140) proves a significant influence of the operation process on the port oil terminal safety what is also clearly expressed in the resilience indicators to operation process impact (4.142)–(4.144).

## References

1. Ferreira F, Pacheco A (2007) Comparison of level-crossing times for Markov and semi-Markov processes. Stat Probab Lett 77(2):151–157
2. Glynn PW, Haas PJ (2006) Laws of large numbers and functional central limit theorems for generalized semi-Markov processes. Stoch Model 22(2):201–231
3. Grabski F (2002) Semi-Markov models of systems reliability and operations analysis. System Research Institute, Polish Academy of Science (in Polish)
4. Kołowrocki K (2005) Reliability of large systems. Elsevier Amsterdam—Boston—Heidelberg—London—New York—Oxford—Paris—San Diego—San Francisco—Singapore—Sydney—Tokyo
5. Kołowrocki K (2014) Reliability of large and complex systems (2nd edn), Elsevier Amsterdam—Boston—Heidelberg—London—New York—Oxford—Paris—San Diego—San Francisco—Singapore—Sydney—Tokyo
6. Kołowrocki K, Kuligowska E, Soszyńska-Budny J, Torbicki M (2018) Oil transport in port Part 1 Port oil piping transportation system safety and resilience impacted by its operation process. J Pol Saf Reliab Assoc, Summer Safety Reliab Seminars 9(2):11–28
7. Kolowrocki K, Soszynska J (2009) Safety and risk evaluation of Stena Baltica ferry in variable operation conditions. Electron J Reliab Risk Anal: Theory App 2(4):168–180
8. Kolowrocki K, Soszynska J (2010) Reliability modeling of a port oil transportation system's operation processes. Int J of Perform Eng 6(1):77–87
9. Kołowrocki K, Soszyńska-Budny J (2011) Reliability and safety of complex technical systems and processes. Modeling—identification—prediction—optimization. Springer, London, Dordrecht, Heildeberg, New York
10. Kołowrocki K, Soszyńska-Budny J (2015) Modelling safety of multistate systems with ageing components. In: Proceedings of international conference of numerical analysis and applied mathematics—ICNAAM 2015 Symposium/Workshop On Safety Of Critical Infrastructures (ISBN: 978-0-7354-1392-4; https://doi.org/10.1063/1.4952224)
11. Kołowrocki K, Soszyńska-Budny J (2016) Modelling critical infrastructure operation process including operating environment threats. J Pol Saf Reliab Assoc, Summer Safety Reliab Seminars 7(3):81–88
12. Kołowrocki K Soszyńska-Budny J (2018) Critical infrastructure safety indicators. In: International conference on industrial engineering and engineering management IEEM 2018
13. Kołowrocki K Soszyńska-Budny J (2018) Critical infrastructure impacted by operation safety and resilience indicators. In: International conference on industrial engineering and engineering management IEEM 2018
14. Kołowrocki K, Soszyńska-Budny J (2019) Safety and resilience indicators of critical infrastructure impacted by operation application to port oil terminal examination Trans Nav 13(4):761–769. https://doi.org/10.12716/1001.13.04.08

15. Kołowrocki K, Soszyńska-Budny J, Torbicki M (2018) Critical infrastructure impacted by operation and climate change safety and resilience indicators. In: International conference on industrial engineering and engineering management IEEM 2018
16. Limnios N, Oprisan G (2005) Semi-Markov processes and reliability. Birkhauser, Boston
17. Mercier S (2008) Numerical bounds for semi-Markovian quantities and application to reliability. Methodol Comput in Appl Probab 10(2):179–198

# Chapter 5
# Modelling Critical Infrastructure Safety Impacted by Climate-Weather Change Process

The safety of critical infrastructure impacted by the climate-weather change process at its operating area is considered. The climate-weather change process at the critical infrastructure operating area and its parameters are defined and its characteristics are determined. The safety and resilience indicators of the critical infrastructure and its assets related to the climate-weather change process at its operating area impact are proposed. The safety of the port oil terminal critical infrastructure impacted by climate-weather change process at its operating area is examined and its safety and resilience indicators are determined.

## 5.1 Climate-Weather Change Process at Critical Infrastructure Operating Area

### 5.1.1 *States of Climate-Weather Change Process*

To define the climate-weather states at the fixed area, we distinguish $a$, $a \in N$, parameters that define (describe) the climate-weather states in this area and mark the values they can take by $w_1, w_2, \ldots, w_a$. Further, we assume that the possible values of the $i$-th parameter $w_i$, $i = 1, 2, \ldots, a$, can belong to the interval $< b_i, d_i)$, $i = 1, 2, \ldots, a$. We divide each of the intervals $< b_i, d_i)$, $i = 1, 2, \ldots, a$, into $n_i$, $n_i \in N$, disjoint subintervals [18]

$$< b_{i1}, d_{i1}), < b_{i2}, d_{i2}), \ldots, < b_{in_j}, d_{in_j}),$$



J. Soszyńska-Budny, *Safety Analysis of Critical Infrastructure*,
Lecture Notes in Intelligent Transportation and Infrastructure,
https://doi.org/10.1007/978-3-030-65339-2_5

such that

$$< b_{i1}, d_{i1}) \cup < b_{i2}, d_{i2}) \cup \cdots \cup < b_{in_i}, d_{in_i}) = < b_i, d_i), \; d_{ij_i} = b_{ij_i+1}, \\ j_i = 1, 2, \ldots, n_i - 1, \; i = 1, 2, \ldots, a.$$

Thus, the points ($w_1$, $w_2$, …, $w_a$) describing the values of the climate-weather parameters are the points from the set of the $a$ dimensional space of the descartes product

$$< b_1, d_1) \times \; < b_2, d_2) \times \cdots \times < b_a, d_a)$$

that is composed of the $a$ dimensional space domains of the form [18]

$$< b_{1j_1}, d_{1j_1}) \times \; < b_{2j_2}, d_{2j_2}) \times \cdots \times < b_{aj_a}, d_{aj_a}),$$

where $j_i$ = 1,2,… $n_i$, $i = 1, 2, \ldots, a$, called the climate-weatcher states and $w_i$, $i$ = 1, 2, …, $a$, can takes values from one of the interwals

$$< b_{i1}, d_{i1}), < b_{i2}, d_{i2}), \ldots < b_{in_j}, d_{in_j}), i = 1, 2, \ldots, a.$$

The domains of the above form called the climate-weather states of the climate-weather change process are numerated from 1 up to the value $w = n_1 \cdot n_2 \cdot \ldots \cdot n_a$ that is the number of all possible climate-weather states and marked by $c_1, c_2, \ldots, c_w$.

The interpretation of the states of the climate-weather change process in the case they are defined by $a = 2$ parameters is given in Fig. 5.1. In this case, we have $w = n_1 \cdot n_2$ climate-weather states of the climate-weather change process represented in Fig. 5.1 by the squares marked by $c_1, c_2, \ldots, c_w$ [18].

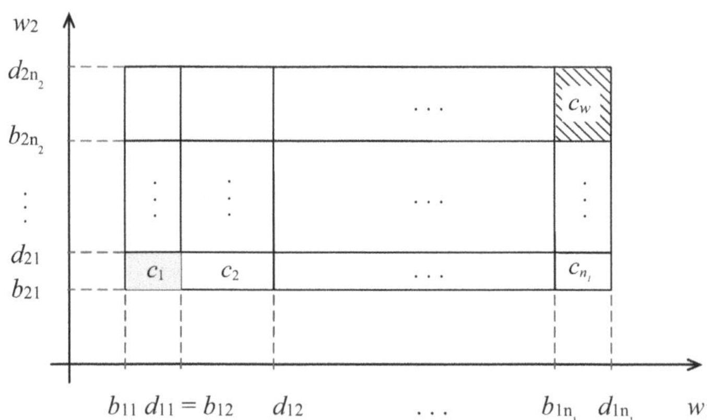


**Fig. 5.1** Interpretation of the climate-weather change process two dimensional climate-weather states [18]

In this particular case, the climate-weather change process can take values ($w_1$,$w_2$) from the climate-weather states defined by the domains

$$< b_{1j_1}, d_{1j_1}) \times < b_{2j_2}, d_{2j_2}),$$

where $j_i = 1, 2, \ldots n_i$, $i = 1, 2$, in the way such that $w_1$ can take value from one of the interwals

$$< b_{11}, d_{11}), \ < b_{12}, d_{12}), \ldots, < b_{1n_1}, d_{1n_1})$$

and $w_2$ can take value from one of the interwals

$$< b_{21}, d_{21}), \ < b_{22}, d_{22}), \ldots, < b_{2n_2}, d_{2n_2})$$

and marked by

$c_1, c_2, \ldots, c_w$,

where

$$w = n_1 \cdot n_2$$

is the number of all possible climate-weather states.

### 5.1.2 *Semi-Markov Model of Climate-Weather Change Process*

To model the climate-weather change process for the critical infrastructure operating area, we assume that the climate-weather process in this area is taking $w$, $w \in N$, different climate-weather states $c_1, c_2, \ldots, c_w$. Further, we define the climate-weather change process $C(t)$, $t \in < 0, +\infty)$, with discrete operation states from the set $\{c_1, c_2, \ldots, c_w\}$.

Assuming that the climate-weather change process $C(t)$ is a semi-Markov process [3, 10, 22] it can be described by the following climate-weather change process parameters (C-WCPP), that can be identified either statistically using the methods given in [10, 16] or evaluated approximately by experts [18]:

- the number of climate-weather states (C-WCPP1)$w$;
- the climate-weather states (C-WCPP2) $\{c_1, c_2, \ldots, c_w\}$;
- the vector

$$[q_l(0)]_{1 \times w} = [q_1(0), q_2(0), \ldots, q_w(0)],$$

of the initial probabilities (C-WCPP3) of the climate-weather change process $C(t)$ staying at particular climate-weather states $c_l$ at the moment $t = 0$

$$q_l(0) = P(C(0) = c_l), l = 1, 2, \ldots, w;$$

- the matrix

$$[q_{lk}]_{w\times w} = \begin{bmatrix} q_{11}\ q_{12} \cdots q_{1w} \\ q_{21}\ q_{22} \cdots q_{2w} \\ \cdots \\ q_{w1}\ q_{w2} \cdots q_{ww} \end{bmatrix},$$

of probabilities of transition (C-WCPP4) of the climate-weather change process $C(t)$ between the climate-weather states $c_l$ and $c_k$

$$q_{lk}, l,\ k = 1, 2, \ldots, w; l \neq k, q_{ll} = 0, l = 1, 2, \ldots, w;$$

- the matrix

$$[N_{lk}]_{w\times w} = \begin{bmatrix} N_{11} & N_{12} & \ldots & N_{1w} \\ N_{21} & N_{22} & \ldots & N_{2w} \\ \cdots & & & \\ N_{w1} & N_{w2} & \ldots & N_{ww} \end{bmatrix},$$

of mean values of the climate-weather change process $C(t)$ conditional sojourn times (C-WCPP5) $C_{lk}$ at the climate-weather state $c_l$ when the next state is $c_k$

$$N_{lk} = E[C_{lk}] = \int_0^\infty t dC_{lk}(t) = \int_0^\infty t c_{lk}(t)dt, l,\ k = 1, 2, \ldots, w, l \neq k,$$
$$N_{ll} = 0, l = 1, 2, \ldots, w, \tag{5.1}$$

where

$$C_{lk}(t) = P(\theta_{lk} < t), t \geq 0, l,\ k = 1, 2, \ldots, w, l \neq k,$$

are conditional distribution functions of the critical infrastructure climate-weather change process $C(t)$ conditional sojourn times $\theta_{lk}$ at the climate-weather states corresponding to conditional density functions

$$c_{lk}(t) = \frac{dC_{lk}(t)}{dt}, t \geq 0, l, k = 1, 2, \ldots, w, l \neq k.$$

Assuming that we have identified the above parameters (C-WCPP1-5) of the climate-weather change process semi-Markov model, we can predict this process basic characteristics (C-WCPC), that can be either calculated analytically or evaluated approximately by experts [10, 16–19]:

- the vector

$$[N_l]_{1\times w} = [N_1, N_2, \ldots, N_w], \tag{5.2}$$

of mean values of the climate-weather change process $C(t)$ critical infrastructure at the operating area unconditional sojourn times $C_l$, $l = 1, 2, \ldots, w$, at the climate-weather states $c_l$ $l = 1, 2, \ldots, w$, (C-WCPC1)

$$N_l = E[C_l] = \sum_{k=1}^{v} q_{lk} N_{lk}, l = 1, 2, \ldots, w, \tag{5.3}$$

where $N_{lk}$ are defined by the Formula (5.1);

- the vector

$$[q_l]_{1\times w} = [q_1, q_2, \ldots, q_w],$$

of limit values of the climate-weather change process $C(t)$ transient probabilities

$$q_l(t) = P(C(t) = c_l), t \in < 0, +\infty), l = 1, 2, \ldots, w,$$

at the particular climate-weather states (C-WCPC2) [18]

$$q_l = \lim_{t\to\infty} q_l(t) = \frac{\pi_l N_l}{\sum_{k=1}^{w} \pi_k N_k}, l = 1, 2, \ldots, w, \tag{5.4}$$

where $N_l$, $l = 1, 2, \ldots, w$, are given by (5.3), while the steady probabilities $\pi_l$ of the vector $[\pi_l]_{1\times w}$ satisfy the system of equations

$$\begin{cases} = [\pi_l][q_{lk}] \\ \sum_{k=1}^{v} \pi_k = 1. \end{cases} \tag{5.5}$$

In the case of a periodic climate-weather change process, the limit transient probabilities $q_l$, $l = 1, 2, \ldots, w$, at the climate-weather states determined by (5.4), are the long term proportions of the climate-weather change process $C(t)$ sojourn times at the particular climate-weather states $c_l$, $l = 1, 2, \ldots, w$.

Another interesting characteristic of the climate-weather change process $C(t)$ possible to obtain is

- the vector of the mean values (C-WCPC3)

$$[\hat{N}_l]_{1 \times w} = [\hat{N}_1, \hat{N}_2, \ldots, \hat{N}_w],$$

of the total sojourn times $\hat{C}_l$, $l = 1, 2, \ldots, w$, of the climate-weather change process $C(t)$ at the critical infrastructure operating area at the particular climate-weather states $c_l$, $l = 1, 2, \ldots, w$, during the fixed time $C$. It is well known, [10, 16–18], that the climate-weather change process total sojourn times $\hat{C}_l$ at the particular climate-weather states $c_l$ for sufficiently large time $C$ have approximately normal distributions with the mean values given by

$$\hat{N}_l = E[\hat{C}_l] = q_l C, l = 1, 2, \ldots, w, \tag{5.6}$$

where $q_l$ are given by (5.4).

## 5.2 Safety and Resilience Indicators of Critical Infrastructure Impacted by Climate-Weather Change Process

We denote by $[T^2(u)]^{(l)}$, $u = 1, 2, \ldots, z$,$l = 1, 2, \ldots, w$, the critical infrastructure conditional lifetime in the safety state subset $\{u, u+1, \ldots, z\}$, $u = 1, 2, \ldots, z$, while the climate-weather change process $C(t)$, $t \in < 0, \infty)$, at the critical infrastructure operating area is at the climate-weather state $c_l$, $l = 1, 2, \ldots, w$, and the conditional safety function of the critical infrastructure related to the climate-weather change process at its operating area $C(t)$, $t \in < 0, \infty)$, by the vector [13, 14, 18, 19]

$$[\boldsymbol{S}^2(t, \cdot)]^{(l)} = [1, [\boldsymbol{S}^2(t, 1)]^{(l)}, \ldots, [\boldsymbol{S}^2(t, z)]^{(l)}], t \in < 0, \infty), l = 1, 2, \ldots, w, \tag{5.7}$$

with the coordinates defined by

$$[\boldsymbol{S}^2(t, u)]^{(l)} = P([T^2(u)]^{(l)} > t | Z(t) = z_l) \tag{5.8}$$

for $t \in < 0, \infty)$,$u = 1, 2, \ldots, z$,$l = 1, 2, \ldots, w$.

The safety function $[\boldsymbol{S}^2(t,u)]^{(l)}$, $t \in <0,\infty)$, $u = 1,2,\ldots,z$, $l = 1,2,\ldots,w$, is the conditional probability that the critical infrastructure impacted by the climate-weather change process $C(t)$, $t \in <0,\infty)$, lifetime $[T^2(u)]^{(l)}$, $u = 1,2,\ldots,z$, $l = 1,2,\ldots,w$, in the safety state subset $\{u, u+1,\ldots,z\}$, $u = 1,2,\ldots,z$, is greater than $t$, while the climate-weather change process $C(t)$, $t \in <0,\infty)$, is at the climate-weather state $c_l$, $l = 1,2,\ldots,w$,[18].

Next, we denote by $T^2(u)$, $u = 1,2,\ldots,z$, the critical infrastructure impacted by the climate-weather change process $C(t)$, $t \in <0,\infty)$, unconditional lifetime in the safety state subset $\{u, u+1,\ldots,z\}$, $u = 1,2,\ldots,z$, and the unconditional safety function (SafI1) of the critical infrastructure impacted by the climate-weather change process $C(t)$,$t \in <0,\infty)$, by the vector [18]

$$\boldsymbol{S}^2(t,\cdot) = [1, \boldsymbol{S}^2(t,1),\ldots,\boldsymbol{S}^2(t,z)],\, t \in <0,\infty), \tag{5.9}$$

with the coordinates defined by

$$\boldsymbol{S}^2(t,u) = P(T^2(u) > t),\, t \in <0,\infty),\, u = 1,2,\ldots,z. \tag{5.10}$$

In the case when the critical infrastructure operation time $\theta$ is large enough, the coordinates of the unconditional safety function of the critical infrastructure related to the climate-weather change process $C(t)$, $t \in <0,\infty)$, defined by (5.10), are given by [18]

$$\boldsymbol{S}^2(t,u) \cong \sum_{l=1}^{w} q_l [\boldsymbol{S}^2(t,u)]^{(l)} \; for \; t \in <0,\infty),\, u = 1,2,\ldots,z, \tag{5.11}$$

where $[\boldsymbol{S}^2(t,u)]^{(l)}$, $u = 1,2,\ldots,z$, $l = 1,2,\ldots,w$, are the coordinates of the critical infrastructure related to the climate-weather change process $C(t)$, $t \in <0,\infty)$, conditional safety functions defined by (5.7), (5.8) and $q_l$, $l = 1,2,\ldots,w$, are the climate-weather change process $C(t)$,$t \in <0,\infty)$, at the critical infrastructure operating area limit transient probabilities at the climate-weather states $c_l$, $l = 1,2,\ldots,w$, given by (5.4).

If $r$ is the critical safety state, then the second safety indicator of the critical infrastructure impacted by the climate-weather change process $C(t)$, $t \in <0,\infty)$, the risk function (SafI2)

$$\boldsymbol{r}^2(t) = P(s(t) < r | s(0) = z) = P\left(T^2(r) \le t\right),\, t \in <0,\infty), \tag{5.12}$$

is defined as a probability that the critical infrastructure impacted by the climate-weather change process $C(t)$, $t \in <0,\infty)$, is in the subset of safety states worse than the critical safety state $r$, $r \in \{1, \ldots, z\}$, while it was in the best safety state $z$ at the moment $t = 0$ and given by [13, 14, 18, 19]

$$\boldsymbol{r}^2(t) = 1 - \boldsymbol{S}^2(t, r), t \in< 0, \infty), \tag{5.13}$$

where $\boldsymbol{S}^2(t, r)$ is the coordinate of the critical infrastructure impacted by the climate-weather change process $C(t)$, $t \in< 0, \infty)$, unconditional safety function given by (5.11) for $u = r$.

The graph of the critical infrastructure risk function $\boldsymbol{r}^2(t)$, $t \in< 0, \infty)$, defined by (5.13), is the safety indicator called the fragility curve (SafI3) of the critical infrastructure impacted by the climate-weather change process $C(t)$,$t \in< 0, \infty)$.

Other practically useful safety and resilience indicators of the critical infrastructure impacted by climate-weather change process $C(t)$,$t \in< 0, \infty)$ at its operating area are:

- the mean value of the critical infrastructure unconditional lifetime $T^2(r)$ up to exceeding critical safety state $r$ (SafI4) given by [18]

$$\boldsymbol{\mu}^2(r) = \int_0^\infty [\boldsymbol{S}^2(t, r)]dt \cong \sum_{l=1}^{w} q_l[\boldsymbol{\mu}^2(r)]^{(l)}, \tag{5.14}$$

where $[\boldsymbol{\mu}^2(r)]^{(l)}$ are the mean values of the critical infrastructure conditional lifetimes $[T^2(r)]^{(l)}$ in the safety state subset $\{r, r+1, \ldots, z\}$ at the climate-weather state $c_l$, $l = 1, 2, \ldots, w$, given by

$$[\boldsymbol{\mu}^2(r)]^{(l)} = \int_0^\infty [\boldsymbol{S}^2(t, r)]^{(l)}dt, \; l = 1, 2, \ldots, w, \tag{5.15}$$

and $[\boldsymbol{S}^2(t, r)]^{(l)}, t \in< 0, \infty), l = 1, 2, \ldots, w$, are defined by (5.7), (5.8) and $q_l, l = 1, 2, \ldots, w$, are given by (5.4);

- the standard deviation of the critical infrastructure lifetime $T^2(r)$ up to the exceeding the critical safety state $r$(SafI5) given by [18]

$$\boldsymbol{\sigma}^2(r) = \sqrt{\boldsymbol{n}^2(r) - [\boldsymbol{\mu}^2(r)]^2}, \tag{5.16}$$

where

$$\boldsymbol{n}^2(r) = 2\int_0^\infty tS^2(t, r)dt, \tag{5.17}$$

and $\boldsymbol{S}^2(t, r)$, $t \in< 0, \infty)$ is defined by (5.11) for $u = r$ and $\boldsymbol{\mu}^2(r)$ is given by (5.14);

- the moment $\boldsymbol{\tau}^2$ of exceeding acceptable value of critical infrastructure risk function level $\delta$ (SafI6) given by [18]

$$\boldsymbol{\tau}^2 = \boldsymbol{r}^{2^{-1}}(\delta), \tag{5.18}$$

where $\boldsymbol{r}^{2-1}(\delta)$ is the inverse function of the risk function $\boldsymbol{r}^2(t)$ given by (5.13);

- the mean lifetimes of the critical infrastructure in the safety state subsets $\{u, u+1, \ldots, z\}$, $u = 1, 2, \ldots, z$, (SafI7), given by [18]

$$\boldsymbol{\mu}^2(u) = \int_0^\infty [\boldsymbol{S}^2(t,u)]dt \cong \sum_{l=1}^{w} p_l[\boldsymbol{\mu}^2(u)]^{(l)}, u = 1, 2, \ldots, z, \tag{5.19}$$

where $[\boldsymbol{\mu}^2(u)]^{(l)}, u = 1, 2, \ldots, z, l = 1, 2, \ldots, w$, are the mean values of the critical infrastructure conditional lifetimes $[T^2(u)]^{(l)}, u = 1, 2, \ldots, z, l = 1, 2, \ldots, w$, in the safety state subsets $\{u, u+1, \ldots, z\}$ at the climate-weather state $c_l$, $l = 1, 2, \ldots, w$, given by

$$[\boldsymbol{\mu}^2(u)]^{(l)} = \int_0^\infty [\boldsymbol{S}^2(t,u)]^{(l)}dt, u = 1, 2, \ldots, z, l = 1, 2, \ldots, w, \tag{5.20}$$

and $[\boldsymbol{S}^2(t,u)]^{(l)}, t \in <0, \infty), u = 1, 2, \ldots, z, l = 1, 2, \ldots, w$, are defined by (5.7), (5.8) and $q_l, l = 1, 2, \ldots, w$, are given by (5.4);

- the standard deviations of the critical infrastructure lifetimes in the safety state subsets $\{u, u+1, \ldots, z\}$, $u = 1, 2, \ldots, z$, (SafI8), given by [18]

$$\boldsymbol{\sigma}^2(u) = \sqrt{\boldsymbol{n}^2(u) - [\boldsymbol{\mu}^2(u)]^2}, u = 1, 2, \ldots, z \tag{5.21}$$

where

$$\boldsymbol{n}^2(u) = 2\int_0^\infty t\boldsymbol{S}^2(t,u)dt, u = 1, 2, \ldots, Z, \tag{5.22}$$

and $\boldsymbol{S}^2(t,u), t \in <0, \infty), u = 1, 2, \ldots, z$, are given by (5.11);

- the mean lifetimes $\overline{\boldsymbol{\mu}}^2(u)$, $u = 1, 2, \ldots, z$, of the critical infrastructure in the particular safety states (SafI9) [18]

$$\overline{\boldsymbol{\mu}}^2(u) = \boldsymbol{\mu}^2(u) - \boldsymbol{\mu}^2(u+1), u = 0, 1, \ldots, z-1, \overline{\boldsymbol{\mu}}^2(z) = \boldsymbol{\mu}^2(z); \quad (5.23)$$

where $\boldsymbol{\mu}^2(u), u = 1, 2, \ldots, z$, are given by (5.20);

- the intensities of degradation of the critical infrastructure/the intensities of critical infrastructure departure from the safety state subset $\{u, u+1, \ldots, z\}$, $u =$ 1, 2, …, $z$, (SafI10), i.e. the coordinates of the vector [18]

$$\boldsymbol{\lambda}^2(t, \cdot) = [0, \boldsymbol{\lambda}^2(t, 1), \ldots, \boldsymbol{\lambda}^2(t, z)], t \in < 0, +\infty), \quad (5.24)$$

where

$$\boldsymbol{\lambda}^2(t, u) = \frac{-\frac{d\boldsymbol{S}^2(t,u)}{dt}}{\boldsymbol{S}^2(t, u)}, t \in < 0, +\infty), u = 1, 2, \ldots, z; \quad (5.25)$$

- the coefficients of climate-weather change process impact on the critical infrastructure intensities of degradation/the coefficients of climate-weather change process impact on critical infrastructure intensities of departure from the safety state subset $\{u, u+1, \ldots, z\}$ (ResI1), i.e. the coordinates of the vector [18]

$$\boldsymbol{\rho}^2(t, \cdot) = [0, \boldsymbol{\rho}^2(t, 1), \ldots, \boldsymbol{\rho}^2(t, z)], t \in < 0, +\infty), \quad (5.26)$$

where

$$\boldsymbol{\lambda}^2(t, u) = \boldsymbol{\rho}^2(t, u) \cdot \boldsymbol{\lambda}^0(t, u), t \in < 0, +\infty), u = 1, 2, \ldots, z, \quad (5.27)$$

i.e.

$$\boldsymbol{\rho}^2(t, u) = \frac{\boldsymbol{\lambda}^2(t, u)}{\boldsymbol{\lambda}^0(t, u)}, t \in < 0, +\infty), u = 1, 2, \ldots, z, \quad (5.28)$$

and $\boldsymbol{\lambda}^{\mathbf{0}}(t, u)$, $t \in < 0, +\infty)$, $u = 1, 2, \ldots, z$, are the intensities of degradation of the critical infrastructure without of climate-weather change process impact, defined by (3.14), i.e. the coordinates of the vector

$$\boldsymbol{\lambda}^{\mathbf{0}}(t, \cdot) = [0, \boldsymbol{\lambda}^{\mathbf{0}}(t, 1), \ldots, \boldsymbol{\lambda}^{\mathbf{0}}(t, z)], t \in < 0, +\infty), \quad (5.29)$$

and $\boldsymbol{\lambda}^{\mathbf{2}}(t, u)$, $t \in < 0, +\infty)$, $u = 1, 2, \ldots, z$, are the intensities of degradation of the critical infrastructure impacted by the climate-weather change process, defined by (5.25), i.e. the coordinates of the vector

$$\boldsymbol{\lambda}^2(t, \cdot) = [0, \boldsymbol{\lambda}^2(t, 1), \ldots, \boldsymbol{\lambda}^2(t, z)], t \in < 0, +\infty); \quad (5.30)$$

- the indicator of critical infrastructure resilience to climate-weather change process impact (ResI2) defined by

$$\boldsymbol{RI}^2(t,r) = \frac{1}{\boldsymbol{\rho}^2(t,r)}, t \in< 0, +\infty), \tag{5.31}$$

where $\boldsymbol{\rho}^2(t,\ r), t \in< 0, +\infty)$, is the coefficients of climate-weather change process impact on the critical infrastructure intensities of degradation given by (5.28) for $u = r$.

In the case, the critical infrastructure have the picewise exponential safety functions, i.e.

$$\boldsymbol{S}^2(t,\ \cdot) = [0, \boldsymbol{S}^2(t,1), \ldots, \boldsymbol{S}^2(t,z)], t \in< 0, +\infty), \tag{5.32}$$

where

$$\boldsymbol{S}^2(t,u) = \exp[-\boldsymbol{\lambda}^2(u)t], t \in< 0, +\infty), \boldsymbol{\lambda}^2(u) \geq 0, u = 1, 2, \ldots, z, \tag{5.33}$$

the critical infrastructure safety and resilience indicators defined by (5.24)–(5.31) take following forms:

- the intensities of degradation of the critical infrastructure impacted by the climate-weather change process impact, (SafI10), i.e. the coordinates of the vector

$$\boldsymbol{\lambda}^2(\cdot) = [0, \boldsymbol{\lambda}^2(1), \ldots, \boldsymbol{\lambda}^2(z)], \tag{5.34}$$

are constant and

$$\boldsymbol{\lambda}^2(u) = \frac{1}{\boldsymbol{\mu}^2(u)}, u = 1, 2, \ldots, z, \tag{5.35}$$

where $\boldsymbol{\mu}^2(u), u = 1, 2, \ldots, z$, is the mean value of the critical infrastructure lifetime in the safety state subset $\{u,u + 1,\ldots,z\}$, $u = 1, 2, \ldots, z$;

- the coefficients of the climate-weather change process impact on the critical infrastructure intensities of degradation/the coefficients of the climate-weather change process impact on critical infrastructure intensities of departure from the safety state subset $\{u, u + 1, \ldots, z\}$, (ResI1), i.e. the coordinates of the vector

$$\boldsymbol{\rho}^2(\cdot) = [0, \boldsymbol{\rho}^2(1), \ldots, \boldsymbol{\rho}^2(z)], \tag{5.36}$$

where

$$\rho^2(u) = \frac{\boldsymbol{\lambda}^2(u)}{\boldsymbol{\lambda}^0(u)} = \frac{\boldsymbol{\mu}^0(u)}{\boldsymbol{\mu}^2(u)}, u = 1, 2, \ldots, z, \tag{5.37}$$

and $\boldsymbol{\lambda}^{\mathbf{0}}(u)$, $u = 1, 2, \ldots, z$, are the intensities of degradation of the critical infrastructure without of climate-weather change process impact, defined by (3.18), i.e. the coordinates of the vector

$$\boldsymbol{\lambda}^{\mathbf{0}}(u) = [0, \boldsymbol{\lambda}^{\mathbf{0}}(1), \ldots, \boldsymbol{\lambda}^{\mathbf{0}}(z)] \tag{5.38}$$

and $\boldsymbol{\lambda}^{\mathbf{2}}(u)$, $u = 1, 2, \ldots, z$, are the intensities of degradation of the critical infrastructure related to the climate-weather process impact, defined by (5.35), i.e. the coordinates of the vector

$$\boldsymbol{\lambda}^2(\cdot) = [0, \boldsymbol{\lambda}^2(1), \ldots, \boldsymbol{\lambda}^2(z)]; \tag{5.39}$$

- the indicator of critical infrastructure resilience to climate-weather change process impact (ResI2) defined by

$$\boldsymbol{RI}^2(r) = \frac{1}{\rho^2(r)}, t \in < 0, +\infty), \tag{5.40}$$

where $\rho^2(r)$ is the coefficient of climate-weather change process impact on the critical infrastructure intensities of degradation given by (5.37) for $u = r$.

## 5.3 Critical Infrastructure Assets Safety Parameters

We assume that the changes of the climate-weather change process $C(t)$ states at the system operating area have an influence on the critical infrastructure asset $A_i$ $i = 1, 2, \ldots, n$, safety parameters.

We denote by $[T_i^2(u)]^{(l)}$, $u = 1, 2, \ldots, z, l = 1, 2, \ldots, w$, $i = 1, 2, \ldots, n$, the critical infrastructure asset $A_i, i = 1, 2, \ldots, n$, conditional lifetime in the safety state subset $\{u, u+1, \ldots, z\}$ while the climate-weather change process at the critical infrastructure operating area is at the climate-weather state $c_l$, $l = 1, 2, \ldots, w$, and we define its conditional safety function (SafI1) by the vector [18]

$$[S_i^2(t, \cdot)]^{(l)} = [1, [S_i^2(t, 1)]^{(l)}, \ldots, [S_i^2(t, z)]^{(l)}, t \in < 0, \infty), l = 1, 2, \ldots, w, i = 1, 2, \ldots, n, ] \tag{5.41}$$

with the coordinates defined by

$$[S_i^2(t,u)]^{(l)} = P([T_i^2(u)]^{(l)} > t | C(t) = c_l) \tag{5.42}$$

for $t \in <0, \infty)$, $u = 1, 2, \ldots, z$, $l = 1, 2, \ldots, w$, $i = 1, 2, \ldots, n$.

The safety function $[S_i^2(t,u)]^{(l)}$, $t \in <0, \infty)$, $u = 1, 2, \ldots, z$, $l = 1, 2, \ldots, w$, $i = 1, 2, \ldots, n$, is the conditional probability that the asset $A_i$ lifetime $[T_i^2(u)]^{(l)}$, $u = 1, 2, \ldots, z$, $l = 1, 2, \ldots, w$, $i = 1, 2, \ldots, n$, in the safety state subset $\{u, u+1, \ldots, z\}$ is greater than $t$, while the climate-weather change process at the critical infrastructure operating area is at the climate-weather state $c_l$,$l = 1, 2, \ldots, w$.

The conditional safety functions $[S_i^2(t,u)]^{(l)}$, $t \in <0, \infty)$, $u = 1, 2, \ldots, z$, $l = 1, 2, \ldots, w$, $i = 1, 2, \ldots, n$, defined by (5.42) are called the coordinates of the asset $A_i$, $i = 1, 2, \ldots, n$, conditional safety function $[S_i^2(t,\cdot)]^{(l)}$, $t \in <0, \infty)$, $l = 1, 2, \ldots, w$, $i = 1, 2, \ldots, n$, while the climate-weather change process $C(t)$ is at the climate-weather state $c_l$,$l = 1, 2, \ldots, w$, given by (5.41). Thus, the relationship between the conditional distribution function $[F_i^2(t,u)]^{(l)}$, $t \in <0, \infty)$, $u = 1, 2, \ldots, z$, $l = 1, 2, \ldots, w$, $i = 1, 2, \ldots, n$, of the asset $A_i$, $i = 1, 2, \ldots, n$, lifetime $[T_i^2(u)]^{(l)}$, $u = 1, 2, \ldots, z$, $l = 1, 2, \ldots, w$, $i = 1, 2, \ldots, n$, in the safety state subset $\{u, u+1, \ldots, z\}$, $u = 1, 2, \ldots, z$, and the coordinate $[S_i^2(t,u)]^{(l)}$, $t \in <0, \infty)$, $u = 1, 2, \ldots, z$, $l = 1, 2, \ldots, w$, $i = 1, 2, \ldots, n$, of its conditional safety function is given by

$$[F_i^2(t,u)]^{(l)} = P([T_i^2(u)]^{(l)} \le t), = 1 - P([T_i^2(u)]^{(l)} > t) = 1 - [S_i^2(t,u)]^{(l)}, t \in <0, \infty), \tag{5.43}$$

$$u = 1, 2, \ldots, z, l = 1, 2, \ldots, w, \quad i = 1, 2, \ldots, n.$$

The function

$$[r_i^2(t)]^{(l)} = 1 - [S_i^2(t,r)]^{(l)}, t \in <0, \infty), l = 1, 2, \ldots, w, i = 1, 2, \ldots, n, \tag{5.44}$$

is the asset $A_i$, $i = 1, 2, \ldots, n$, the conditional risk function (SafI2) and its graph is the asset $A_i$, $i = 1, 2, \ldots, n$, fragility curve (SafI3) while the climate-weather change process at the critical infrastructure operating area is at the climate-weather state $c_l$,$l = 1, 2, \ldots, w$.

Moreover, the conditional mean lifetimes, (SafI4, SafI7), of the asset $A_i$ in the safety state subsets $\{u, u+1, \ldots, z\}$, $u = 1, 2, \ldots, z$, while the climate-weather change process at the critical infrastructure operating area is at the climate-weather state $c_l$,$l = 1, 2, \ldots, w$, are given by

$$[\mu_i^2(u)]^{(l)} = \int_0^\infty [S_i^2(t,u)]^{(l)} dt, u = 1, 2, \ldots, z, l = 1, 2, \ldots, w, i = 1, 2, \ldots, n, \tag{5.45}$$

and the conditional standard deviations, (SafI5, SafI8), of the asset $A_i$ lifetimes in the safety state subsets $\{u, u+1, \ldots, z\}$, $u = 1, 2, \ldots, z$, while the climate-weather change process at the critical infrastructure operating area is at the climate-weather state $c_l$, $l = 1, 2, \ldots, w$, are given by

$$[\sigma_i^2(u)]^{(l)} = \sqrt{[n_i^2(u)]^{(l)} - [[\mu_i^2(u)]^{(l)}]^2}, u = 1, 2, \ldots, z, l = 1, 2, \ldots, w, i = 1, 2, \ldots, n, \tag{5.46}$$

where

$$[n_i^2(u)]^{(l)} = 2\int_0^\infty t[S_i^2(t,u)]^{(l)}dt, u = 1, 2, \ldots, z, l = 1, 2, \ldots, w, i = 1, 2, \ldots, n. \tag{5.47}$$

Next, it can be justified that the conditional mean lifetimes $[\overline{\mu}_i^2(u)]^{(l)}$, $u = 1, 2, \ldots, z$, $l = 1, 2, \ldots, w$,
$i = 1, 2, \ldots, n$, of the asset $A_i$, $1, 2, \ldots, n$, in the particular safety states, (SafI9), while the climate-weather change process at the critical infrastructure operating area is at the climate-weather state $c_l$,$l = 1, 2, \ldots, w$, are

$$[\overline{\mu}_i^2(u)]^{(l)} = [\mu_i^2(u)]^{(l)} - [\mu_i^2(u+1)]^{(l)}, u = 1, 2, \ldots, z-1, [\overline{\mu}_i^2(z)]^{(l)} = [\mu_i^2(z)]^{(l)}, \tag{5.48}$$

$$u = 1, 2, \ldots, z, l = 1, 2, \ldots, w, i = 1, 2, \ldots, n.$$

In the case, when the critical infrastructure assets $A_i$, $i = 1, 2, \ldots, n$, lifetimes at the climate-weather states $c_l$,$l = 1, 2, \ldots, w$, have the piecewise exponential safety functions, the coordinates (5.42) of the vector (5.41) are given by

$$\begin{aligned} [S_i^2(t,u)]^{(l)} &= P([T_i^2(u)]^{(l)} > t | C(t) = c_l) \\ &= \exp[-[\lambda_i^2(u)]^{(l)} t], t \in <0, \infty), \\ u &= 1, 2, \ldots, z, l = 1, 2, \ldots, w, \quad i = 1, 2, \ldots, n. \end{aligned} \tag{5.49}$$

Existing in (5.49) the intensities of degradation of the critical infrastructure asset $A_i$, $i = 1, 2, \ldots, n$, with the climate-weather change process at the critical infrastructure operating area impact at the climate-weather states $c_l$,$l = 1, 2, \ldots, w$, (SafI10), i.e. the coordinates of the vector

$$[\lambda_i^2(\cdot)]^{(l)} = [0, [\lambda_i^2(1)]^{(l)}, \ldots, [\lambda_i^2(z)]^{(l)}], t \in <0, +\infty), l = 1, 2, \ldots, w, i = 1, 2, \ldots, n, \tag{5.50}$$

are constant and given by

$$[\lambda_i^2(u)]^{(l)} = \frac{1}{[\mu_i^2(u)]^{(l)}},$$
$$u = 1, 2, \ldots, z, u = 1, 2, \ldots, z, \quad l = 1, 2, \ldots, w, \quad i = 1, 2, \ldots, n, \tag{5.51}$$

and moreover

$$[\lambda_i^2(u)]^{(l)} = [\rho_i^2(u)]^{(l)} \cdot \lambda_i^0(u),$$
$$u = 1, 2, \ldots, z, u = 1, 2, \ldots, z, l = 1, 2, \ldots, w, i = 1, 2, \ldots, n, \tag{5.52}$$

where $\lambda_i^0(u)$ are the intensities of degradation of the critical infrastructure asset $A_i$, $i = 1, 2, \ldots, n$, without the climate-weather change process at the critical infrastructure operating area impact, i.e. the coordinate of the vector

$$\lambda_i^0(\cdot) = [0, \lambda_i^0(1), \ldots, \lambda_i^0(z)], i = 1, 2, \ldots, n, \tag{5.53}$$

and $[\rho_i^2(u)]^{(l)}$, $u = 1, 2, \ldots, z$, $l = 1, 2, \ldots, w$, $i = 1, 2, \ldots, n$, are the coefficients of the climate-weather change process at the critical infrastructure operating area impact on the critical infrastructure asset $A_i$, $i = 1, 2, \ldots, n$, intensities of degradation at the climate-weather states $c_l$,$l = 1, 2, \ldots, w$, (ResI1), i.e. the coordinate of the vector

$$[\rho_i^2(\cdot)]^{(l)} = [0, [\rho_i^2(1)]^{(l)}, \ldots, [\rho_i^2(z)]^{(l)}], l = 1, 2, \ldots, w, i = 1, 2, \ldots, n. \tag{5.54}$$

where by (5.52)

$$[\rho_i^2(u)]^{(l)} = \frac{[\lambda_i^2(u)]^{(l)}}{\lambda_i^0(u)} = \frac{\mu_i^0(u)}{[\mu_i^2(u)]^{(l)}}, u = 1, 2, \ldots, z, l = 1, 2, \ldots, w, i = 1, 2, \ldots, n. \tag{5.55}$$

The indicators of critical infrastructure asset $A_i$, $i = 1, 2, \ldots, n$, resilience to the climate-weather change process impact (ResI2) at the critical infrastructure climate-weather states $c_l$, $l = 1, 2, \ldots, w$, can be defined by

$$[RI_i^2(r)]^{(l)} = \frac{1}{[\rho_i^2(r)]^{(l)}}, t \in < 0, +\infty), l = 1, 2, \ldots, w, i = 1, 2, \ldots, n, \tag{5.56}$$

where $[\rho_i^2(r)]^{(l)}$, $l = 1, 2, \ldots, w$, $i = 1, 2, \ldots, n$, are the coefficients of the climate-weather change process at the critical infrastructure operating area impact on the critical infrastructure assets intensities of degradation at the climate-weather states $c_l$, $l = 1, 2, \ldots, w$, given by (5.55) for $u = r$.

The conditional safety function (5.41), (5.42), the conditional risk function (5.44), the conditional intensities' of degradation (5.51), (5.52) and the coefficients of the operation process impact on the critical infrastructure assets intensities of degradation (5.55) are the assets of the critical infrastructure main safety and resilience indicators impacted by the climate-weather change process at its operating area.

## 5.4 Application

We consider the port oil terminal critical infrastructure impacted by the climate-weather change process at its operating area. This port oil terminal main technical assets are presented in Chaps. 1 and 3. The considered port oil terminal placed at the Baltic seaside is described and its safety indicators without concidering outside impacts are determined in Chap. 3. In this section, the parameters and characteristics of the climate-weather change process at considered port oil terminal operating area are evaluated. Next, the safety analysis of the port oil terminal critical infrastructure impacted by the climate-weather change process at its operating area is performed and its safety indicators are evaluated.

### *5.4.1 Description of Climate-Weather Change Processes at Port Oil Terminal Operating Area*

The main asstet $A_1$ of the port oil terminal critical infrastructure, the port oil piping transportation system operating area is divided into two parts, the sea water operating area and the land operating area. We distinguish two different climate weather-change processes for these two fixed operation areas:

- the climate weather change process $C^1(t)$, $t \geq 0$, at the sea water operating area presented in Figs. 5.2 and 5.3, where the measurement points 1–3 of the weather parameters are marked;
- the climate-weather change process $C^2(t)$, $t \geq 0$, at the seaside operating area presented in Fig. 5.2, where the measurement point 4 of the the weather parameters is marked.

The weather statistical data sets from measurement points 1–3 after their successful data uniformity testing [18, 21] were joined into one weather data set and analyzed commonly.

Fig. 5.2 The port oil piping transportation system operating area [22]

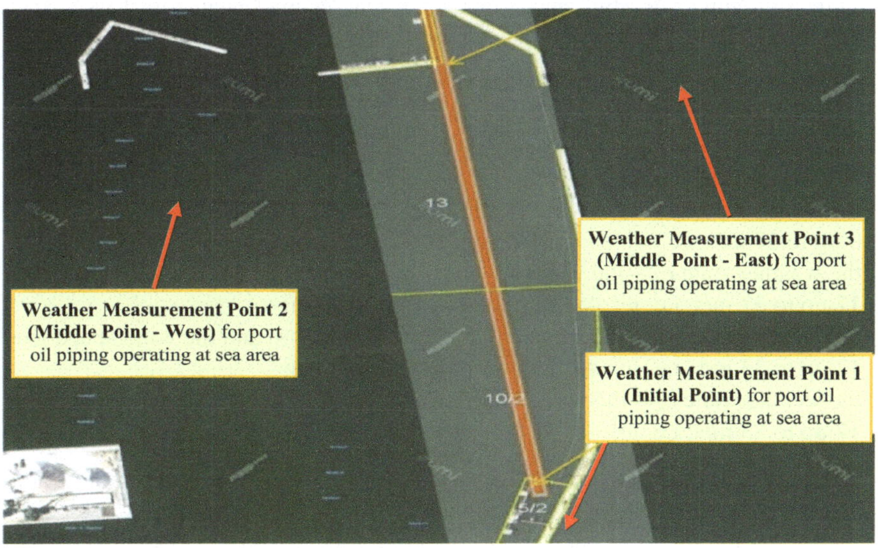


Fig. 5.3 The port oil piping transportation system sea wather area [22]

### 5.4.2 Parameters and Characteristics of Climate-Weather Change Process at Port Oil Terminal Critical Infrastructure Water Operating Area

To define the climate-weather change process $C^1(t)$, $t \geq 0$, at port oil terminal critical infrastructure water operating area, we distinguish $a^1 = 2$ its parameters, the water wave height and the wind speed and we marked the values they can take respectively by $w_1^1$ and $w_2^1$. We assume that the value of $w_1^1$ of the water wave height measured in metres ($m$) belongs to the interval $< b_1^1, d_1^1) = < 0, 14$ m) and the value $w_2^1$ of the wind speed measured in metres per second ($m/s$) belongs to the interval $< b_2^1, d_2^1) = < 0, 33$ m/s).

We devid the interval <0, 14) of the value $w_1^1$ of the water wave height into the $n_1^1 = 3$ subintervals:

$< 0, 2); < 2, 5); < 5, 14)$.

We devided the interval $< 0, 33)$ of the value $w_2^1$ of the wind speed into the $n_3^1 = 2$ subintervals:

$< 0, 17); < 17, 33)$.

Now we can define $w^1 = n_1^1 \cdot n_2^1 = 3 \cdot 2 = 6$ climate-weater states of the climate-weather change process $C^1(t)$, $t \geq 0$, marked by: $c_1^1$, $c_2^1$,$c_3^1$,$c_4^1$, $c_5^1$, $c_6^1$.

Thus, the climate-weather states of the climate-weather change proces $C^1(t)$ at port oil terminal critical infrastructure water operating area, are defined as follows:

- the weather state $c_1^1$—the wave height from 0 up to 2 m and the wind speed from 0 up to 17 m/s, i.e. $w_1^1 \in< 0, 2)$ and $w_2^1 \in< 0, 17)$;
- the weather state $c_2^1$—the wave height from 2 up to 5 m and the wind speed from 0 up to 17 m/s, i.e. $w_1^1 \in< 2, 5)$ and $w_2^1 \in< 0, 17)$;
- the weather state $c_3^1$—the wave height from 5 up to 14 m and the wind speed from 0 up to 17 m/s, i.e. $w_1^1 \in< 5, 14)$ and $w_2^1 \in< 0, 17)$;
- the weather state $c_4^1$—the wave height from 0 up to 2 m and the wind speed from 17 up to 33 m/s, i.e. $w_1^1 \in< 0, 2)$ and $w_2^1 \in< 17, 33)$;
- the weather state $c_5^1$—the wave height from 2 up to 5 m and the wind speed from 17 up to 33 m/s, i.e. $w_1^1 \in< 2, 5)$ and $w_2^1 \in< 17, 33)$;
- the weather state $c_6^1$—the wave height from 5 up to 14 m and the wind speed from 17 up to 33 m/s; i.e. $w_1^1 \in< 5, 14)$ and $w_2^1 \in< 17, 33)$;

The vector of the initial probabilities $q_l^1(0)$, $l = 1, 2, \ldots, 6$, of the climate-weather change process $C^1(t)$, $t \geq 0$, staying at the particular climate-weather states $c_l^1$, $l = 1, 2, \ldots, 6$, at the initial moment $t = 0$, fixed on the basis of the statistical data for February, collected at the measurement points 1–3 is [22]:

$$\left[q_l^1(0)\right]_{1\times 6} = [0.718, 0.225, 0.004, 0, 0.035, 0.018]. \tag{5.57}$$

The matrix of the probabilities $q_{lk}^1$, $l, k = 1, 2, \ldots, 6$, of transitions of the climate-weather change process $C^1(t)$, $t \geq 0$, between the climate-weather states $c_l^1$ and $c_k^1$, $l, k = 1, 2, \ldots, 6$, fixed on the basis of the statistical data for February, collected at the measurement points 1–3 is [22]:

$$[q^1_{lk}]_{6\times 6} = \begin{bmatrix} 0 & 0.99 & 0 & 0 & 0.01 & 0 \\ 0.82 & 0 & 0 & 0 & 0.18 & 0 \\ 0 & 0.86 & 0 & 0 & 0 & 0.14 \\ 0 & 0 & 0 & 0 & 1 & 0 \\ 0.02 & 0.65 & 0.06 & 0 & 0 & 0.27 \\ 0 & 0.11 & 0.74 & 0 & 0.15 & 0 \end{bmatrix}. \tag{5.58}$$

The matrix of the mean values $N^1_{lk}$, $l,k = 1, 2, \ldots, 6$, of the conditional sojourn times $C^1_{lk}$, $l,k = 1, 2, \ldots, 6$, of the climate-weather change process $C^1(t)$, $t \geq 0$, at the climate-weather states $c^1_l$, $l = 1, 2, \ldots, 6$, when the next climate-weather state is $c^1_k$, $k = 1, 2, \ldots, 6$, fixed in [22] is given by

$$[N^1_{lk}]_{6\times 6} = \begin{bmatrix} 0 & 317.84 & 0 & 0 & 12 & 0 \\ 18.01 & 0 & 0 & 0 & 17.09 & 0 \\ 0 & 3.5 & 0 & 0 & 0 & 6 \\ 0 & 0 & 0 & 0 & 3 & 0 \\ 3 & 8.27 & 6 & 0 & 0 & 10.94 \\ 0 & 7 & 14.4 & 0 & 7.5 & 0 \end{bmatrix}. \tag{5.59}$$

Further, applying (5.5), we get the following system of equations

$$\begin{cases} [\pi_1, \pi_2, \pi_3, \pi_4, \pi_5, \pi_6] = [\pi_1, \pi_2, \pi_3, \pi_4, \pi_5, \pi_6][q^1_{lk}]_{6x6} \\ \pi_1 + \pi_2 + \pi_3 + \pi_4 + \pi_5 + \pi_6 = 1 \end{cases},$$

where $[q^1_{lk}]_{6\times 6}$ is given by (5.58), equivalent to the system of equations

$$\begin{cases} 0.82\pi_2 + 0.02\pi_5 = \pi_1 \\ 0.99\pi_1 + 0.86\pi_3 + 0.65\pi_5 + 0.11\pi_6 = \pi_2 \\ 0.06\pi_5 + 0.74\pi_6 = \pi_3 \\ 0 = \pi_4 \\ 0.01\pi_1 + 0.18\pi_2 + \pi_4 + 0.15\pi_6 = \pi_5 \\ 0.14\pi_3 + 0.27\pi_5 = \pi_6 \\ \pi_1 + \pi_2 + \pi_3 + \pi_4 + \pi_5 + \pi_6 = 1, \end{cases}$$

reasulting in its following solution

$$\pi_1 = 0.3850, \pi_2 = 0.4673, \pi_3 = 0.0267, \pi_4 = 0, \pi_5 = 0.0923, \pi_6 = 0.0287. \tag{5.60}$$

Next, using (5.3), we obtain the evaluations of the mean values of the climate-weather change process $C^1(t)$, unconditional sojourn times at the particular climate-weather states:

$$
\begin{aligned}
N_1^1 &= 0.99 \cdot 317.84 + 12 \cdot 0.01 \cong 314.78,\\
N_2^1 &= 0.82 \cdot 18.01 + 0.18 \cdot 17.09 \cong 17.84,\\
N_3^1 &= 0.86 \cdot 3.5 + 0.14 \cdot 6 \cong 3.85,\\
N_4^1 &= 1 \cdot 3 = 3,\\
N_5^1 &= 0.02 \cdot 3 + 0.65 \cdot 8.27 + 0.06 \cdot 6 + 0.27 \cdot 10.94 \cong 8.75,\\
N_6^1 &= 0.11 \cdot 7 + 0.74 \cdot 14.4 + 0.15 \cdot 7.5 \cong 12.55.
\end{aligned}
\tag{5.61}
$$

Using (5.4) and (5.60), (5.61), the vector of the limit values of the climate-weather change process $C^1(t)$, $t \geq 0$, transient probabilities at the climate-weather states $c_l^1$, $l = 1, 2, \ldots, 6$, for February is given by

$$
\left[q_l^1\right]_{1\times 6} = [0.926, 0.064, 0.001, 0, 0.006, 0.003]. \tag{5.62}
$$

The vector of the mean values of the total sojourn times $\hat{C}_l^1$, $l = 1, 2, \ldots, w$, of the climate-weather change process $C^1(t)$ at port oil terminal critical infrastructure water operating area at the particular climate-weather states $c_l^1$, $l = 1, 2, \ldots, 6$ during the fixed operation time $C^1 = 1$ year $= 365$ days is

$$
[\hat{N}_l^1]_{1\times 6} = [337.99; 23.36; 0.365; 0; 2.19; 1.095],
$$

where $\hat{N}_l = E[\hat{C}_l^1] = q_l^1 C^1$, $l = 1, 2, \ldots, 6$, are determined according to (5.6) and (5.62) in the following way

$$
\begin{aligned}
&\hat{N}_1^1 = E[\hat{C}_1^1] = 0.926 \text{ year } = 337.99 \text{ days}, \hat{N}_2^1 = E[\hat{C}_2^1] = 0.064 \text{ year } = 23.36 \text{ days},\\
&\hat{N}_3^1 = E[\hat{C}_3^1] = 0.001 \text{ year } = 0.365 \text{ days}, \hat{N}_4^1 = E[\hat{C}_4^1] = 0 \text{ year } = 0 \text{ days},\\
&\hat{N}_5^1 = E[\hat{C}_5^1] = 0.006 \text{ year } = 2.19 \text{ days}, \hat{N}_6^1 = E[\hat{C}_6^1] = 0.003 \text{ year } = 1.095 \text{ days}.
\end{aligned}
$$

### 5.4.3 Parameters and Characteristics of Climate-Weather Change Process at Port Oil Terminal Critical Infrastructure Seaside Operating Area

To define the weather change process $C^2(t)$, $t \geq 0$, at port oil terminal critical infrastructure seaside operating area we distinguish $a^2 = 2$ its parameters, the air temperature and the soil temperature and we marked the values they can take respectively by $w_1^2$ and $w_2^2$. We assume that the value $w_1^2$ of the air temperature measured in Celsius degrees (°C) belongs to the interval $< b_1^2, d_1^2) = < -25, 35$ °C) and the value $w_2^2$ of the soil temperature measured in Celsius degrees (°C) belongd to the interval $< b_2^2, d_2^2) = < -15, 37$ °C).

We devide the interval $< -25, 35$ °C) of the value $w_1^2$ of the air temperature into the $n_1^2 = 4$ subintervals:

$$< -25, -15); < -15, 5); < 5, 25); < 25, 35).$$

We devide the interval $< -15, 37$ °C) of the value $w_2^2$ of the soil temperature into the $n_2^2 = 4$ subintervals:

$$< -30, -5); < -5, 5); < 5, 20); < 20, 37).$$

Now we can define $w^2 = n_1^2 \cdot n_2^2 = 4 \cdot 4 = 16$ states of the weather change process $C^2(t)$, $t \geq 0$, and marked them by: $c_1^2, c_2^2, c_3^2, c_4^2, c_5^2, c_6^2, c_7^2, c_8^2, c_9^2, c_{10}^2, c_{11}^2, c_{12}^2, c_{13}^2, c_{14}^2, c_{15}^2, c_{16}^2$.

Thus, the climate-weather states of the climate-weather change proces $C^2(t)$ at port oil terminal critical infrastructure seaside operating area are defined as follows:

- the weather state $c_1^2$—the air temperature from −25 up to −15 °C and the soil temperature from −30 up to −5 °C, i.e. $w_1^2 \in< -25, -15)$ and $w_2^2 \in< -30, -5)$;
- the weather state $c_2^2$—the air temperature from −15 up to 5 °C and the soil temperature from −30 up to −5 °C, i.e. $w_1^2 \in< -15, 5)$ and $w_2^2 \in< -30, -5)$;
- the weather state $c_3^2$—the air temperature from 5 up to 25 °C and the soil temperature from −30 up to −5 °C, i.e. $w_1^2 \in< 5, 25)$ and $w_2^2 \in< -30, -5)$;
- the weather state $c_4^2$—the air temperature from 25 up to 35 °C and the soil temperature from −30 up to −5 °C, i.e. $w_1^2 \in< 25, 35)$ and $w_2^2 \in< -30, -5)$;
- the weather state $c_5^2$—the air temperature from −25 up to −15 °C and the soil temperature from −5 up to 5 °C, i.e. $w_1^2 \in< -25, -15)$ and $w_2^2 \in< -5, 5)$;
- the weather state $c_6^2$—the air temperature from −15 up to 5 °C and the soil temperature from −5 up to 5 °C, i.e. $w_1^2 \in< -15, 5)$ and $w_2^2 \in< -5, 5)$;
- the weather state $c_7^2$—the air temperature from 5 up to 25 °C and the soil temperature from −5 up to 5 °C, i.e. $w_1^2 \in< 5, 25)$ and $w_2^2 \in< -5, -5)$;
- the weather state $c_8^2$—the air temperature from 25 up to 35 °C and the soil temperature from −5 up to 5 °C, i.e. $w_1^2 \in< 25, 35)$ and $w_2^2 \in< -5, 5)$;

- the weather state $c_9^2$—the air temperature from −25 up to −15 °C and the soil temperature from 5 up to 20 °C, i.e. $w_1^2 \in < -25, -15)$ and $w_2^2 \in < 5, 20)$;
- the weather state $c_{10}^2$—the air temperature from −15 up to 5 °C and the soil temperature from 5 up to 20 °C, i.e. $w_1^2 \in < -15, 5)$ and $w_2^2 \in < 5, 20)$;
- the weather state $c_{11}^2$—the air temperature from 5 up to 25 °C and the soil temperature from 5 up to 20 °C; i.e. $w_1^2 \in < 5, 25)$ and $w_2^2 \in < 5, 20)$;
- the weather state $c_{12}^2$—the air temperature from 25 up to 35 °C and the soil temperature from 5 up to 20 °C, i.e. $w_1^2 \in < 25, 35)$ and $w_2^2 \in < 5, 20)$;
- the weather state $c_{13}^2$—the air temperature from −25 up to −15 °C and the soil temperature from 20 up to 37 °C, i.e. $w_1^2 \in < -25, -15)$ and $w_2^2 \in < 20, 37)$;
- the weather state $c_{14}^2$—the air temperature from −15 up to 5 °C and the soil temperature from 20 up to 37 °C, i.e. $w_1^2 \in < -15, 5)$ and $w_2^2 \in < 20, 37)$;
- the weather state $c_{15}^2$—the air temperature from 5 up to 25 °C and the soil temperature from 20 up to 37 °C, i.e. $w_1^2 \in < 5, 25)$ and $w_2^2 \in < 20, 37)$;
- the weather state $c_{16}^2$—the air temperature from 25 up to 35 °C and the soil temperature from 20 up to 37 °C; i.e. $w_1^2 \in < 25, 35)$ and $w_2^2 \in < 20, 37)$.

The vector of the initial probabilities $q_b^2$, $b = 1, 2, \ldots, 16$, of the climate-weather change process $C^2(t)$, $t \geq 0$, staying at the particular climate-weather states $c_b^2$, $b = 1, 2, \ldots, 16$, at the initial moment $t = 0$, fixed on the basis of the statistical data for February, collected at the measurement point 4 is [22]:

$$\left[q_l^2(0)\right]_{1\times 16} = [0.012, 0.118, 0, 0, 0, 0.784, 0.035, 0, 0, 0, 0.051, 0, 0, 0, 0, 0]. \tag{5.63}$$

The matrix of the probabilities $q_{lk}^2$, $l, k = 1, 2, \ldots, 16$, of transitions of the climate-weather change process $C^2(t)$, $t \geq 0$, between the climate-weather states $c_l^2$ and $c_k^2$, $l,k = 1, 2, \ldots, 16$, fixed on the basis of the statistical data for February, collected at the measurement point 4 is [22]:

$$[q_{lk}^2]_{16\times 16} = \begin{bmatrix} 0 & 1 & 0 & 0 & 0 & 0 & 0 & 0 & 0 & 0 & 0 & 0 & 0 & 0 & 0 & 0 \\ 0.1 & 0 & 0 & 0 & 0 & 0.9 & 0 & 0 & 0 & 0 & 0 & 0 & 0 & 0 & 0 & 0 \\ 0 & 0 & 0 & 0 & 0 & 0 & 0 & 0 & 0 & 0 & 0 & 0 & 0 & 0 & 0 & 0 \\ 0 & 0 & 0 & 0 & 0 & 0 & 0 & 0 & 0 & 0 & 0 & 0 & 0 & 0 & 0 & 0 \\ 0 & 0 & 0 & 0 & 0 & 0 & 0 & 0 & 0 & 0 & 0 & 0 & 0 & 0 & 0 & 0 \\ 0 & 0.18 & 0 & 0 & 0 & 0 & 0.22 & 0 & 0 & 0.47 & 0.13 & 0 & 0 & 0 & 0 & 0 \\ 0 & 0 & 0 & 0 & 0 & 0.72 & 0 & 0 & 0 & 0 & 0.28 & 0 & 0 & 0 & 0 & 0 \\ 0 & 0 & 0 & 0 & 0 & 0 & 0 & 0 & 0 & 0 & 0 & 0 & 0 & 0 & 0 & 0 \\ 0 & 0 & 0 & 0 & 0 & 0 & 0 & 0 & 0 & 0 & 0 & 0 & 0 & 0 & 0 & 0 \\ 0 & 0 & 0 & 0 & 0 & 0.55 & 0.02 & 0 & 0 & 0 & 0.43 & 0 & 0 & 0 & 0 & 0 \\ 0 & 0 & 0 & 0 & 0 & 0.2 & 0.69 & 0 & 0 & 0.11 & 0 & 0 & 0 & 0 & 0 & 0 \\ 0 & 0 & 0 & 0 & 0 & 0 & 0 & 0 & 0 & 0 & 0 & 0 & 0 & 0 & 0 & 0 \\ 0 & 0 & 0 & 0 & 0 & 0 & 0 & 0 & 0 & 0 & 0 & 0 & 0 & 0 & 0 & 0 \\ 0 & 0 & 0 & 0 & 0 & 0 & 0 & 0 & 0 & 0 & 0 & 0 & 0 & 0 & 0 & 0 \\ 0 & 0 & 0 & 0 & 0 & 0 & 0 & 0 & 0 & 0 & 0 & 0 & 0 & 0 & 0 & 0 \\ 0 & 0 & 0 & 0 & 0 & 0 & 0 & 0 & 0 & 0 & 0 & 0 & 0 & 0 & 0 & 0 \end{bmatrix}. \tag{5.64}$$

The matrix of the mean values $N_{lk}^2$, $l$, $k = 1, 2, \ldots, 16$, of the conditional sojourn times $C_{lk}^2$, $l,k = 1, 2, \ldots, 16$, of the climate-weather change process $C^2(t)$, $t \geq 0$, at the climate-weather states $c_l^2$, $l = 1, 2, \ldots, 16$, when the next climate-weather state is $c_k^2$, $k = 1, 2, \ldots, 16$, fixed in [22], is given by

$$[N_{lk}^2]_{16\times 16} = \begin{bmatrix} 0 & 3.57 & 0 & 0 & 0 & 0 & 0 & 0 & 0 & 0 & 0 & 0 & 0 & 0 & 0 & 0 \\ 19.57 & 0 & 0 & 0 & 0 & 43.84 & 0 & 0 & 0 & 0 & 0 & 0 & 0 & 0 & 0 & 0 \\ 0 & 0 & 0 & 0 & 0 & 0 & 0 & 0 & 0 & 0 & 0 & 0 & 0 & 0 & 0 & 0 \\ 0 & 0 & 0 & 0 & 0 & 0 & 0 & 0 & 0 & 0 & 0 & 0 & 0 & 0 & 0 & 0 \\ 0 & 0 & 0 & 0 & 0 & 0 & 0 & 0 & 0 & 0 & 0 & 0 & 0 & 0 & 0 & 0 \\ 0 & 42.05 & 0 & 0 & 0 & 0 & 57.85 & 0 & 0 & 66.56 & 29.79 & 0 & 0 & 0 & 0 & 0 \\ 0 & 0 & 0 & 0 & 0 & 2.82 & 0 & 0 & 0 & 0 & 2 & 0 & 0 & 0 & 0 & 0 \\ 0 & 0 & 0 & 0 & 0 & 0 & 0 & 0 & 0 & 0 & 0 & 0 & 0 & 0 & 0 & 0 \\ 0 & 0 & 0 & 0 & 0 & 0 & 0 & 0 & 0 & 0 & 0 & 0 & 0 & 0 & 0 & 0 \\ 0 & 0 & 0 & 0 & 0 & 1.97 & 4 & 0 & 0 & 0 & 1.86 & 0 & 0 & 0 & 0 & 0 \\ 0 & 0 & 0 & 0 & 0 & 3.81 & 12.63 & 0 & 0 & 3.11 & 0 & 0 & 0 & 0 & 0 & 0 \\ 0 & 0 & 0 & 0 & 0 & 0 & 0 & 0 & 0 & 0 & 0 & 0 & 0 & 0 & 0 & 0 \\ 0 & 0 & 0 & 0 & 0 & 0 & 0 & 0 & 0 & 0 & 0 & 0 & 0 & 0 & 0 & 0 \\ 0 & 0 & 0 & 0 & 0 & 0 & 0 & 0 & 0 & 0 & 0 & 0 & 0 & 0 & 0 & 0 \\ 0 & 0 & 0 & 0 & 0 & 0 & 0 & 0 & 0 & 0 & 0 & 0 & 0 & 0 & 0 & 0 \\ 0 & 0 & 0 & 0 & 0 & 0 & 0 & 0 & 0 & 0 & 0 & 0 & 0 & 0 & 0 & 0 \end{bmatrix}. \tag{5.65}$$

Further, applying (5.5), we get the system of equations

$$\begin{cases} [\pi_1, \pi_2, \ldots, \pi_{16}] = [\pi_1, \pi_2, \ldots, \pi_{16}][q_{lk}^2]_{16x16} \\ \pi_1 + \pi_2 + \cdots + \pi_{16} = 1 \end{cases},$$

where $[q_{lk}^2]_{16\times16}$ is given by (5.64), equivalent to the system of equations

$$\begin{cases} 0.1\pi_2 = \pi_1 \\ \pi_1 + 0.18\pi_6 = \pi_2 \\ 0 = \pi_3 \\ 0 = \pi_4 \\ 0 = \pi_5 \\ 0.9\pi_2 + 0.72\pi_7 + 0.55\pi_{10} + 0.2\pi_{11} = \pi_6 \\ 0.22\pi_2 + 0.02\pi_{10} + 0.69\pi_{11} = \pi_7 \\ 0 = \pi_8 \\ 0 = \pi_9 \\ 0.47\pi_6 + 0.11\pi_{11} = \pi_{10} \\ 0.13\pi_6 + 0.28\pi_7 + 0.43\pi_{10} = \pi_{11} \\ 0 = \pi_{12} \\ 0 = \pi_{13} \\ 0 = \pi_{14} \\ 0 = \pi_{15} \\ 0 = \pi_{16} \\ \pi_1 + \pi_2 + \pi_3 + \pi_4 + \pi_5 + \pi_6 + \pi_7 + \pi_8 + \pi_9 + \pi_{10} + \pi_{11} + \pi_{12} + \pi_{13} + \pi_{14} + \pi_{15} + \pi_{16} = 1, \end{cases}$$

reasulting in its following solution

$$\pi_1 = -0.0069, \pi_2 = 0.0699, \pi_3 = 0, \pi_4 = 0, \pi_5 = 0, \pi_6 = 0.3496, \pi_7 = 0.2066,$$
$$\pi_8 = 0, \pi_9 = 0, \pi_{10} = 0.1844, \pi_{11} = 0.1826, \pi_{12} = 0, \pi_{13} = 0, \pi_{14} = 0,$$
$$\pi_{15} = 0, \pi_{16} = 0. \tag{5.66}$$

Next, using (5.3), we obtain the evaluations of the mean values of the climate-weather change process $C^2(t)$, unconditional sojourn times at the particular climate-weather states:

$$N_1^2 = 1 \cdot 3.57 = 3.57,$$
$$N_2^2 = 0.1 \cdot 19.57 + 0.9 \cdot 43.84 \cong 41.41,$$
$$N_3^2 = 0,$$
$$N_4^2 = 0,$$
$$N_5^2 = 0,$$
$$N_6^2 = 0.18 \cdot 42.05 + 0.22 \cdot 57.85 + 0.47 \cdot 66.56 + 0.13 \cdot 29.79 \cong 55.45,$$
$$N_7^2 = 0.72 \cdot 2,82 + 0.28 \cdot 2 \cong 2.59,$$
$$N_8^2 = 0,$$

$$
\begin{aligned}
N_6^2 &= 0.18 \cdot 42.05 + 0.22 \cdot 57.85 + 0.47 \cdot 66.56 + 0.13 \cdot 29.79 \cong 55.45, \\
N_7^2 &= 0.72 \cdot 2,82 + 0.28 \cdot 2 \cong 2.59, \\
N_8^2 &= 0, \\
N_9^2 &= 0, \\
N_{10}^2 &= 0.55 \cdot 1.97 + 0.02 \cdot 4 + 0.43 \cdot 1.86 \cong 1.96, \\
N_{11}^2 &= 0.2 \cdot 3.81 + 0.69 \cdot 12.63 + 0.11 \cdot 3.11 \cong 9.82, \\
N_{12}^2 &= 0, \\
N_{13}^2 &= 0, \\
N_{14}^2 &= 0, \\
N_{15}^2 &= 0, \\
N_{16}^2 &= 0.
\end{aligned} \tag{5.67}
$$

Using (5.4) and (5.66), (5.67), the vector of the limit values of the climate-weather change process $C^2(t)$, $t \geq 0$, transient probabilities at the climate-weather states $c_l^2$, $l = 1, 2, \ldots, 16$, for February is given by

$$
\left[ q_b^2 \right]_{1\times 16} == [0.001, 0.116, 0, 0, 0, 0.776, 0.021, 0, 0, 0.014, 0.072, 0, 0, 0, 0, 0]. \tag{5.68}
$$

The vector of the mean values of the total sojourn times $\hat{C}_l^2$, $l = 1, 2, \ldots, 16$, of the climate-weather change process $C^2(t)$ at port oil terminal critical infrastructure seaside operating area at the particular climate-weather states $c_l^2$, $l = 1, 2, \ldots, 16$ during the fixed operation time $C^2 = 1$ year $= 365$ days is

$$
[\hat{N}_l^2]_{1\times 16} = [337.99; 23.36; 0.365; 0; 2.19; 1.095],
$$

where $\hat{N}_l = E[\hat{C}_l^1] = q_l^1 C^1$, $l = 1, 2, \ldots, 6$, are determined according to (5.6) and (5.68) in the following way

$$
\begin{aligned}
&\hat{N}_1^2 = E[\hat{C}_1^2] = 0.001 \text{ year } = 0.365 \text{ days}, \ \hat{N}_2^2 = E[\hat{C}_2^2] = 0.116 \text{ year } = 42.34 \text{ days}, \\
&\hat{N}_3^2 = E[\hat{C}_3^2] = 0 \text{ year } = 0 \text{ days}, \ \hat{N}_4^2 = E[\hat{C}_4^2] = 0 \text{ year } = 0 \text{ days}, \\
&\hat{N}_5^2 = E\left[\hat{C}_5^2\right] = 0 \text{ year } = 0 \text{ days}, \ \hat{N}_6^2 = E\left[\hat{C}_6^2\right] = 0.776 \text{ year } = 283.24 \text{ days}, \\
&\hat{N}_7^2 = E\left[\hat{C}_7^2\right] = 0.021 \text{ year } = 7.67 \text{ days}, \ \hat{N}_8^2 = E\left[\hat{C}_8^2\right] = 0 \text{ year } = 0 \text{ days}, \\
&\hat{N}_9^2 = E\left[\hat{C}_9^2\right] = 0 \text{ year } = 0 \text{ days}, \ \hat{N}_{10}^2 = E\left[\hat{C}_{10}^2\right] = 0.014 \text{ year } = 5.11 \text{ days}, \\
&\hat{N}_{11}^2 = E\left[\hat{C}_{11}^2\right] = 0.072 \text{ year } = 26.28 \text{ days}, \ \hat{N}_{12}^2 = E\left[\hat{C}_{12}^2\right] = 0 \text{ year } = 0 \text{ days}, \\
&\hat{N}_{13}^2 = E\left[\hat{C}_{13}^2\right] = 0 \text{ year } = 0 \text{ days}, \ \hat{N}_{14}^2 = E\left[\hat{C}_{14}^2\right] = 0 \text{ year } = 0 \text{ days},
\end{aligned}
$$

$$\hat{N}_{15}^2 = E\left[\hat{C}_{15}^2\right] = 0 \text{ year } = 0 \text{ days}, \ \hat{N}_{16}^2 = E\left[\hat{C}_{16}^2\right] = 0 \text{ year } = 0 \text{ days}.$$

### 5.4.4 Parameters and Characteristics of Joint Climate-Weather Change Process at Port Oil Terminal Critical Infrastructure Water and Seaside Operating Area

We define the climate-weather change process $C(t), t \geq 0$, at port oil terminal critical infrastructure joining its climate-weather change process $C^1(t), t \geq 0$, at water operating area, and its climate-weather change process $C^2(t), t \geq 0$, at seaside operating area, defined in Sect. 5.4.2 and Sect. 5.4.3 respectively, according to the following agreement.

*Agreement 5.1.*

The rule for numbering of the climate-weather states of the joint climate weather-change process $C(t), t \geq 0$, is as follows.

The joint climate-weather change process $C(t)$ is at the climate-weather state

$$c_{(i-1)w_2+j}, i = 1, 2, \ldots, w^1, j = 1, 2, \ldots, w^2,$$

i.e. at the climate-weather state

$$c_{(i-1)16+j}, i = 1, 2, \ldots, 6, j = 1, 2, \ldots, 16,$$

when the climate-weather change process $C^1(t)$ is at the climate-weather state

$$c_i^1, i = 1, 2, \ldots, 6$$

and the climate-weather change process $C^2(t)$ is in the climate-weather state

$$c_j^2, j = 1, 2, \ldots, 16.$$

Thus, the joint climate-weather change process $C(t), t \geq 0$, has

$$w = w^1 \cdot w^2 = 6 \cdot 16 = 96$$

climate-weather states.

Further in accordance with Agreement 5.1 and definitions of the climate-weather states of climate-weather change process $C^1(t)$, defined in Sect. 5.4.2, and climate-weather change process $C^2(t)$, defined in Sect. 5.4.3, we define the joint climate weather-change process $C(t)$, $t \geq 0$, with 96 following climate-weather states:

- the climate-weather state $c_1$—the climate-weather change process $C^1(t)$ is at the climate-weather state $c_1^1$ and the climate-weather change process $C^2(t)$ is at the climate-weather state $c_1^2$;
- the climate-weather state $c_2$—the climate-weather change process $C^1(t)$ is at the climate-weather state $c_1^1$ and the climate-weather change process $C^2(t)$ is at the climate-weather state $c_2^2$;
- the climate-weather state $c_3$—the climate-weather change process $C^1(t)$ is at the climate-weather state $c_1^1$ and the climate-weather change process $C^2(t)$ is at the climate-weather state $c_3^2$;
- the climate-weather state $c_4$—the climate-weather change process $C^1(t)$ is at the climate-weather state $c_1^1$ and the climate-weather change process $C^2(t)$ is at the climate-weather state $c_4^2$;
- the climate-weather state $c_5$—the climate-weather change process $C^1(t)$ is at the climate-weather state $c_1^1$ and the climate-weather change process $C^2(t)$ is at the climate-weather state $c_5^2$;
- the climate-weather state $c_6$—the climate-weather change process $C^1(t)$ is at the climate-weather state $c_1^1$ and the climate-weather change process $C^2(t)$ is at the climate-weather state $c_6^2$;
- the climate-weather state $c_7$—the climate-weather change process $C^1(t)$ is at the climate-weather state $c_1^1$ and the climate-weather change process $C^2(t)$ is at the climate-weather state $c_7^2$;
- the climate-weather state $c_8$—the climate-weather change process $C^1(t)$ is at the climate-weather state $c_1^1$ and the climate-weather change process $C^2(t)$ is at the climate-weather state $c_8^2$;
- the climate-weather state $c_9$—the climate-weather change process $C^1(t)$ is at the climate-weather state $c_1^1$ and the climate-weather change process $C^2(t)$ is at the climate-weather state $c_9^2$;
- the climate-weather state $c_{10}$—the climate-weather change process $C^1(t)$ is at the climate-weather state $c_1^1$ and the climate-weather weather change process $C^2(t)$ is at the climate-weather state $c_{10}^2$;
- the climate-weather state $c_{11}$—the climate-weather change process $C^1(t)$ is at the climate-weather state $c_1^1$ and the climate-weather change process $C^2(t)$ is at the climate-weather state $c_{11}^2$;
- the climate-weather state $c_{12}$—the climate-weather change process $C^1(t)$ is at the climate-weather state $c_1^1$ and the climate-weather change process $C^2(t)$ is at the climate-weather state $c_{12}^2$;
- the climate-weather state $c_{13}$—the climate-weather change process $C^1(t)$ is at the climate-weather state $c_1^1$ and the climate-weather change process $C^2(t)$ is at the climate-weather state $c_{13}^2$;

- the climate-weather state $c_{14}$—the climate-weather change process $C^1(t)$ is at the climate-weather state $c_1^1$ and the climate-weather change process $C^2(t)$ is at the climate-weather state $c_{14}^2$;
- the climate-weather state $c_{15}$—the climate-weather change process $C^1(t)$ is at the climate-weather state $c_1^1$ and the climate-weather change process $C^2(t)$ is at the climate-weather state $c_{15}^2$;
- the climate-weather state $c_{16}$—the climate-weather change process $C^1(t)$ is at the climate-weather state $c_1^1$ and the climate-weather change process $C^2(t)$ is at the climate-weather state $c_{16}^2$;
- the climate-weather state $c_{17}$—the climate-weather change process $C^1(t)$ is at the climate-weather state $c_2^1$ and the climate-weather change process $C^2(t)$ is at the climate-weather state $c_1^2$;
- the climate-weather state $c_{18}$—the climate-weather change process $C^1(t)$ is at the climate-weather state $c_2^1$ and the climate-weather change process $C^2(t)$ is at the climate-weather state $c_2^2$;
- the climate-weather state $c_{19}$—the climate-weather change process $C^1(t)$ is at the climate-weather state $c_2^1$ and the climate-weather change process $C^2(t)$ is at the climate-weather state $c_3^2$;
- the climate-weather state $c_{20}$—the climate-weather change process $C^1(t)$ is at the climate-weather state $c_2^1$ and the climate-weather change process $C^2(t)$ is at the climate-weather state $c_4^2$;
- the climate-weather state $c_{21}$—the climate-weather change process $C^1(t)$ is at the climate-weather state $c_2^1$ and the climate-weather change process $C^2(t)$ is at the climate-weather state $c_5^2$;
- the climate-weather state $c_{22}$—the climate-weather change process $C^1(t)$ is at the climate-weather state $c_2^1$ and the climate-weather change process $C^2(t)$ is at the climate-weather state $c_6^2$;
- the climate-weather state $c_{23}$—the climate-weather change process $C^1(t)$ is at the climate-weather state $c_2^1$ and the climate-weather change process $C^2(t)$ is at the climate-weather state $c_7^2$;
- the climate-weather state $c_{24}$—the climate-weather change process $C^1(t)$ is at the climate-weather state $c_2^1$ and the climate-weather change process $C^2(t)$ is at the climate-weather state $c_8^2$;
- the climate-weather state $c_{25}$—the climate-weather change process $C^1(t)$ is at the climate-weather state $c_2^1$ and the climate-weather change process $C^2(t)$ is at the climate-weather state $c_9^2$;
- the climate-weather state $c_{26}$—the climate-weather change process $C^1(t)$ is at the climate-weather state $c_2^1$ and the climate-weather change process $C^2(t)$ is at the climate-weather state $c_{10}^2$;
- the climate-weather state $c_{27}$—the climate-weather change process $C^1(t)$ is at the climate-weather state $c_2^1$ and the climate-weather change process $C^2(t)$ is at the climate-weather state $c_{11}^2$;
- the weather state $c_{28}$—the climate-weather change process $C^1(t)$ is at the climate-weather state $c_2^1$ and the climate-weather change process $C^2(t)$ is at the climate-wether state $c_{12}^2$;

- the climate-weather state $c_{29}$—the climate-weather change process $C^1(t)$ is at the climate-weather state $c_2^1$ and the climate-weather change process $C^2(t)$ is at the climate-weather state $c_{13}^2$;
- the climate-weather state $c_{30}$—the climate-weather change process $C^1(t)$ is at the climate-weather state $c_2^1$ and the climate-weather change process $C^2(t)$ is at the climate-weather state $c_{14}^2$;
- the climate-weather state $c_{31}$—the climate-weather change process $C^1(t)$ is at the climate-weather state $c_2^1$ and the climate-weather change process $C^2(t)$ is at the climate-weather state $c_{15}^2$;
- the climate-weather state $c_{32}$—the climate-weather change process $C^1(t)$ is at the climate-weather state $c_2^1$ and the climate-weather change process $C^2(t)$ is at the climate-weather state $c_{16}^2$;
- the climate-weather state $c_{33}$—the climate-weather change process $C^1(t)$ is at the climate-weather state $c_3^1$ and the climate-weather change process $C^2(t)$ is at the climate-weather state $c_1^2$;
- the climate-weather state $c_{34}$—the climate-weather change process $C^1(t)$ is at the climate-weather state $c_3^1$ and the climate-weather change process $C^2(t)$ is at the climate-weather state $c_2^2$;
- the climate-weather state $c_{35}$—the climate-weather change process $C^1(t)$ is at the climate-weather state $c_3^1$ and the climate-weather change process $C^2(t)$ is at the climate-weather state $c_3^2$;
- the climate-weather state $c_{36}$—the climate-weather change process $C^1(t)$ is at the climate-weather state $c_3^1$ and the climate-weather change process $C^2(t)$ is at the climate-weather state $c_4^2$;
- the climate-weather state $c_{37}$—the climate-weather change process $C^1(t)$ is at the climate-weather state $c_3^1$ and the climate-weather change process $C^2(t)$ is at the climate-weather state $c_5^2$;
- the climate-weather state $c_{38}$—the climate-weather change process $C^1(t)$ is at the climate-weather state $c_3^1$ and the climate-weather change process $C^2(t)$ is at the climate-weather state $c_6^2$;
- the climate-weather state $c_{39}$—the climate-weather change process $C^1(t)$ is at the climate-weather state $c_3^1$ and the climate-weather change process $C^2(t)$ is at the climate-weather state $c_7^2$;
- the climate-weather state $c_{40}$—the climate-weather change process $C^1(t)$ is at the climate-weather state $c_3^1$ and the climate-weather change process $C^2(t)$ is at the climate-weather state $c_8^2$;
- the climate-weather state $c_{41}$—the climate-weather change process $C^1(t)$ is at the climate-weather state $c_3^1$ and the climate-weather change process $C^2(t)$ is at the climate-weather state $c_9^2$;
- the climate-weather state $c_{42}$—the climate-weather change process $C^1(t)$ is at the climate-weather state $c_3^1$ and the climate-weather change process $C^2(t)$ is at the climate-weather state $c_{10}^2$;
- the climate-weather state $c_{43}$—the climate-weather change process $C^1(t)$ is at the climate-weather state $c_3^1$ and the climate-weather change process $C^2(t)$ is at the climate-weather state $c_{11}^2$;

- the climate-weather state $c_{44}$—the climate-weather change process $C^1(t)$ is at the climate-weather state $c_3^1$ and the climate-weather change process $C^2(t)$ is at the climate-weather state $c_{12}^2$;
- the climate-weather state $c_{45}$—the climate-weather change process $C^1(t)$ is at the climate-weather state $c_3^1$ and the climate-weather change process $C^2(t)$ is at the climate-weather state $c_{13}^2$;
- the climate-weather state $c_{46}$—the climate-weather change process $C^1(t)$ is at the climate-weather state $c_3^1$ and the climate-weather change process $C^2(t)$ is at the climate-weather state $c_{14}^2$;
- the climate-weather state $c_{47}$—the climate-weather change process $C^1(t)$ is at the climate-weather state $c_3^1$ and the climate-weather change process $C^2(t)$ is at the climate-weather state $c_{15}^2$;
- the climate-weather state $c_{48}$—the climate-weather change process $C^1(t)$ is at the climate-weather state $c_3^1$ and the climate-weather change process $C^2(t)$ is at the climate-weather state $c_{16}^2$;
- the climate-weather state $c_{49}$—the climate-weather change process $C^1(t)$ is at the climate-weather state $c_4^1$ and the climate-weather change process $C^2(t)$ is at the climate-weather state $c_1^2$;
- the climate-weather state $c_{50}$—the climate-weather change process $C^1(t)$ is at the climate-weather state $c_4^1$ and the climate-weather change process $C^2(t)$ is at the climate-weather state $c_2^2$;
- the climate-weather state $c_{51}$—the climate-weather change process $C^1(t)$ is at the climate-weather state $c_4^1$ and the climate-weather change process $C^2(t)$ is at the climate-weather state $c_3^2$;
- the climate-weather state $c_{52}$—the climate-weather change process $C^1(t)$ is at the climate-weather state $c_4^1$ and the climate-weather change process $C^2(t)$ is at the climate-weather state $c_4^2$;
- the climate-weather state $c_{53}$—the climate-weather change process $C^1(t)$ is at the climate-weather state $c_4^1$ and the climate-weather change process $C^2(t)$ is at the climate-weather state $c_5^2$;
- the climate-weather state $c_{54}$—the climate-weather change process $C^1(t)$ is at the climate-weather state $c_4^1$ and the climate-weather change process $C^2(t)$ is at the climate-weather state $c_6^2$;
- the climate-weather state $c_{55}$—the climate-weather change process $C^1(t)$ is at the climate-weather state $c_4^1$ and the climate-weather change process $C^2(t)$ is at the climate-weather state $c_7^2$;
- the climate-weather weather state $c_{56}$—the climate-weather change process $C^1(t)$ is at the climate-weather state $c_4^1$ and the climate-weather change process $C^2(t)$ is at the climate-weather state $c_8^2$;
- the climate-weather state $c_{57}$—the climate-weather change process $C^1(t)$ is at the climate-weather state $c_4^1$ and the climate-weather change process $C^2(t)$ is at the climate-weather state $c_9^2$;
- the climate-weather state $c_{58}$—the climate-weather change process $C^1(t)$ is at the climate-weather state $c_4^1$ and the climate-weather change process $C^2(t)$ is at the climate-weather state $c_{10}^2$;

- the climate-weather state $c_{59}$—the climate-weather change process $C^1(t)$ is at the climate-weather state $c_4^1$ and the climate-weather change process $C^2(t)$ is at the climate-weather state $c_{11}^2$;
- the climate-weather state $c_{60}$—the climate-weather change process $C^1(t)$ is at the climate-weather state $c_4^1$ and the climate-weather change process $C^2(t)$ is at the climate-weather state $c_{12}^2$;
- the climate-weather state $c_{61}$—the climate-weather change process $C^1(t)$ is at the climate-weather state $c_4^1$ and the climate-weather change process $C^2(t)$ is at the climate-weather state $c_{13}^2$;
- the climate-weather state $c_{62}$—the climate-weather change process $C^1(t)$ is at the climate-weather state $c_4^1$ and the climate-weather change process $C^2(t)$ is at the climate-weather state $c_{14}^2$;
- the climate-weather state $c_{63}$—the climate-weather change process $C^1(t)$ is at the climate-weather state $c_4^1$ and the climate-weather change process $C^2(t)$ is at the climate-weather state $c_{15}^2$;
- the climate-weather state $c_{64}$—the climate-weather change process $C^1(t)$ is at the climate-weather state $c_4^1$ and the climate-weather change process $C^2(t)$ is at the climate-weather state $c_{16}^2$;
- the climate-weather state $c_{65}$—the climate-weather change process $C^1(t)$ is at the climate-weather state $c_5^1$ and the climate-weather change process $C^2(t)$ is at the climate-weather state $c_1^2$;
- the climate-weather state $c_{66}$—the climate-weather change process $C^1(t)$ is at the climate-weather state $c_5^1$ and the climate-weather change process $C^2(t)$ is at the climate-weather state $c_2^2$;
- the climate-weather state $c_{67}$—the climate-weather change process $C^1(t)$ is at the climate-weather state $c_5^1$ and the climate-weather change process $C^2(t)$ is at the climate-weather state $c_3^2$;
- the climate-weather state $c_{68}$—the climate-weather change process $C^1(t)$ is at the climate-weather state $c_5^1$ and the climate-weather change process $C^2(t)$ is at the climate-weather state $c_4^2$;
- the climate-weather state $c_{69}$—the climate-weather change process $C^1(t)$ is at the climate-weather state $c_5^1$ and the climate-weather change process $C^2(t)$ is at the climate-weather state $c_5^2$;
- the climate-weather state $c_{70}$—the climate-weather change process $C^1(t)$ is at the climate-weather state $c_5^1$ and the climate-weather change process $C^2(t)$ is at the climate-weather state $c_6^2$;
- the climate-weather state $c_{71}$—the climate-weather change process $C^1(t)$ is at the climate-weather state $c_5^1$ and the climate-weather change process $C^2(t)$ is at the climate-weather state $c_7^2$;
- the climate-weather state $c_{72}$—the climate-weather change process $C^1(t)$ is at the climate-weather state $c_5^1$ and the climate-weather change process $C^2(t)$ is at the climate-weather state $c_8^2$;
- the climate-weather state $c_{73}$—the climate-weather change process $C^1(t)$ is at the climate-weather state $c_5^1$ and the climate-weather change process $C^2(t)$ is at the climate-weather state $c_9^2$;

- the climate-weather state $c_{74}$—the climate-weather change process $C^1(t)$ is at the climate-weather state $c_5^1$ and the climate-weather change process $C^2(t)$ is at the climate-weather state $c_{10}^2$;
- the climate-weather state $c_{75}$—the climate-weather change process $C^1(t)$ is at the climate-weather state $c_5^1$ and the climate-weather change process $C^2(t)$ is at the climate-weather state $c_{11}^2$;
- the climate-weather state $c_{76}$—the climate-weather change process $C^1(t)$ is at the climate-weather state $c_5^1$ and the climate-weather change process $C^2(t)$ is at the climate-weather state $c_{12}^2$;
- the climate-weather state $c_{77}$—the climate-weather change process $C^1(t)$ is at the climate-weather state $c_5^1$ and the climate-weather change process $C^2(t)$ is at the climate-weather state $c_{13}^2$;
- the climate-weather state $c_{78}$—the climate-weather change process $C^1(t)$ is at the climate-weather state $c_5^1$ and the climate-weather change process $C^2(t)$ is at the climate-weather state $c_{14}^2$;
- the climate-weather state $c_{79}$—the climate-weather change process $C^1(t)$ is at the climate-weather state $c_5^1$ and the climate-weather change process $C^2(t)$ is at the climate-weather state $c_{15}^2$;
- the climate-weather state $c_{80}$—the climate-weather change process $C^1(t)$ is at the climate-weather state $c_5^1$ and the climate-weather change process $C^2(t)$ is at the climate-weather state $c_{16}^2$;
- the climate-weather state $c_{81}$—the climate-weather change process $C^1(t)$ is at the climate-weather state $c_6^1$ and the climate-weather change process $C^2(t)$ is at the climate-weather state $c_1^2$;
- the climate-weather state $c_{82}$—the climate-weather change process $C^1(t)$ is at the climate-weather state $c_6^1$ and the climate-weather change process $C^2(t)$ is at the climate-weather state $c_2^2$;
- the climate-weather state $c_{83}$—the climate-weather change process $C^1(t)$ is at the climate-weather state $c_6^1$ and the climate-weather change process $C^2(t)$ is at the climate-weather state $c_3^2$;
- the climate-weather state $c_{84}$—the climate-weather change process $C^1(t)$ is at the climate-weather state $c_6^1$ and the climate-weather change process $C^2(t)$ is at the climate-weather state $c_4^2$;
- the climate-weather state $c_{85}$—the climate-weather change process $C^1(t)$ is at the climate-weather state $c_6^1$ and the climate-weather change process $C^2(t)$ is at the climate-weather state $c_5^2$;
- the climate-weather state $c_{86}$—the climate-weather change process $C^1(t)$ is at the climate-weather state $c_6^1$ and the climate-weather change process $C^2(t)$ is at the climate-weather state $c_6^2$;
- the climate-weather state $c_{87}$—the climate-weather change process $C^1(t)$ is at the climate-weather state $c_6^1$ and the climate-weather change process $C^2(t)$ is at the climate-weather state $c_7^2$;
- the climate-weather state $c_{88}$—the climate-weather change process $C^1(t)$ is at the climate-weather state $c_6^1$ and the climate-weather change process $C^2(t)$ is at the climate-weather state $c_8^2$;

- the climate-weather state $c_{89}$—the climate-weather change process $C^1(t)$ is at the climate-weather state $c_6^1$ and the climate-weather change process $C^2(t)$ is at the climate-weather state $c_9^2$;
- the climate-weather state $c_{90}$—the climate-weather change process $C^1(t)$ is at the climate-weather state $c_6^1$ and the climate-weather change process $C^2(t)$ is at the climate-weather state $c_{10}^2$;
- the climate-weather state $c_{91}$—the climate-weather change process $C^1(t)$ is at the climate-weather state $c_6^1$ and the climate-weather change process $C^2(t)$ is at the climate-weather state $c_{11}^2$;
- the climate-weather state $c_{92}$—the climate-weather change process $C^1(t)$ is at the climate-weather state $c_6^1$ and the climate-weather change process $C^2(t)$ is at the climate-weather state $c_{12}^2$;
- the climate-weather state $c_{93}$—the climate-weather change process $C^1(t)$ is at the climate-weather state $c_6^1$ and the climate-weather change process $C^2(t)$ is at the climate-weather state $c_{13}^2$;
- the climate-weather state $c_{94}$—the climate-weather change process $C^1(t)$ is at the climate-weather state $c_6^1$ and the climate-weather change process $C^2(t)$ is at the climate-weather state $c_{14}^2$;
- the climate-weather state $c_{95}$—the climate-weather change process $C^1(t)$ is at the climate-weather state $c_6^1$ and the climate-weather change process $C^2(t)$ is at the climate-weather state $c_{15}^2$;
- the climate-weather state $c_{96}$—the climate-weather change process $C^1(t)$ is at the climate-weather state $c_6^1$ and the climate-weather change process $C^2(t)$ is at the climate-weather state $c_{16}^2$.

Next, considering the Agreement 5.1, we determine the limit values $q_b$, $b = 1, 2, \ldots, 96$, of the joint climate-weather change process $C(t)$, transient probabilities

$$q_b(t) = (P(C(t) = c_b), t \geq 0,$$

at the weather states $c_b$, $b = 1, 2, \ldots, 96$, from the following formulae

$$q_{(i-1)w^2+j} = q_i^1 \cdot q_{j/i}^2, i = 1, 2, \ldots, w^1, j = 1, 2, \ldots, w^2, w^1 = 6, w^2 = 16,$$

i.e.

$$q_{(i-1)16+j} = q_i^1 \cdot q_{j/i}^2, i = 1, 2, \ldots, 6, j = 1, 2, \ldots, 16,$$

where $q_{j/i}^2$, are the conditional limit probabilities that the climate-weather change process $C^2(t)$ is at the climate- weather state $c_j^2$, $j = 1, 2, \ldots, 16$, while the climate-weather change process $C^1(t)$ is at the climate-weather state $c_i^1$, $i = 1, 2, \ldots, 6$.

Assuming the independence of the processes $C^1(t)$ and $C^2(t)$, i.e. considering the expression

$$P((C^1(t) = c_i^1) \cap (C^2(t) = c_j^2)) = P(C^1(t) = c_i^1) \cdot P(C^2(t) = c_j^2),$$
$$t \geq 0, i = 1, 2, \ldots, 6, j = 1, 2, \ldots, 16,$$

which implies that

$$q_b = \lim_{t \to \infty} P(C(t) = c_b) = \lim_{t \to \infty} P(C^1(t) = c_i^1) \cdot \lim_{t \to \infty} P(C^2(t) = c_j^2)$$
$$= q_i^1 \cdot q_j^2, i = 1, 2, \ldots, 6, j = 1, 2, \ldots, 16,$$

from the above formulae, we get

$$q_{(i-1)16+j} = q_i^1 \cdot q_j^2, i = 1, 2, \ldots, 6, j = 1, 2, \ldots, 16,$$

and consequently, the limit transient probabilities of the joint climate-weather change process $C(t)$, $t \geq 0$, are as follows:

- for $i = 1$, we have

$$q_{0 \cdot 16+j} = q_1^1 \cdot q_j^2, j = 1, 2, \ldots, 16,$$

and particularly

$$q_1 = q_1^1 \cdot q_1^2 = 0.926 \cdot 0.001 = 0.000926,$$
$$q_2 = q_1^1 \cdot q_2^2 = 0.926 \cdot 0.116 = 0.107416,$$
$$q_3 = q_1^1 \cdot q_3^2 = 0.926 \cdot 0 = 0,$$
$$q_4 = q_1^1 \cdot q_4^2 = 0.926 \cdot 0 = 0,$$
$$q_5 = q_1^1 \cdot q_5^2 = 0.926 \cdot 0 = 0,$$
$$q_6 = q_1^1 \cdot q_6^2 = 0.926 \cdot 0.776 = 0.718576,$$
$$q_7 = q_1^1 \cdot q_7^2 = 0.926 \cdot 0.021 = 0.019446,$$
$$q_8 = q_1^1 \cdot q_8^2 = 0.926 \cdot 0 = 0,$$
$$q_9 = q_1^1 \cdot q_9^2 = 0.926 \cdot 0 = 0,$$
$$q_{10} = q_1^1 \cdot q_{10}^2 = 0.926 \cdot 0.014 = 0.012964,$$
$$q_{11} = q_1^1 \cdot q_{11}^2 = 0.926 \cdot 0.072 = 0.066672,$$
$$q_{12} = q_1^1 \cdot q_{12}^2 = 0.926 \cdot 0 = 0,$$
$$q_{13} = q_1^1 \cdot q_{13}^2 = 0.926 \cdot 0 = 0,$$

$$
\begin{aligned}
q_{14} &= q_1^1 \cdot q_{14}^2 = 0.926 \cdot 0 = 0, \\
q_{15} &= q_1^1 \cdot q_{15}^2 = 0.926 \cdot 0 = 0, \\
q_{16} &= q_1^1 \cdot q_{16}^2 = 0.926 \cdot 0 = 0; \\
q_{16+j} &= q_2^1 \cdot q_j^2, \ j = 1, 2, \ldots, 16,
\end{aligned} \tag{5.69}
$$

- for $i = 2$, we have

$$q_{16+j} = q_2^1 \cdot q_j^2, \ j = 1, 2, \ldots, 16,$$

and particularly

$$
\begin{aligned}
q_{17} &= q_2^1 \cdot q_1^2 = 0.064 \cdot 0.001 = 0.000064, \\
q_{18} &= q_2^1 \cdot q_2^2 = 0.064 \cdot 0.116 = 0.007424, \\
q_{19} &= q_2^1 \cdot q_3^2 = 0.064 \cdot 0 = 0, \\
q_{20} &= q_2^1 \cdot q_4^2 = 0.064 \cdot 0 = 0, \\
q_{21} &= q_2^1 \cdot q_5^2 = 0.064 \cdot 0 = 0, \\
q_{22} &= q_2^1 \cdot q_6^2 = 0.064 \cdot 0.776 = 0.049664, \\
q_{23} &= q_2^1 \cdot q_7^2 = 0.064 \cdot 0.021 = 0.001344, \\
q_{24} &= q_2^1 \cdot q_8^2 = 0.064 \cdot 0 = 0, \\
q_{17} &= q_2^1 \cdot q_1^2 = 0.064 \cdot 0 = 0, \\
q_{26} &= q_2^1 \cdot q_{10}^2 = 0.064 \cdot 0.014 = 0.000896, \\
q_{27} &= q_2^1 \cdot q_{11}^2 = 0.064 \cdot 0.072 = 0.004608, \\
q_{28} &= q_2^1 \cdot q_{12}^2 = 0.064 \cdot 0 = 0, \\
q_{29} &= q_2^1 \cdot q_{13}^2 = 0.064 \cdot 0 = 0, \\
q_{30} &= q_2^1 \cdot q_{14}^2 = 0.064 \cdot 0 = 0, \\
q_{31} &= q_2^1 \cdot q_{15}^2 = 0.064 \cdot 0 = 0, \\
q_{32} &= q_2^1 \cdot q_{16}^2 = 0.064 \cdot 0 = 0;
\end{aligned} \tag{5.70}
$$

- for $i = 3$, we have

$$q_{2.16+j} = q_3^1 \cdot q_j^2, \ j = 1, 2, \ldots, 16,$$

and particularly

$$\begin{aligned}
q_{33} &= q_3^1 \cdot q_1^2 = 0.001 \cdot 0.001 = 0.000001,\\
q_{34} &= q_3^1 \cdot q_2^2 = 0.001 \cdot 0.116 = 0.000116,\\
q_{35} &= q_3^1 \cdot q_3^2 = 0.001 \cdot 0 = 0,\\
q_{36} &= q_3^1 \cdot q_4^2 = 0.001 \cdot 0 = 0,\\
q_{37} &= q_3^1 \cdot q_5^2 = 0.001 \cdot 0 = 0,\\
q_{38} &= q_3^1 \cdot q_6^2 = 0.001 \cdot 0.776 = 0.000776,\\
q_{39} &= q_3^1 \cdot q_7^2 = 0.001 \cdot 0.021 = 0.000021,\\
q_{40} &= q_3^1 \cdot q_8^2 = 0.001 \cdot 0 = 0,\\
q_{41} &= q_3^1 \cdot q_9^2 = 0.001 \cdot 0 = 0,\\
q_{42} &= q_3^1 \cdot q_{10}^2 = 0.001 \cdot 0.014 = 0.000014,\\
q_{43} &= q_3^1 \cdot q_{11}^2 = 0.001 \cdot 0.072 = 0.000072,\\
q_{44} &= q_3^1 \cdot q_{12}^2 = 0.001 \cdot 0 = 0,\\
q_{45} &= q_3^1 \cdot q_{13}^2 = 0.001 \cdot 0 = 0,\\
q_{46} &= q_3^1 \cdot q_{14}^2 = 0.001 \cdot 0 = 0,\\
q_{47} &= q_3^1 \cdot q_{15}^2 = 0.001 \cdot 0 = 0,\\
q_{48} &= q_3^1 \cdot q_{16}^2 = 0.001 \cdot 0 = 0,\\
q_{48} &= q_3^1 \cdot q_{16}^2 = 0.001 \cdot 0 = 0;
\end{aligned} \tag{5.71}$$

- for $i = 4$, we have

$$q_{316+j} = q_3^1 \cdot q_j^2,\ j = 1, 2, \ldots, 16,$$

and particularly

$$\begin{aligned}
q_{49} &= q_4^1 \cdot q_1^2 = 0 \cdot 0.001 = 0,\\
q_{50} &= q_4^1 \cdot q_2^2 = 0 \cdot 0.116 = 0,\\
q_{51} &= q_4^1 \cdot q_3^2 = 0 \cdot 0 = 0,\\
q_{52} &= q_4^1 \cdot q_4^2 = 0 \cdot 0 = 0,\\
q_{53} &= q_4^1 \cdot q_5^2 = 0 \cdot 0 = 0,\\
q_{54} &= q_4^1 \cdot q_6^2 = 0 \cdot 0.776 = 0,\\
q_{55} &= q_4^1 \cdot q_7^2 = 0 \cdot 0.021 = 0,\\
q_{56} &= q_4^1 \cdot q_8^2 = 0 \cdot 0 = 0,\\
q_{57} &= q_4^1 \cdot q_9^2 = 0 \cdot 0 = 0,\\
q_{58} &= q_4^1 \cdot q_{10}^2 = 0 \cdot 0.014 = 0,\\
q_{59} &= q_4^1 \cdot q_{11}^2 = 0 \cdot 0.072 = 0,\\
q_{60} &= q_4^1 \cdot q_{12}^2 = 0 \cdot 0 = 0,\\
q_{61} &= q_4^1 \cdot q_{13}^2 = 0 \cdot 0 = 0,\\
q_{62} &= q_4^1 \cdot q_{14}^2 = 0 \cdot 0 = 0,\\
q_{63} &= q_4^1 \cdot q_{15}^2 = 0 \cdot 0 = 0,\\
q_{64} &= q_4^1 \cdot q_{16}^2 = 0 \cdot 0 = 0;
\end{aligned} \tag{5.72}$$

- for $i = 5$, we have

$$q_{416+j} = q_5^1 \cdot q_j^2, \; j = 1, 2, \ldots, 16,$$

and particularly

$$\begin{aligned}
q_{65} &= q_5^1 \cdot q_1^2 = 0.006 \cdot 0.001 = 0.000006,\\
q_{66} &= q_5^1 \cdot q_2^2 = 0.006 \cdot 0.116 = 0.000696,\\
q_{67} &= q_5^1 \cdot q_3^2 = 0.006 \cdot 0 = 0,\\
q_{68} &= q_5^1 \cdot q_4^2 = 0.006 \cdot 0 = 0,\\
q_{69} &= q_5^1 \cdot q_5^2 = 0.006 \cdot 0 = 0,\\
q_{70} &= q_5^1 \cdot q_6^2 = 0.006 \cdot 0.776 = 0.004656,\\
q_{71} &= q_5^1 \cdot q_7^2 = 0.006 \cdot 0.021 = 0.000126,\\
q_{72} &= q_5^1 \cdot q_8^2 = 0.006 \cdot 0 = 0,\\
q_{73} &= q_5^1 \cdot q_9^2 = 0.006 \cdot 0 = 0,\\
q_{74} &= q_5^1 \cdot q_{10}^2 = 0.006 \cdot 0.014 = 0.000084,\\
q_{75} &= q_5^1 \cdot q_{11}^2 = 0.006 \cdot 0.072 = 0.000432,\\
q_{76} &= q_5^1 \cdot q_{12}^2 = 0.006 \cdot 0 = 0,\\
q_{77} &= q_5^1 \cdot q_{13}^2 = 0.006 \cdot 0 = 0,\\
q_{78} &= q_5^1 \cdot q_{14}^2 = 0.006 \cdot 0 = 0,\\
q_{79} &= q_5^1 \cdot q_{15}^2 = 0.006 \cdot 0 = 0,\\
q_{80} &= q_5^1 \cdot q_{16}^2 = 0.006 \cdot 0 = 0;
\end{aligned} \tag{5.73}$$

- for $i = 6$, we have

$$q_{516+j} = q_6^1 \cdot q_j^2, \; j = 1, 2, \ldots, 16,$$

and particularly

$$\begin{aligned}
q_{81} &= q_6^1 \cdot q_1^2 = 0.003 \cdot 0.001 = 0.000003,\\
q_{82} &= q_6^1 \cdot q_2^2 = 0.003 \cdot 0.116 = 0.000348,\\
q_{83} &= q_6^1 \cdot q_3^2 = 0.003 \cdot 0 = 0,\\
q_{84} &= q_6^1 \cdot q_4^2 = 0.003 \cdot 0 = 0,\\
q_{85} &= q_6^1 \cdot q_5^2 = 0.003 \cdot 0 = 0,\\
q_{86} &= q_6^1 \cdot q_6^2 = 0.003 \cdot 0.776 = 0.002328,\\
q_{87} &= q_6^1 \cdot q_7^2 = 0.003 \cdot 0.021 = 0.000063,\\
q_{88} &= q_6^1 \cdot q_8^2 = 0.003 \cdot 0 = 0,\\
q_{89} &= q_6^1 \cdot q_9^2 = 0.003 \cdot 0 = 0,\\
q_{90} &= q_6^1 \cdot q_{10}^2 = 0.003 \cdot 0.014 = 0.000042,
\end{aligned}$$

$$
\begin{aligned}
q_{91} &= q_6^1 \cdot q_{11}^2 = 0.003 \cdot 0.072 = 0.000216, \\
q_{92} &= q_6^1 \cdot q_{12}^2 = 0.003 \cdot 0 = 0, \\
q_{93} &= q_6^1 \cdot q_{13}^2 = 0.003 \cdot 0 = 0, \\
q_{94} &= q_6^1 \cdot q_{14}^2 = 0.003 \cdot 0 = 0, \\
q_{95} &= q_6^1 \cdot q_{15}^2 = 0.003 \cdot 0 = 0, \\
q_{96} &= q_6^1 \cdot q_{16}^2 = 0.003 \cdot 0 = 0.
\end{aligned} \tag{5.74}
$$

The mean values of the total sojourn times $\hat{C}_l$, $l = 1, 2, \ldots, 96$, of the joint climate-weather change process $C(t)$ at port oil terminal critical infrastructure operating area at the particular climate-weather states $c_l$, $l = 1, 2, \ldots, 96$ during the fixed operation time $C = 1$ year $= 365$ days, determined according to (5.6) and (5.74) are

$$
\begin{aligned}
&\hat{N}_1 = E[\hat{C}_1] = 0.000926 \text{ year } = 0.338 \text{ days }, \hat{N}_2 = E[\hat{C}_2] = 0.107416 \text{ year } = 39.207 \text{ days}, \\
&\hat{N}_6 = E[\hat{C}_6] = 0.0718576 \text{ year } = 26.228 \text{ days}, \hat{N}_7 = E[\hat{C}_7] = 0.019446 \text{ year } = 7.098 \text{ days}, \\
&\hat{N}_{10} = E[\hat{C}_{10}] = 0.000064 \text{ year } = 0.023 \text{ days}, \hat{N}_{11} = E[\hat{C}_{11}] = 0.066672 \text{ year } = 24.335 \text{ days}, \\
&\hat{N}_{17} = E[\hat{C}_{17}] = 0.000064 \text{ year } = 0.023 \text{ days}, \hat{N}_{18} = E[\hat{C}_{18}] = 0.007424 \text{ year } = 2.709 \text{ days}, \\
&\hat{N}_{22} = E[\hat{C}_{22}] = 0.049664 \text{ year } = 18.127 \text{ days}, \hat{N}_{23} = E[\hat{C}_{23}] = 0.001344 \text{ year } = 0.490 \text{ days}, \\
&\hat{N}_{26} = E[\hat{C}_{26}] = 0.000896 \text{ year } = 0.327 \text{ days}, \hat{N}_{27} = E[\hat{C}_{27}] = 0.004608 \text{ year} = 1.682 \text{ days}, \\
&\hat{N}_{33} = E[\hat{C}_{33}] = 0.000001 \text{ year } = 0.000365 \text{ days}, \hat{N}_{34} = E[\hat{C}_{34}] = 0.000116 \text{ year } = 0.042 \text{ days}, \\
&\hat{N}_{38} = E[\hat{C}_{38}] = 0.000776 \text{ year } = 0.283 \text{ days}, \hat{N}_{39} = E[\hat{C}_{39}] = 0.000021 \text{ year } = 0.008 \text{ days}, \\
&\hat{N}_{42} = E[\hat{C}_{43}] = 0.000014 \text{ year} = 0,00511 \text{ days}, \hat{N}_{43} = E[\hat{C}_{43}] = 0.000072 \text{ year } = 0.026 \text{ days}, \\
&\hat{N}_{65} = E[\hat{C}_{65}] = 0.000006 \text{ year } = 0.002 \text{ days}, \hat{N}_{66} = E[\hat{C}_{66}] = 0.000696 \text{ year } = 0.254 \text{ days}, \\
&\hat{N}_{70} = E[\hat{C}_{70}] = 0.004656 \text{ year } = 1.699 \text{ days}, \hat{N}_{71} = E[\hat{C}_{71}] = 0.000126 \text{ year } = 0.046 \text{ days}, \\
&\hat{N}_{74} = E[\hat{C}_{74}] = 0.000084 \text{ year } = 0.031 \text{ days}, \hat{N}_{75} = E[\hat{C}_{75}] = 0.000432 \text{ year } = 0.158 \text{ days}, \\
&\hat{N}_{81} = E[\hat{C}_{81}] = 0.000003 \text{ year } = 0.001 \text{ days}, \hat{N}_{82} = E[\hat{C}_{82}] = 0.000348 \text{ year } = 0.127 \text{ days}, \\
&\hat{N}_{86} = E[\hat{C}_{86}] = 0.002328 \text{ year } = 0.849 \text{ days}, \hat{N}_{87} = E[\hat{C}_{87}] = 0.000063 \text{ year } = 0.023 \text{ days}, \\
&\hat{N}_{90} = E[\hat{C}_{90}] = 0.000042 \text{ year } = 0.015 \text{ days}, \hat{N}_{91} = E[\hat{C}_{91}] = 0.000216 \text{ year } = 0.079 \text{ days},
\end{aligned}
$$

and remaining mean values are equal zero.

### 5.4.5 Port Oil Terminal Critical Infrastructure Safety Parameters Impacted by Climate-Weather Change Process

In this Section, we accept assumptions and agreements such as in Sects. 3.3.2 and 3.3.3.

Next, we assume that the coefficients of the climate-weather change process impact on the port oil terminal critical infrastructure intensities of ageing at the climate-weather states $c_l, l = 1, 2, \ldots, 96$, of the climate-weather change process $C(t), t \geq 0$, are as follows [17, 19, 22]:

- for asset $A_1$

$$\left[\rho_1^2(1)\right]^{(i)} = 1.05, \left[\rho_1^2(2)\right]^{(i)} = 1.05$$
$$l = 1, 2, 3, 4, 5, 8, 9, 12, 13, 14, 15, 16, 17, 18, 19, 20, 21, 24, 25, 28, 29, 30, 31, 32, \quad (5.75)$$

$$\left[\rho_1^2(1)\right]^{(l)} = 1.08, \left[\rho_1^2(2)\right]^{(l)} = 1.09$$
$$l = 33, 34, 35, 36, 37, 40, 41, 44, 45, 46, 47, 48, 49, 50, 51, 52, 53, 56, 57, 60, 61, 62, 63, 64, 65, 66, 67, 68, 69, 72, 73, 76, 77, 78, 79, 80, 81, 82, 83, 84, 85, 88, 89, 92, 93, 94, 95, 96 \quad (5.76)$$

$$[\rho_1^2(1)]^{(l)} = 1.03, [\rho_1^2(2)]^{(l)} = 1.04,$$
$$l = 38, 39, 42, 43, 54, 55, 58, 59, 70, 71, 74, 75, 86, 87, 90, 91, \quad (5.77)$$

$$\left[\rho_1^2(1)\right]^{(i)} = 1, \left[\rho_1^2(2)\right]^{(l)} = 1,$$
$$l = 6, 7, 10, 11, 22, 23, 26, 27; \quad (5.78)$$

- for asset $A_2$–$A_9$

$$[\rho_i^2(1)]^{(l)} = 1, [\rho_i^2(2)]^{(l)} = 1, i = 2, 3, 4, 5, 6, 7, 8, 9,$$
$$l = 6, 7, 10, 11, 22, 23, 26, 27, 38, 39, 42, 43, 54, 55, 58, 59, 70, 71, 74, 75, 86, 87, 90, 91,$$

$$\left[\rho_i^2(1)\right]^{(i)} = 1.10, \left[\rho_i^2(2)\right]^{(i)} = 1.10, i = 2, 3, 4, 5, 6, 7, 8, 9,$$
$$l = 1, 2, 3, 4, 5, 8, 9, 12, 13, 14, 15, 16, 17, 18, 19, 20, 21, 24, 25, 28, 29, 30, 31, 32, 33, 34, 35, 36, 37, 40, 41, 44, 45, 46, 47, 48, 49, 50, 51, 52, 53, 56, 57, 60, 61, 62, 63, 64, 65, 66, 67, 68, 69, 72, 73, 76, 77, 78, 79, 80, 81, 82, 83, 84, 85, 88, 89, 92, 93, 94, 95, 96. \quad (5.79)$$

Considering above assumption similarly to Sect. 3.3.5, we assume that the port oil terminal critical infrastructure assets $A_i$, $i = 1, 2, \ldots, 9$, at the climate-weather change process $C(t)$ states $c_l, l = 1, 2, \ldots, 96$, conditional safety functions (5.41), (5.42) are picewice exponential with the coordinates $[S_i^2(t, u)]^{(l)}$, $t \in < 0, \infty)$, $u = 1, 2$, $l = 1, 2, \ldots, 96$, $i = 1, 2, \ldots, 9$, defined by (5.49). Whereas, the intensities of ageing $[\lambda_i^2(u)]^{(l)}, u = 1, 2$, $l = 1, 2, \ldots, 96$, $i = 1, 2, \ldots, 9$, for the port oil terminal critical infrastructure assets $A_i$, $i = 1, 2, \ldots, 9$, at the climate-weather states $c_l, l = 1, 2, \ldots, 96$, existing in (5.49) are given by (5.52), where $\lambda_i^0(u)$, $u = 1, 2$, $i = 1, 2, \ldots, 9$, are the intensities of ageing of the port oil terminal critical infrastructure assets $A_i$, $i = 1, 2, \ldots, 9$, without considering climate-weather change process impact and $[\rho_i^2(u)]^{(l)}$, $u = 1, 2$, $l = 1, 2, \ldots, 96$, $i = 1, 2, \ldots, 9$, are the coefficients of climate-weather change process impact on the port oil terminal critical infrastructure assets $A_i$, $i = 1, 2, \ldots, 9$, intensities of ageing at the climate-weather states $c_l, l = 1, 2, \ldots, 96$.

After that, from (5.52), considering (3.96), (3.99) and (5.75)–(5.79), it follows that the intensities of assets departure from the safety states subset {1, 2}, {2}, related to the climate-weather change process impact on their safety at the particular climate-weather states $c_l$, $l = 1, 2, \ldots, 96$, are:

- for asset $A_1$

$$[\lambda_1^2(1)]^{(l)} = [\rho_1^2(1)]^{(l)} \cdot \lambda_1^0(1) = 1.05 \cdot 0.015873 = 0.0166666,$$
$$[\lambda_1^2(2)]^{(l)} = [\rho_1^2(2)]^{(l)} \cdot \lambda_1^0(2) = 1.05 \cdot 0.021739 = 0.0228259,$$
$$l = 1, 2, 3, 4, 5, 8, 9, 12, 13, 14, 15, 16, 17, 18, 19, 20, 21, 24, 25, 28, 29, 30, 31, 32, \quad (5.80)$$

$$[\lambda_1^2(1)]^{(l)} = [\rho_1^2(1)]^{(l)} \cdot \lambda_1^0(1) = 1.08 \cdot 0.015873 = 0.0171428,$$
$$[\lambda_1^2(2)]^{(l)} = [\rho_1^2(2)]^{(l)} \cdot \lambda_1^0(2) = 1.09 \cdot 0.021739 = 0.0236955,$$
$$l = 33, 34, 35, 36, 37, 40, 41, 44, 45, 46, 47, 48, 49, 50, 51, 52, 53, 56, 57, 60, 61, 62, 63, 64, 65, 66, 67, 68, 69, 72, 73, 76, 77, 78, 79, 80, 81, 82, 83, 84, 85, 88, 89, 92, 93, 94, 95, 96, \quad (5.81)$$

$$[\lambda_1^2(1)]^{(l)} = [\rho_1^2(1)]^{(l)} \cdot \lambda_1^0(1) = 1.03 \cdot 0.015873 = 0.013491,$$
$$[\lambda_1^2(2)]^{(l)} = [\rho_1^2(2)]^{(l)} \cdot \lambda_1^0(2) = 1.04 \cdot 0.021739 = 0.0226085,$$
$$l = 38, 39, 42, 43, 54, 55, 58, 59, 70, 71, 74, 75, 86, 87, 90, 91, \quad (5.82)$$

$$[\lambda_1^2(1)]^{(l)} = [\rho_1^2(1)]^{(l)} \cdot \lambda_1^0(1) = 1. \cdot 0.015873 = 0.015873,$$
$$[\lambda_1^2(2)]^{(l)} = [\rho_1^2(2)]^{(l)} \cdot \lambda_1^0(2) = 1 \cdot 0.021739 = 0.021739,$$
$$l = 6, 7, 10, 11, 22, 23, 26, 27, \quad (5.83)$$

- for asset $A_2$–$A_9$

$$[\lambda_i^2(1)]^{(l)} = [\rho_i^2(1)]^{(l)} \cdot \lambda_i^0(1) = 1 \cdot 0.0125 = 0.0125,$$
$$[\lambda_i^2(2)]^{(l)} = [\rho_i^2(2)]^{(l)} \cdot \lambda_i^0(2) = 1 \cdot 0.02 = 0.02, i = 2, 3, 4, 5, 6, 7, 8, 9,$$
$$l = 6, 7, 10, 11, 22, 23, 26, 27, 38, 39, 42, 43, 54, 55, 58, 59, 70, 71, 74, 75, 86, 87, 90, 91, \quad (5.84)$$

$$[\lambda_i^2(1)]^{(l)} = [\rho_i^2(1)]^{(l)} \cdot \lambda_i^0(1) = 1.10 \cdot 0.0125 = 0.01375,$$

$$\left[\lambda_i^2(2)\right]^{(l)} = \left[\rho_i^2(2)\right]^{(l)} \cdot \lambda_i^0(2) = 1.10 \cdot 0.02, 0.022, i = 2, 3, 4, 5, 6, 7, 8, 9,$$

$$\begin{aligned} l = {} & 1, 2, 3, 4, 5, 8, 9, 12, 13, 14, 15, 16, 17, 18, 19, 20, 21, 24, 25, \\ & 28, 29, 30, 31, 32, 33, 34, 35, 36, 37, 40, 41, 44, 45, 46, 47, \\ & 48, 49, 50, 51, 52, 53, 56, 57, 60, 61, 62, 63, 64, 65, 66, 67, \\ & 68, 69, 72, 73, 76, 77, 78, 79, 80, 81, 82, 83, 84, 85, \\ & 88, 89, 92, 93, 94, 95, 96, \end{aligned} \tag{5.85}$$

### 5.4.6 *Safety and Resilience Indicators of Port Oil Terminal Critical Infrastructure Impacted by Climate-Weather Change Process*

Considering that the coordinates $[S_i^2(t, u)]^{(l)}$, $t \geq 0$, $u = 1, 2, i = 1, 2, \ldots 9$, $l = 1, 2, \ldots, 96$, of the conditional safety functions (5.41), (5.42) for the port oil terminal critical infrastructure assets $A_i$, $i = 1, 2, \ldots 9$, at the climate-weather states $c_l$, $l = 1, 2, \ldots, 96$, are of the form (5.49), i.e. they are given by the vectors

$$\left[S_i^2(t, \cdot)\right]^{(l)} = \left[1, \left[S_i^2(t, 1)\right]^{(l)}, \left[S_i^2(t, 2)\right]^{(l)}\right],$$
$$t \geq 0, u = 1, 2, i = 1, 2, \ldots 9, l = 1, 2, \ldots, 96,$$

with the coordinates

$$[S_i^2(t, u)]^{(l)} = \exp[-[\lambda_i^2(u)]^{(l)} t], t \geq 0, u = 1, 2, i = 1, 2, \ldots 9, \quad l = 1, 2, \ldots, 96,$$

and the intensities of ageing $[\lambda_i^2(u)]^{(l)}$, $t \geq 0$, $u = 1, 2, i = 1, 2, \ldots 9$, $l = 1, 2, \ldots, 96$, at the climate-weather states $c_l, l = 1, 2, \ldots, 96$, are given respectively by (5.80)–(5.85), we get their coordinates following final forms at the particular climate-weather states $c_l, l = 1, 2, \ldots, 96$:

- for asset $A_1$

$$\left[S_1^2(1)\right]^{(l)} = \exp[-0.0166666t], \left[S_1^2(2)\right]^{(l)} = \exp[-0.0228259t]$$
$$l = 1, 2, 3, 4, 5, 8, 9, 12, 13, 14, 15, 16, 17, 18, 19, 20, 21, 24, 25, 28, 29, 30, 31, 32, \tag{5.86}$$

$$\left[S_1^2(1)\right]^{(l)} = \exp[-0.0171428t], \left[S_1^2(2)\right]^{(l)} = \exp[-0.0236955t]$$
$$l = 33, 34, 35, 36, 37, 40, 41, 44, 45, 46, 47, 48, 49, 50, 51, 52, 53, 56, 57, 60, 61, 62, 63, 64, 65, 66, 67, 68, 69, 72, 73, 76, 77, 78, 79, 80, 81, 82, 83, 84, 85, 88, 89, 92, 93, 94, 95, 96, \quad (5.87)$$

$$\left[S_1^2(1)\right]^{(l)} = \exp[-0.013491t], \left[S_1^2(2)\right]^{(l)} = \exp[-0.0226085t]$$
$$l = 38, 39, 42, 43, 54, 55, 58, 59, 70, 71, 74, 75, 86, 87, 90, 91, \quad (5.88)$$

$$\left[S_1^2(1)\right]^{(l)} = \exp[-0.015873t], \left[S_1^2(2)\right]^{(l)} = \exp[-0.021739t], \quad l = 6, 7, 10, 11, 22, 23, 26, 27; \quad (5.89)$$

- for asset $A_2$ - $A_9$

$$\left[S_i^2(1)\right]^{(l)} = \exp[-0.0125t], \left[S_i^2(2)\right]^{(l)} = \exp[-0.02t], i = 2, 3, 4, 5, 6, 7, 8, 9,$$
$$l = 6, 7, 10, 11, 22, 23, 26, 27, 38, 39, 42, 43, 54, 55, 58, 59, 70, 71, 74, 75, 86, 87, 90, 91, \quad (5.90)$$

$$\left[S_i^2(1)\right]^{(l)} = \exp[-0.01375t], \left[S_i^2(2)\right]^{(l)} = \exp[-0.022t], i = 2, 3, 4, 5, 6, 7, 8, 9,$$
$$l = 1, 2, 3, 4, 5, 8, 9, 12, 13, 14, 15, 16, 17, 18, 19, 20, 21, 24, 25, 28, 29, 30, 31, 32, 33, 34, 35, 36, 37, 40, 41, 44, 45, 46, 47, 48, 49, 50, 51, 52, 53, 56, 57, 60, 61, 62, 63, 64, 65, 66, 67, 68, 69, 72, 73, 76, 77, 78, 79, 80, 81, 82, 83, 84, 85, 88, 89, 92, 93, 94, 95, 96. \quad (5.91)$$

Since the oil terminal critical infrastructure is a three-state (z = 2) series system, then considering (5.86)–(5.91) and applying (2.8), (2.9) its conditional safety functions at the particular climate-weather states $c_l, l = 1, 2, ..., 96$, are given by:

$$\left[\mathbf{S}^2(t,\cdot)\right]^{(l)} = \left[1, \quad \left[\mathbf{S}^2(t,1)\right]^{(l)}, \left[\mathbf{S}^2(t,2)\right]^{(l)}\right] t \geq 0, \quad (5.92)$$

where

$$\begin{aligned}[\mathbf{S}^2(t,1)]^{(l)} &= \prod_{i=1}^{9} [S_i^2(t,1)]^{(l)} \\ &= [S_1^2(t,1)]^{(l)} \cdot [S_2^2(t,1)]^{(l)} \cdot [S_3^2(t,1)]^{(l)} \cdot [S_4^2(t,1)]^{(l)} \cdot [S_5^2(t,1)]^{(l)} \\ &\cdot [S_6^2(t,1)]^{(l)} \cdot [S_7^2(t,1)]^{(l)} \cdot [S_8^2(t,1)]^{(l)} \cdot [S_9^2(t,1)]^{(l)} \\ &= \exp[-0.0166666t]\exp[-0.01375t]\exp[-0.01375t]\exp[-0.01375t] \\ &\exp[-0.01375t]\exp[-0.01375t]\exp[-0.01375t]\exp[-0.01375t] \\ &\exp[-0.01375t] = \exp[-0.1266666t],\end{aligned} \quad (5.93)$$

$$[S^2(t,2)]^{(l)} = \prod_{i=1}^{9}[S_i^2(t,2)]^{(l)}$$
$$= [S_1^2(t,2)]^{(l)} \cdot [S_2^2(t,2)]^{(l)} \cdot [S_3^2(t,2)]^{(l)} \cdot [S_4^2(t,2)]^{(l)} \cdot [S_5^2(t,2)]^{(l)}$$
$$\cdot [S_6^2(t,2)]^{(l)} \cdot [S_7^2(t,2)]^{(l)} \cdot [S_8^2(t,2)]^{(l)} \cdot [S_9^2(t,2)]^{(l)}$$
$$= \exp[-0.0228259t]\exp[-0.022t]\exp[-0.022t]\exp[-0.022t]\exp[-0.022t]$$
$$\exp[-0.022t]\exp[-0.022t]\exp[-0.022t]\exp[-0.022t]$$
$$= \exp[-0.1988259t], \tag{5.94}$$

for the climate-weather states $c_l$, $l = 1,2, 3, 4, 5, 8, 9, 12, 13, 14, 15, 16, 17, 18, 19, 20, 21, 24, 25, 28, 29, 30, 31, 32$;

$$[\mathbf{S}^2(t,\cdot)]^{(l)} = [1, [\mathbf{S}^2(t,1)]^{(l)}, [\mathbf{S}^2(t,2)]^{(l)}]t \geq 0, \tag{5.95}$$

where

$$[S^2(t,1)]^{(l)} = \prod_{i=1}^{9}[S_i^2(t,1)]^{(l)}\,[S_1^2(t,1)]^{(l)} \cdot [S_2^2(t,1)]^{(l)} \cdot [S_3^2(t,1)]^{(l)}$$
$$\cdot [S_4^2(t,1)]^{(l)} \cdot [S_5^2(t,1)]^{(l)} \cdot [S_5^2(t,1)]^{(l)} \cdot [S_7^2(t,1)]^{(l)} \cdot [S_8^2(t,1)]^{(l)}$$
$$\cdot [S_9^2(t,1)]^{(l)} = \exp[-0.0171428t]\exp[-0.01375t]\exp[-0.01375t]$$
$$\exp[-0.01375t]\exp[-0.01375t]\exp[-0.01375t]\exp[-0.01375t]$$
$$\exp[-0.01375t]\exp[-0.01375t] = \exp[-0.1271428t], \tag{5.96}$$

$$[S^2(t,2)]^{(l)} = \prod_{i=1}^{9}[S_i^2(t,2)]^{(l)}$$
$$[S_1^2(t,2)]^{(l)} \cdot [S_2^2(t,2)]^{(l)} \cdot [S_3^2(t,2)]^{(l)} \cdot [S_4^2(t,2)]^{(l)} \cdot [S_5^2(t,2)]^{(l)} \cdot [S_6^2(t,2)]^{(l)}$$
$$\cdot [S_7^2(t,2)]^{(l)} \cdot [S_8^2(t,2)]^{(l)} \cdot [S_9^2(t,2)]^{(l)}$$
$$= \exp[-0.0236955t]\exp[-0.022t]\exp[-0.022t]\exp[-0.022t]\exp[-0.022t]$$
$$\exp[-0.022t]\exp[-0.022t]\exp[-0.022t]\exp[-0.022t]$$
$$= \exp[-0.1996955t], \tag{5.97}$$

for the climate-weather states $c_l$, $l = 33, 34, 35, 36, 37, 40, 41, 44, 45, 46, 47, 48, 49, 50, 51, 52, 53, 56, 57, 60, 61, 62, 63, 64, 65, 66, 67, 68, 69, 72, 73, 76, 77, 78, 79, 80, 81, 82, 83, 84, 85, 88, 89, 92, 93, 94, 95, 96$;

$$\left[\mathbf{S}^2(t,\cdot)\right]^{(l)} = \left[1,\ \left[\mathbf{S}^2(t,1)\right]^{(l)}, \left[\mathbf{S}^2(t,2)\right]^{(l)}\right]t \geq 0, \tag{5.98}$$

where

$$[S^2(t,1)]^{(l)} = \prod_{i=1}^{9} [S_i^2(t,1)]^{(l)}$$
$$[S_1^2(t,1)]^{(l)} \cdot S_2^0(t,1) \cdot S_3^0(t,1) \cdot S_4^0(t,1) \cdot S_5^0(t,1) \cdot S_6^0(t,1)$$
$$\cdot\ S_7^0(t,1) \cdot S_8^0(t,1) \cdot S_9^0(t,1)$$
$$= \exp[-0.0163491t]\exp[-0.0125t]\exp[-0.0125t]$$
$$\exp[-0.0125t]\exp[-0.0125t]\exp[-0.0125t]\exp[-0.0125t]\exp[-0.0125t]\exp[-0.0125t]$$
$$= \exp[-0.1163491t], \quad (5.99)$$

$$[S^2(t,2)]^{(l)} = \prod_{i=1}^{9} [S_i^2(t,2)]^{(l)}$$
$$= [S_1^2(t,2)]^{(l)} \cdot S_2^0(t,2) \cdot S_3^0(t,2) \cdot S_4^0(t,2) \cdot S_5^0(t,2) \cdot S_6^0(t,2)$$
$$\cdot\ S_7^0(t,2) \cdot S_8^0(t,2) \cdot S_9^0(t,2)$$
$$= \exp[-0.0226085t]\exp[-0.02t]\exp[-0.02t]\exp[-0.02t]\exp[-0.02t]$$
$$\exp[-0.02t]\exp[-0.02t]\exp[-0.02t]\exp[-0.02t]$$
$$= \exp[-0.1966944t]; \quad (5.100)$$

for the climate-weather states $c_l$, $l = 38, 39, 42, 43, 54, 55, 58, 59, 70, 71, 74, 75, 86, 87, 90, 91$;

$$[\boldsymbol{S}^2(t,\cdot)]^{(l)} = [1, [\boldsymbol{S}^2(t,1)]^{(l)}, [\boldsymbol{S}^2(t,2)]^{(l)}]t \geq 0, \quad (5.101)$$

where

$$[S^2(t,1)]^{(l)} = \prod_{i=1}^{9} [S_i^2(t,1)]^{(l)}$$
$$S_1^0(t,1) \cdot S_2^0(t,1) \cdot S_3^0(t,1) \cdot S_4^0(t,1) \cdot S_5^0(t,1) \cdot S_6^0(t,1)$$
$$\cdot\ S_7^0(t,1) \cdot S_8^0(t,1) \cdot S_9^0(t,1)$$
$$= \exp[-0.015873t]\exp[-0.0125t]\exp[-0.0125t]$$
$$\exp[-0.0125t]\exp[-0.0125t]\exp[-0.0125t]\exp[-0.0125t]\exp[-0.0125t]\exp[-0.0125t]$$
$$= \exp[-0.115873t], \quad (5.102)$$

$$[S^2(t,2)]^{(l)} = \prod_{i=1}^{9} [S_i^2(t,2)]^{(l)}$$
$$= S_1^0(t,2) \cdot S_2^0(t,2) \cdot S_3^0(t,2) \cdot S_4^0(t,2) \cdot S_5^0(t,2) \cdot S_6^0(t,2)$$
$$\cdot\ S_7^0(t,2) \cdot S_8^0(t,2) \cdot S_9^0(t,2)$$
$$= \exp[-0.021739t]\exp[-0.02t]\exp[-0.02t]\exp[-0.02t]\exp[-0.02t]$$
$$\exp[-0.02t]\exp[-0.02t]\exp[-0.02t]\exp[-0.02t]$$
$$= \exp[-0.181739t], \quad (5.103)$$

for the climate-weather states $c_l$, $l = 6, 7, 10, 11, 22, 23, 26, 27$.

Hence, applying (5.45), the expected values of the port oil terminal critical infrastructure lifetimes in the safety state subsets $\{1, 2\}$, $\{2\}$, at the climate-weather states $c_l$, $l = 1, 2, ..., 96$, respectively are:

$$[\boldsymbol{\mu}^2(1)]^{(l)} = \int_0^\infty [\boldsymbol{S}^2(t, 1)]^{(l)} dt = \int_0^\infty \exp[-0.1266666t]dt \cong 7.89 \text{ years},$$

$$[\boldsymbol{\mu}^2(2)]^{(l)} = \int_0^\infty [\boldsymbol{S}^2(t, 2)]^{(l)} dt = \int_0^\infty \exp[-0.1988259t]dt \cong 5.03 \text{ years}, \quad (5.104)$$

for $l = 1,2, 3, 4, 5, 8, 9, 12, 13, 14, 15, 16, 17, 18, 19, 20, 21, 24, 25, 28, 29, 30, 31, 32$;

$$[\boldsymbol{\mu}^2(1)]^{(l)} = \int_0^\infty [\boldsymbol{S}^2(t, 1)]^{(l)} dt = \int_0^\infty \exp[-0.1271428t]dt \cong 7.87 \text{ years},$$

$$[\boldsymbol{\mu}^2(2)]^{(l)} = \int_0^\infty [\boldsymbol{S}^2(t, 2)]^{(l)} dt = \int_0^\infty \exp[-0.1996955t]dt \cong 5.01 \text{ years}, \quad (5.105)$$

for the climate-weather states $c_l$, $l = 33, 34, 35, 36, 37, 40, 41, 44, 45, 46, 47, 48, 49, 50, 51, 52, 53, 56, 57, 60, 61, 62, 63, 64, 65, 66, 67, 68, 69, 72, 73, 76, 77, 78, 79, 80, 81, 82, 83, 84, 85, 88, 89, 92, 93, 94, 95, 96$;

$$[\boldsymbol{\mu}^2(1)]^{(l)} = \int_0^\infty [\boldsymbol{S}^2(t, 1)]^{(l)} dt = \int_0^\infty \exp[-0.1163491t]dt \cong 8.59 \text{years},$$

$$[\boldsymbol{\mu}^2(2)]^{(l)} = \int_0^\infty [\boldsymbol{S}^2(t, 2)]^{(l)} dt = \int_0^\infty \exp[-0.1826085t]dt \cong 5.48 \text{ years}, \quad (5.106)$$

for the climate-weather states $c_l$, $l = 38, 39, 42, 43, 54, 55, 58, 59, 70, 71, 74, 75, 86, 87, 90, 91$;

$$[\boldsymbol{\mu}^2(1)]^{(l)} = \int_0^\infty [\boldsymbol{S}^2(t, 1)]^{(l)} dt \int_0^\infty \exp[-0.115873t]dt \cong 8.63 \text{ years},$$

$$[\boldsymbol{\mu}^2(2)]^{(l)} = \int_0^\infty [\boldsymbol{S}^2(t, 2)]^{(l)} dt = \int_0^\infty \exp[-0.181739t]dt \cong 5.50 \text{ years}, \quad (5.107)$$

for the climate-weather states $c_l$, $l = 6, 7, 10, 11, 22, 23, 26, 27$.

From the results (5.69)–(5.74) and (5.92)–(5.103), applying (5.11), the port oil terminal critical infrastructure unconditional safety function (SafI1) is given by

$$\boldsymbol{S}^2(t,\ \cdot) = [1, \boldsymbol{S}^2(t, 1)\boldsymbol{S}^2(t, 2)], t \geq 0, \tag{5.108}$$

where

$$\begin{aligned}
\boldsymbol{s}^2(t, 1) &\cong \sum_{l=1}^{96} q_l [\boldsymbol{S}^2(t, 1)]^{(l)} \\
&= 0.000926 \exp[-0.1266666t] + 0.107416 \exp[-0.1266666t] \\
&+ 0 \exp[-0.1266666t] + 0 \exp[-0.1266666t] \\
&+ 0 \exp[-0.1266666t] + 0.718576 \exp[-0.115873t] + 0.019446 \exp[-0.115873t] \\
&+ 0 \exp[-0.1266666t] + 0 \exp[-0.1266666t] + 0.012964 \exp[-0.115873t] \\
&+ 0.066672 \exp[-0.115873t] + 0 \exp[-0.1266666t] \\
&+ 0 \exp[-0.1266666t] + 0 \exp[-0.1266666t] \\
&+ 0 \exp[-0.1266666t] + 0 \exp[-0.1266666t] \\
&+ 0.000064 \exp[-0.1266666t] + 0.007424 \exp[-0.1266666t] \\
&+ 0 \exp[-0.1266666t] + 0 \exp[-0.1266666t] \\
&+ 0 \exp[-0.1266666t] + 0.049664 \exp[-0.115873t] \\
&+ 0.001344 \exp[-0.115873t] + 0 \exp[-0.1266666t] \\
&+ 0 \exp[-0.1266666t] + 0.000896 \exp[-0.115873t] \\
&+ 0.004608 \exp[-0.115873t] + 0 \exp[-0.1266666t] \\
&+ 0 \exp[-0.1266666t] + 0 \exp[-0.1266666t] + 0 \exp[-0.1266666t] + 0 \exp[-0.1266666t] \\
&+ 0.000001 \exp[-0.1271428t] + 0.000116 \exp[-0.1271428t] \\
&+ 0 \exp[-0.1271428t] + 0 \exp[-0.1271428t] \\
&+ 0 \exp[-0.1271428t] + 0.000776 \exp[-0.1163491] + 0 \exp[-0.1163491t] \\
&+ 0 \exp[-0.1271428t] + 0 \exp[-0.1271428t] + 0.000014 \exp[-0.1163491t] \\
&+ 0.000072 \exp[-0.1163491t] + 0 \exp[-0.1271428t] + 0 \exp[-0.1271428t] \\
&+ 0 \exp[-0.1271428t] + 0 \exp[-0.1271428t] + 0 \exp[-0.1271428t] + 0 \exp[-0.1271428t] \\
&+ 0 \exp[-0.1271428t] + 0 \exp[-0.1271428t] + 0 \exp[-0.1271428t] \\
&+ 0 \exp[-0.1271428t] + 0 \exp[-0.1163491t] + 0 \exp[-0.1163491t] \\
&+ 0 \exp[-0.1271428t] + 0 \exp[-0.1271428t] + 0 \exp[-0.1163491t] \\
&+ 0 \exp[-0.1163491t] + 0 \exp[-0.1271428t] + 0 \exp[-0.1271428t] \\
&+ 0 \exp[-0.1271428t] + 0 \exp[-0.1271428t] + 0 \exp[-0.1271428t] \\
&+ 0.000006 \exp[-0.1271428t] + 0.000696 \exp[-0.1271428t] \\
&+ 0 \exp[-0.1271428t] + 0 \exp[-0.1271428t] \\
&+ 0 \exp[-0.1271428t] + 0.004656 \exp[-0.1163491t] + 0.000126 \exp[-0.1163491t] \\
&+ 0 \exp[-0.1271428t] + 0 \exp[-0.1271428t] + 0.000084 \exp[-0.1163491t] \\
&+ 0.000432 \exp[-0.1163491t] + 0 \exp[-0.1271428t] \\
&+ 0 \exp[-0.1271428t] + 0 \exp[-0.1271428t] + 0 \exp[-0.1271428t]
\end{aligned}$$

$$+ 0\exp[-0.1271428t] + 0.000003\exp[-0.1271428t] + 0.000348\exp[-0.1271428t]$$
$$+ 0\exp[-0.1271428t] + 0\exp[-0.1271428t] + 0\exp[-0.1271428t]$$
$$+ 0.002328\exp[-0.1163491t] + 0.000063\exp[-0.1163491t] + 0\exp[-0.1271428t]$$
$$+ 0\exp[-0.1271428t] + 0.000042\exp[-0.1163491t]$$
$$+ 0.000216\exp[-0.1163491t] + 0\exp[-0.1271428t]$$
$$+ 0\exp[-0.1271428t] + 0\exp[-0.1271428t] + 0\exp[-0.1271428t]$$
$$+ 0\exp[-0.1271428t] = 0.87417\exp[-0.115873t]$$
$$+ 0.008809\exp[-0.1163491t] + 0.11583\exp[-0.1266666t]$$
$$+ 0.00117\exp[-0.1271428t], \tag{5.109}$$

$$\boldsymbol{S}^2(t,2) \cong \sum_{l=1}^{96} q_l [\boldsymbol{S}^2(t,2)]^{(l)}$$
$$= 0.000926\exp[-0.1988259t] + 0.107416\exp[-0.1988259t]$$
$$+ 0\exp[-0.1988259t] + 0\exp[-0.1988259t]$$
$$+ 0\exp[-0.1988259t] + 0.718576\exp[-0.181739t]$$
$$+ 0.019446\exp[-0.181739t] + 0\exp[-0.1988259t]$$
$$+ 0\exp[-0.1988259t] + 0.012964\exp[-0.181739t]$$
$$+ 0.066672\exp[-0.181739t] + 0\exp[-0.1988259t]$$
$$+ 0\exp[-0.1988259t] + 0\exp[-0.1988259t] + 0\exp[-0.1988259t]$$
$$+ 0\exp[-0.1988259t] + 0.000064\exp[-0.1988259t] + 0.007424\exp[-0.1988259t]$$
$$+ 0\exp[-0.1988259t] + 0\exp[-0.1988259t]$$
$$+ 0\exp[-0.1988259t] + 0.049664\exp[-0.181739t]$$
$$+ 0.001344\exp[-0.181739t] + 0\exp[-0.1988259t]$$
$$+ 0\exp[-0.1988259t] + 0.000896\exp[-0.181739t]$$
$$+ 0.004608\exp[-0.181739t] + 0\exp[-0.1988259t]$$
$$+ 0\exp[-0.1988259t] + 0\exp[-0.1988259t]$$
$$+ 0\exp[-0.1988259t] + 0\exp[-0.1988259t]$$
$$+ 0.000001\exp[-0.1996955t] + 0.000116\exp[-0.1996955t]$$
$$+ 0\exp[-0.1996955t] + 0\exp[-0.1996955t]$$
$$+ 0\exp[-0.1996955t] + 0.000776\exp[-0.1826085t]$$
$$+ 0\exp[-0.1826085t] + 0\exp[-0.1996955t]$$
$$+ 0\exp[-0.1996955t] + 0.000014\exp[-0.1826085t]$$
$$+ 0.000072\exp[-0.1826085t] + 0\exp[-0.1996955t] + 0\exp[-0.1996955t]$$
$$+ 0\exp[-0.1996955t] + 0\exp[-0.1996955t]$$
$$+ 0\exp[-0.1996955t] + 0\exp[-0.1996955t] + 0\exp[-0.1996955t]$$
$$+ 0\exp[-0.1996955t] + 0\exp[-0.1996955t] + 0\exp[-0.1996955tt]$$
$$+ 0\exp[-0.1826085t] + 0\exp[-0.1826085t] + 0\exp[-0.1996955t]$$
$$+ 0\exp[-0.1996955t] + 0\exp[-0.1826085t] + 0\exp[-0.1826085t]$$
$$+ 0\exp[-0.1996955t] + 0\exp[-0.1996955t] + 0\exp[-0.1996955t]$$
$$+ 0\exp[-0.1996955t] + 0\exp[-0.1996955t] + 0.000006\exp[-0.1996955t]$$
$$+ 0.000696\exp[-0.1996955t] + 0\exp[-0.1996955t] + 0\exp[-0.1996955t]$$

$$
\begin{aligned}
&+ 0\exp[-0.1826085t] + 0.004656\exp[-0.1826085t]\\
&+ 0.000126\exp[-0.1996955t] + 0\exp[-0.1996955t] + 0\exp[-0.1826085t]\\
&+ 0.000084\exp[-0.1826085t] + 0.000432\exp[-0.1996955t]\\
&+ 0\exp[-0.1996955t] + 0\exp[-0.1996955t] + 0\exp[-0.1996955t]\\
&+ 0\exp[-0.1996955t] + 0\exp[-0.1996955t] + 0.000003\exp[-0.1996955t]\\
&+ 0.000348\exp[-0.1996955t] + 0\exp[-0.1996955t] + 0\exp[-0.1996955t]\\
&+ 0\exp[-0.1826085t] + 0.002328\exp[-0.1826085]\\
&+ 0.000063\exp[-0.1996955t] + 0\exp[-0.1996955t] + 0\exp[-0.1826085t]\\
&+ 0.000042\exp[-0.1826085t] + 0.000216\exp[-0.1996955t] + 0\exp[-0.1996955t]\\
&+ 0\exp[-0.1996955t] + 0\exp[-0.1996955t] + 0\exp[-0.1996955t] + 0\exp[-0.1996955t]\\
&= 0.87417\exp[-0.181739t] + 0.008809\exp[-0.1826085t]\\
&+ 0.11583\exp[-0.19969557t] + 0.00117\exp[-0.1988259t], \qquad (5.110)
\end{aligned}
$$

The graph of the three-state oil terminal critical infrastructure safety function is shown in Fig. 5.4.

As the critical safety state is $r = 1$, then by (5.13) and (5.109), the port oil terminal critical infrastructure risk function (SafI2), is given by

$$
\begin{aligned}
\boldsymbol{r}^2(t) &= 1 - \boldsymbol{S}^2(t, 1)\\
&= 1 - [0.87417\exp[-0.115873t] + 0.008809\exp[-0.1163491t]\\
&+ 0.11583\exp[-0.1266666t] \qquad (5.111)
\end{aligned}
$$

The graph of the risk function $\boldsymbol{r}^2(t)$ of the oil terminal critical infrastructure, the fragility curve (SafI3), is shown in Fig. 5.5.

The values of the above risk function are given in Table 5.1.

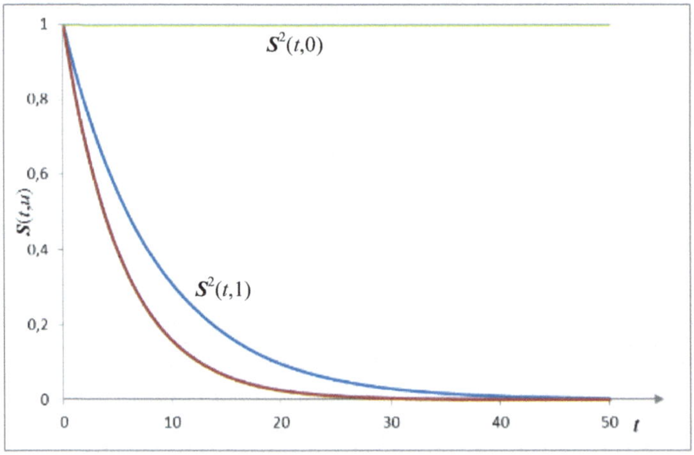


**Fig. 5.4** The graph of the oil terminal critical infrastructure safety function $\boldsymbol{S}^2(t, \cdot)$ coordinates

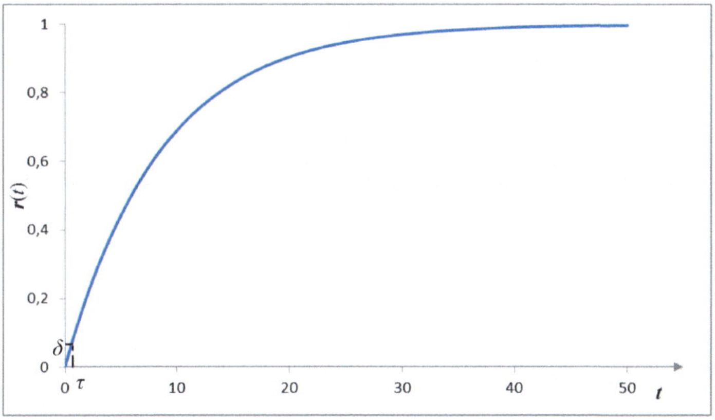


**Fig. 5.5** The graph of the risk function $\boldsymbol{r}^2(t)$ (the fragility curve) of the oil terminal critical infrastructure

**Table. 5.1** The values of the port oil terminal critical infrastructure risk function

| $t$ | $r^2(t) = 1 - S^2(t,1)$ |
|---|---|
| 0.430 | 0.049142 |
| 0.431 | 0.049253 |
| 0.432 | 0.049365 |
| 0.433 | 0.049476 |
| 0.434 | 0.049587 |
| 0.435 | 0.049699 |
| 0.436 | 0.04981 |
| 0.437 | 0.049921 |
| **0.438** | 0.050032 |
| 0.439 | 0.050144 |
| 0.440 | 0.050255 |
| 0.441 | 0.050366 |
| 0.442 | 0.050477 |
| 0.443 | 0.050589 |
| 0.444 | 0.0507 |
| 0.445 | 0.050811 |
| 0.446 | 0.050922 |
| 0.447 | 0.051033 |
| 0.448 | 0.051145 |
| 0.449 | 0.051256 |
| 0.450 | 0.051367 |

From (5.18) and (5.111), and the Table 5.1, the moment when the oil terminal critical infrastructure risk function exceeds a permitted level $\delta = 0.05$ (SafI6), is

$$\tau^2 = r^{2-1}(\delta) \cong 0.438 \text{ years.}$$

Considering (5.69)–(5.74) and (5.109)-(5.110) and applying (5.14), (5.16), (5.17), and (5.19), (5.21)–(5.22), the expected values (SafI4, SafI7) and standard deviations (SafI5, Saf 8) of the port oil terminal critical infrastructure lifetimes in the safety state subsets {1, 2}, {2}, respectively are:

$$\begin{aligned}\boldsymbol{\mu}^2(1) &\cong \int_0^\infty [\boldsymbol{S}^2(t,1)]dt \cong \sum_{l=1}^{96} q_l[\boldsymbol{\mu}^2(1)]^{(l)} \\ &= 0.87417 \cdot 8.63 + 0.008809 \cdot 8.59 + 0.11583 \cdot 7.89 + 0.00117 \cdot 7.87 \\ &\cong 8.54 \text{ years,}\end{aligned} \tag{5.112}$$

$$\begin{aligned}\boldsymbol{\mu}^2(2) \cong \boldsymbol{\mu}^2(1) &\cong \int_0^\infty [\boldsymbol{S}^2(t,2)]dt \cong \sum_{l=1}^{96} q_l[\boldsymbol{\mu}^2(2)]^{(l)} \\ &= 0.87417 \cdot 5.40 + 0.008809 \cdot 5.48 + 0.11583 \cdot 5.03 + 0.00117 \cdot 5.01 \\ &\cong 5.36 \text{ years,}\end{aligned} \tag{5.113}$$

$$\boldsymbol{\sigma}^2(1) \cong 8.55, \boldsymbol{\sigma}^2(2) \cong 5.35 \text{ years.} \tag{5.114}$$

The detailed way of finding the above standard deviations (5.114) is illustrated below.

According to (5.17), (5.22) using (5.109), (5.110)

$$\boldsymbol{n}^2(1) = 2\int_0^\infty t\boldsymbol{S}^2(t,1)dt = 146.10, \tag{5.115}$$

and

$$\boldsymbol{n}^2(2) = 2\int_0^\infty t\boldsymbol{S}^2(t,2)dt = 57.43. \tag{5.116}$$

Hence, after applying (5.16), (5.21) we get

$$\boldsymbol{\sigma}^2(1) = \sqrt{\boldsymbol{n}^2(1) - [\boldsymbol{\mu}^2(1)]^2} = \sqrt{146.10 - [8.54]^2} \cong 8.55,$$
$$\boldsymbol{\sigma}^2(2) = \sqrt{\boldsymbol{n}^2(2) - [\boldsymbol{\mu}^2(2)]^2} = \sqrt{57.43 - [5.36]^2} \cong 5.35.$$

Further, it follows that the mean values of the oil terminal critical infrastructure lifetimes in the particular safety states (SafI9), according (5.23) and using (5.112), (5.113) are:

$$\overline{\mu}^2(1) = \mu^2(1) - \mu^2(2) \cong 3.18, \overline{\mu}^2(2) = \mu^2(2) \cong 5.36 \text{ years}. \tag{5.117}$$

By (5.109), (5.110), applying (5.25), the oil terminal critical infrastructure intensities of ageing (SafI10) are:

$$\begin{aligned}\boldsymbol{\lambda}^2(t,1) &= \frac{-\frac{d\boldsymbol{S}^2(t,1)}{dt}}{\boldsymbol{S}^2(t,1)}\\ &= \{0.115873 \cdot 0.87417 \exp[-0.115873t] + 0.1163491\\ &\cdot 0.008809 \exp[-0.1163491t]\\ &+ 0.1266666 \cdot 0.11583 \exp[-0.1266666t]\\ &+ 0.1271428 \cdot 0.00117 \exp[-0.1271428t]\}\\ &/\{0.87417 \exp[-0.115873t] + 0.008809 \exp[-0.1163491t]\\ &+ 0.11583 \exp[-0.1266666t]\\ &+ 0.00117 \exp[-0.1271428t]\}, t \geq 0,\end{aligned} \tag{5.118}$$

$$\begin{aligned}\boldsymbol{\lambda}^2(t,2) &= \frac{-\frac{d\boldsymbol{S}^2(t,2)}{dt}}{\boldsymbol{S}^2(t,2)}\\ &= \{0.181739 \cdot 0.87417 \exp[-0.181739t] + 0.1826085\\ &\cdot 0.008809 \exp[-0.1826085t]\\ &+ 0.19969557 \cdot 0.11583 \exp[-0.19969557t] + 0.1988259\\ &\cdot 0.00117 \exp[-0.1988259t]\}\\ &/\{0.87417 \exp[-0.181739t] + 0.008809 \exp[-0.1826085t]\\ &+ 0.11583 \exp[-0.19969557t]\\ &+ 0.00117 \exp[-0.1988259t]\}, t \geq 0,\end{aligned} \tag{5.119}$$

From (5.118) and (5.119), the oil terminal critical infrastructure intensities of ageing are:

$$\lim_{t\to\infty} \boldsymbol{\lambda}^2(t,1) \cong \lim_{t} \frac{[0.115873 \cdot 0.87417 \exp[-0.115873t]][1+o(1)]}{[0.87417 \exp[-0.115873t]][1+o(1)]} = 0.115873, \tag{5.120}$$

$$\lim_{t\to\infty} \boldsymbol{\lambda}^2(t,2) \cong \lim_{t} \frac{[0.18173 \cdot 0.87417 \exp[-0.181739t]][1+o(1)]}{[0.87417 \exp[0.181739t]][1+o(1)]} = 0.181739. \tag{5.121}$$

From (5.112) and (5.113), applying (5.35) the approximate mean values of the oil terminal critical infrastructure intensities of ageing (SafI10) are:

$$\boldsymbol{\lambda}^2(t,1) \cong \frac{1}{8.54} \cong 0.11710, \tag{5.122}$$

$$\boldsymbol{\lambda}^2(t,2) \cong \frac{1}{5.36} \cong 0.18657. \tag{5.123}$$

The graphs of the intensities of ageing of the oil terminal critical infrastructure are shown in Fig. 5.6.

Considering (5.120), (5.121), (5.122), (5.123) and the values of the port oil terminal critical infrastructure intensities of ageing without of operation impact, determined in Chap. 3,

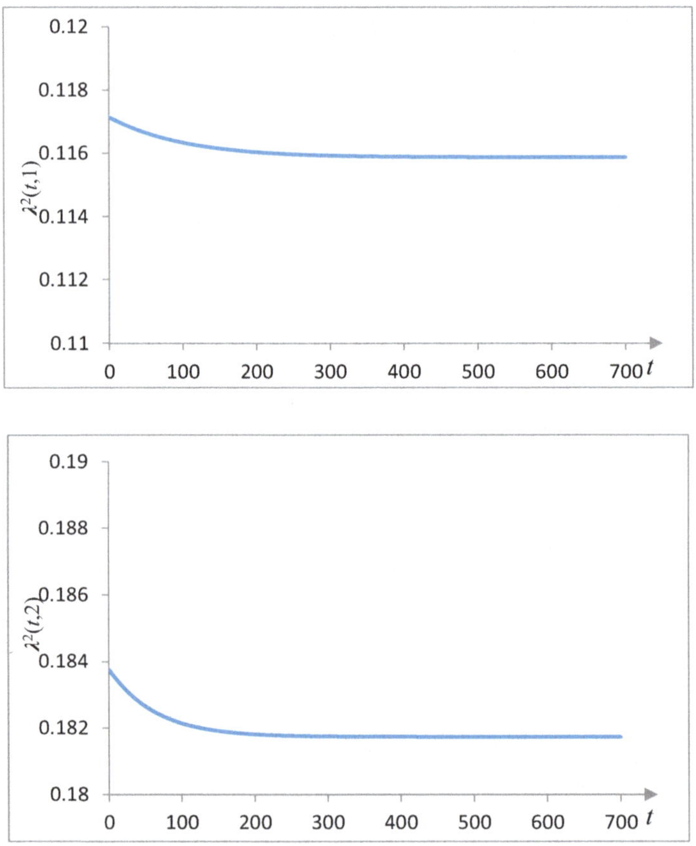


**Fig. 5.6** The graphs of the intensities of ageing of the oil terminal critical infrastructure

$$\boldsymbol{\lambda}^{\mathbf{0}}(t,1) = 0.115873, \boldsymbol{\lambda}^{\mathbf{0}}(t,2) = 0.181739, \tag{5.124}$$

and applying (5.28), the coefficients of the operation process impact on the oil terminal critical infrastructure intensities of ageing (ResI1) are:

$$\boldsymbol{\rho}^2(t,1) = \frac{\boldsymbol{\lambda}^2(t,1)}{\boldsymbol{\lambda}^0(t,1)} \cong \frac{0.11710}{0.115873} \cong 1.01059, \tag{5.125}$$

$$\boldsymbol{\rho}^2(t,1) = \frac{\boldsymbol{\lambda}^2(t,1)}{\boldsymbol{\lambda}^0(t,1)} \cong \frac{0.115873}{0.115873} = 1, \tag{5.126}$$

$$\boldsymbol{\rho}^2(t,2) = \frac{\boldsymbol{\lambda}^2(t,2)}{\boldsymbol{\lambda}^0(t,2)} \cong \frac{0.18657}{0.181739} \cong 1.02658, \tag{5.127}$$

$$\boldsymbol{\rho}^2(t,2) = \frac{\boldsymbol{\lambda}^2(t,2)}{\boldsymbol{\lambda}^0(t,2)} \cong \frac{0.181739}{0.181739} = 1. \tag{5.128}$$

Finally, by (5.31) and (5.126), the port oil terminal critical infrastructure resilence indicator (ResI2), i.e. the coefficient of the port oil terminal critical infrastructure resilience to climate-weather change process impact, is

$$\boldsymbol{RI}^2(t,1) = \frac{1}{\boldsymbol{\rho}^2(t,1)}, \cong 0.990 = 99\%. \quad t \in < 0, +\infty). \tag{5.129}$$

The comparison of safety indicators (3.102)–(3.114) and (5.108)–(5.123) proves a negligible influence of the climate-weather change process on the port oil terminal safety what is also clearly expressed in its resilience indicators to climate weather change process impact given by (5.125)–(5.129).

The comparison of safety indicators (4.124)–(4.136) and (5.108)–(5.123) proves a significantly larger influence of the operation process then the influence of the climate-weather change process on the port oil terminal safety what is also clearly expressed in its resilience indicators to operation process impact given by (4.142)–(4.144) and in its resilience indicators to climate-weather change process impact given by (5.125)–(5.129).

## References


1. Ferreira F, Pacheco A (2007) Comparison of level-crossing times for Markov and semi-Markov processes. Stat Probab Lett 77(2):151–157
2. Glynn PW, Haas PJ (2006) Laws of large numbers and functional central limit theorems for generalized semi-Markov processes. Stoch Model 22(2):201–231
3. Grabski F (2002) Semi-Markov models of systems reliability and operations analysis. System Research Institute, Polish Academy of Science (in Polish)

4. Jakusik E, Kołowrocki K, Kuligowska E, Soszyńska-Budny J, Torbicki M (2016) Modelling climate-weather change process including extreme weather hazards for port oil piping transportation system. J Pol Saf Reliab Assoc Summer Saf Reliab Seminars 7(3):31–40
5. Jakusik E, Kołowrocki K, Kuligowska E, Soszyńska-Budny J, Torbicki M (2016) Identification methods and procedures of climate-weather change process including extreme weather hazards for port oil piping transportation system operating at under water Baltic Sea area. J Pol Saf Reliab Assoc Summer Saf Reliab Seminars 7(3):47–56
6. Jakusik E, Kołowrocki K, Kuligowska E, Soszyńska-Budny J, Torbicki M (2016) Identification methods and procedures of climate-weather change process including extreme weather hazards of port oil piping transportation system operating at land Baltic seaside area. J Pol Saf Reliab Assoc Summer Saf Reliab Seminars 7(3):57–64
7. Kołowrocki K (2005) Reliability of large systems. Elsevier Amsterdam, Boston, Heidelberg, London, New York, Oxford, Paris, San Diego, San Francisco, Singapore, Sydney, Tokyo
8. Kołowrocki K (2014) Reliability of large and complex systems, 2nd Edition, Elsevier Amsterdam, Boston, Heidelberg, London, New York, Oxford, Paris, San Diego, San Francisco, Singapore, Sydney, Tokyo
9. Kołowrocki K, Kuligowska E, Soszyńska-Budny J, Torbicki M (2018) Oil transport in port part 1 port oil piping transportation system safety and resilience impacted by its operation process. J Pol Saf Reliab Assoc Summer Saf Reliab Seminars 9(2):11–28
10. Kołowrocki K, Soszyńska-Budny J (2011) Reliability and safety of complex technical systems and processes. Modeling—identification—prediction—optimization. London, Dordrecht, Heildeberg, New York, Springer
11. Kołowrocki K, Soszyńska-Budny J (2015) Modelling safety of multistate systems with ageing components. In: Proceedings the international conference of numerical analysis and applied mathematics—ICNAAM 2015 symposium/workshop on safety of critical infrastructures, ISBN: 978-0-7354-1392-4, https://doi.org/10.1063/1.4952224
12. Kołowrocki K, Soszyńska-Budny J (2016) Modelling critical infrastructure operation process including operating environment threats. J Pol Saf Reliab Assoc Summer Saf Reliab Seminars 7(3):81–88
13. Kołowrocki K Soszyńska-Budny J (2018) Critical Infrastructure Safety Indicators. International Conference on Industrial Engineering and Engineering Management IEEM 2018
14. Kołowrocki K Soszyńska-Budny J (2018) critical infrastructure impacted by operation safety and resilience indicators. In: International conference on industrial engineering and engineering management, IEEM
15. Kołowrocki K, Soszyńska-Budny J, Torbicki M (2017) Critical infrastructure operating area climate-weather change process including extreme weather hazards. J Pol Saf Reliab Assoc Summer Saf Reliab Seminars 8(2):15–24
16. Kołowrocki K, Soszyńska-Budny J, Torbicki M (2017) Identification methods and procedures of climate-weather change process including extreme weather hazards. J Pol Saf Reliab Assoc Summer Saf Reliab Seminars 8(2):85–106
17. Kołowrocki K, Soszyńska-Budny J, Torbicki M (2017) Identification and prediction of port oil piping transportation system operation process related to climate-weather change. J Pol Saf Reliab Assoc Summer Saf Reliab Seminars 8(2):113–120
18. Kołowrocki K, Soszyńska-Budny J, Torbicki M (2017) Integrated impact model on critical infrastructure safety related to climate-weather change proces including extreme weather hazard. J Pol Saf Reliab Assoc Summer Saf Reliab Seminars 8(4):21–32
19. Kołowrocki K, Soszyńska-Budny J, Torbicki M (2018) Critical infrastructure impacted by operation and climate change safety and resilience indicators. In: International conference on industrial engineering and engineering management, IEEM
20. Limnios N, Oprisan G (2005) Semi-Markov processes and reliability. Birkhauser, Boston
21. Mercier S (2008) Numerical bounds for semi-Markovian quantities and application to reliability. Methodol Comput Appl Probab 10(2):179–198
22. Torbicki M (2019) Safety of a critical network infrastructure exposed to operation and weather condition changes. Ph.D. thesis, Polish Academy of Science, System Research Institute, Warsaw

# Chapter 6
# Modelling Critical Infrastructure Safety Impacted by Operation Process and Climate-Weather Change Process

The safety of critical infrastructure impacted by its operation process and climate-weather change process at its operating area is considered. The critical infrastructure operation process related to climate-weather change process at its operating area and its parameters are defined and its characteristics are determined. The safety and resilience indicators of the critical infrastructure and its assets impacted by the operation process related to climate-weather change process are proposed. The safety of the port oil terminal critical infrastructure impacted by its operation process related to climate-weather change process at its operating area is examined and its safety and resilience indicators are determined.

## 6.1 Critical Infrastructure Operation Process Related to Climate-Weather Change Process at Its Operating Area

We consider the critical infrastructure impacted by its operation process $ZC(t), t \in< 0, \infty)$ related to the climate-weather change process at its operating area in a various way at this process states $zc_{bl}$, $b = 1, 2, \ldots, v$, $l = 1, 2, \ldots, w$. We assume that the changes of the states of operation process related to the climate-weather change process $ZC(t)$, $t \in< 0, \infty)$, at the critical infrastructure operating area have an influence on the critical infrastructure safety structure and on the safety parameters of the critical infrastructure assets $A_i$, $i = 1, 2, \ldots, n$, as well [9, 10, 12, 21].

We assume, as in Chap. 4, that the critical infrastructure during its operation process is taking $v$, $v \in N$, different operation states $z_1, z_2, \ldots z_v$. We define the critical infrastructure operation process $Z(t)$, $t \in< 0, \infty)$, with discrete operation



J. Soszyńska-Budny, *Safety Analysis of Critical Infrastructure*,
Lecture Notes in Intelligent Transportation and Infrastructure,
https://doi.org/10.1007/978-3-030-65339-2_6

states from the set $\{z_1, z_2, \ldots z_v\}$. Moreover, we assume that the critical infrastructure operation process $Z(t)$ is a semi-Markov process that can be described by the following parameters [16]:

- the number of operation states $v$;
- the operation states $\{z_1, z_2, \ldots, z_v\}$;
- the vector $[p_b(0)]_{1xv}$ of the initial probabilities $p_b(0)$, $b = 1, 2, \ldots, v$, of the critical infrastructure operation process $Z(t)$ staying at particular operation state $z_b$, $b = 1, 2, \ldots, v$, at the moment $t = 0$;
- the matrix $[p_{bl}]_{1xv}$ of probabilities $p_{bl}$, $b, l = 1, 2, \ldots, v$, of the critical infrastructure operation process $Z(t)$ transitions between the operation states $z_b$ and $z_l$, $b, l = 1, 2, \ldots, v$;
- the matrix $[H_{bl}]_{1xv}$ of conditional distribution functions $H_{bl}(t)$, $b, l = 1, 2, \ldots, v$, of the critical infrastructure operation process $Z(t)$ conditional sojourn times $\theta_{bl}$, at the operation states $z_b$ under the condition that the next operation state will be $z_l$, $b, l = 1, 2, \ldots, v$.

Further, we assume that we have either calculated analytically using the above parameters of the operation process or evaluated approximately by experts the vector of limit values (OPC1)

$$[p_b]_{1xv} = [p_1, p_2, \ldots p_v],$$

of transient probabilities

$$p_b(t) = P(Z(t) = Z_b), t \in < 0, \infty), b = 1, 2, \ldots, v,$$

of the critical infrastructure operation process $Z(t)$ at the particular operation states $z_b$, $b = 1, 2, \ldots, v$.

Moreover, as in Chap. 5, we assume that the climate-weather change process $C(t)$, $t \in < 0, \infty)$, at the critical infrastructure operating area is taking $w$, $w \in N$, different climate-weather states $c_1, c_2, \ldots, c_w$. We assume that the climate-weather change process $C(t)$ is a semi-Markov process and it can be described by [16]:

- the number of climate-weather states $w$;
- the climate-weather states $\{c_1, c_2, \ldots, c_w\}$;
- the vector $[q_b(0)]_{1\times w}$ of the initial probabilities $q_b(0)$, $b = 1, 2, \ldots, w$, of the climate-weather change process $C(t)$ staying at particular climate-weather states $c_b$, $b = 1, 2, \ldots, w$, at the moment $t = 0$;
- the matrix $[q_{bl}]_{w\times w}$ of the probabilities $q_{bl}$, $b, l = 1, 2, \ldots, w$, of transitions of the climate-weather change process $C(t)$ from the climate-weather states $c_b$ to the climate-weather state $c_l$, $b, l = 1, 2, \ldots, w$;
- the matrix $[C_{bl}(t)]_{wxw}$ of the conditional distribution functions $C_{bl}(t)$, $b, l = 1, 2, \ldots, w$, of the conditional sojourn times $C_{bl}$ at the climate-weather states $c_b$ when its next climate-weather state is $c_l$, $b, l = 1, 2, \ldots, w$.

Further, we assume that we have either calculated analytically using the above parameters of the climate-weather change process or evaluated approximately by experts the vector of limit values (C-WCPC1)

$$[q_l]_{1xw} = [q_1, q_2, \ldots q_w],$$

of transient probabilities

$$q_l(t) = P(C(t) = c_l), t \in< 0, \infty), l = 1, 2, \ldots, w,$$

of the climate-weather change process $C(t)$ at the particular climate-weather states $c_l$, $l = 1, 2, \ldots, w$.

### 6.1.1 Joint Model of Critical Infrastructure Independent Operation Process and Climate-Weather Change Process

Under the assumption that the critical infrastructure operation process $Z(t)$, $t \in< 0, \infty)$, and the climate-weather change process $C(t)$, $t \in< 0, \infty)$, at its operating area are independent, we introduce the joint process of critical infrastructure operation process and climate-weather change process called the critical infrastructure operation process related to climate-weather change marked by [16]

$$ZC(t), ), t \in< 0, \infty), \tag{6.1}$$

and we assume that it can take $vw$, $vw \in N$, different operation and climate-weather states

$$zc_{11}, zc_{12}, \ldots, zc_{vw}, b = 1, 2, \ldots, v, l = 1, 2, \ldots, w, \tag{6.2}$$

We assume that the critical infrastructure operation process related to climate-weather change $ZC(t)$, at the moment $t \in< 0, \infty)$, is at the operation and climate-weather state $zc_{bl}$, $b = 1, 2, \ldots, v$, $l = 1, 2, \ldots, w$, if and only if at that moment $t$, the operation process $Z(t)$ is at the operation states $z_b$, $b = 1, 2, \ldots, v$, and the climate-weather change process $C(t)$ is at the climate-weather state $c_l$, $l = 1, 2, \ldots, w$, at the moment $t \in< 0, \infty)$, what we express as follows [16]:

$$(ZC(t) = zc_{bl})\tau \leftrightarrow ((Z(t) = z_b) \cap (C(t) = c_l)),$$
$$\times t \in< 0, \infty), b = 1, 2, \ldots, v, l = 1, 2, \ldots, w. \tag{6.3}$$

Further, we define the initial probabilities [16]

$$pq_{bl}(0) = P(ZC(0) = zc_{bl}), b = 1, 2, \ldots, v, l = 1, 2, \ldots, w, \tag{6.4}$$

of the critical infrastructure operation process related to climate-weather change $ZC(t)$, at the initial moment $t = 0$ at the operation and climate-weather state $zc_{bl}$, $b = 1, 2, \ldots, v$, $l = 1, 2, \ldots, w$, and this way we have the vector

$$[pq_{bl}(0)]_{1xvw} = [pq_{11}(0), pq_{12}(0), \ldots, pq_{1w}(0); pq_{21}(0), pq_{22}(0), \ldots, pq_{2w}(0); \ldots; pq_{v1}(0), pq_{v2}(0), \ldots, pq_{vw}(0)] \tag{6.5}$$

of the initial probabilities the critical infrastructure operation process related to climate-weather change $ZC(t)$ staying at the particular operation and climate-weather states at the initial moment $t = 0$.

From the assumption that the critical infrastructure operation process $Z(t)$ and climate-weather change process $C(t)$ are independent, it follows that [16]

$$\begin{aligned} pq_{bl}(0) &= P(ZC(0) = zc_{bl}) = P((Z(0) = z_b) \cap (C(0) = c_l)) \\ &= P(Z(0) = z_b) \cdot P(C(0) = c_l) = p_b(0) \cdot q_l(0), \\ &\quad b = 1, 2, \ldots, v, l = 1, 2, \ldots, w, \end{aligned} \tag{6.6}$$

where $p_b(0)$, $b = 1, 2, \ldots, v$, and $q_l(0)$, $l = 1, 2, \ldots, w$, are the critical infrastructure initial probabilities of the operation process and the critical infrastructure initial probabilities of the climate-weather change process at its operating area, respectively introduced in Chaps. 4 and 5.

Hence, the vector of the initial probabilities of the critical infrastructure operation process related to climate-weather change $ZC(t)$ defined by (6.5) take the following form [16]

$$[pq_{bl}(0)]_{1xvw} = [p_i(0)q_j(0)]_{1xvw} = [p_1(0)q_1(0), p_1(0)q_2(0), \ldots, p_1(0)q_w(0); \ldots; p_v(0)q_1(0), p_v(0)q_2(0), \ldots, p_v(0)q_w(0)]. \tag{6.7}$$

Further, we introduce the probabilities

$$pq_{bl\,mn}, b = 1, 2, \ldots, v, l = 1, 2, \ldots, w, m = 1, 2, \ldots, v, n = 1, 2, \ldots, w, \tag{6.8}$$

of the transitions of the critical infrastructure operation process related to climate-weather change $ZC(t)$ between the operation and climate-weather states

$$zc_{bl} \text{ and } zc_{mn}, b = 1, 2, \ldots, v, l = 1, 2, \ldots, w, m = 1, 2, \ldots, v, n = 1, 2, \ldots, w, \tag{6.9}$$

and get their following matrix form [16]

$$[pq_{blmn}]_{vwxvw} = \begin{bmatrix} pq_{11\,11}\, pq_{11\,12} \cdots pq_{11\,1w};\ pq_{11\,21}\, pq_{11\,22} \cdots pq_{11\,2w};\ \ldots;\ pq_{11\,v1}\, pq_{11\,v2} \cdots pq_{11\,vw} \\ pq_{12\,11}\, pq_{12\,12} \cdots pq_{12\,1w};\ pq_{12\,21}\, pq_{12\,22} \cdots pq_{12\,2w};\ \ldots;\ pq_{12\,v1}\, pq_{12\,v2} \cdots pq_{12\,vw} \\ \cdots \\ pq_{vw\,11}\, pq_{vw\,12} \cdots pq_{vw\,1w};\ pq_{vw\,21}\, pq_{vw\,22} \cdots pq_{vw\,2w};\ \ldots;\ pq_{vw\,v1}\, pq_{vw\,v2} \cdots pq_{vw\,vw} \end{bmatrix}. \tag{6.10}$$

From the assumption that the critical infrastructure operation process Z(t) and climate-weather change process $C(t)$ are independent, it follows that [16]

$$pq_{bl\,mn} = p_{bm} q_{ln},\ b = 1, 2, \ldots, v,\ l = 1, 2, \ldots, w,$$
$$m = 1, 2, \ldots, v,\ n = 1, 2, \ldots, w, \tag{6.11}$$

where

$$p_{bm},\ b = 1, 2, \ldots, v,\ m = 1, 2, \ldots, v, \text{ and } q_{ln},\ l = 1, 2, \ldots, w,\ n = 1, 2, \ldots, w, \tag{6.12}$$

are the probabilities of critical infrastructure operation process transitions between the operation states and the probabilities of critical infrastructure climate-weather change process transitions between climate-weather states, respectively defined in Chaps. 4 and 5.

Hence, the matrix of the probabilities of transitions between the critical infrastructure operation process related to climate-weather change $ZC(t)$ defined by (6.1), (6.2) takes the following form [16]

$$[pq_{bl\,mn}]_{vwxvw} = \begin{bmatrix} p_{11}q_{11}\, p_{11}q_{12} \cdots p_{11}q_{1w};\ p_{12}q_{11}\, p_{12}q_{12} \\ p_{11}q_{21}\, p_{11}q_{22} \cdots p_{11}q_{2w};\ p_{12}q_{21}\, p_{12}q_{22} \\ \cdots \\ p_{v1}q_{w1}\, p_{v1}q_{w2} \cdots p_{v1}q_{ww};\ p_{v2}q_{w1}\, p_{v2}q_{w2} \end{bmatrix}$$
$$\begin{matrix} \cdots p_{12}q_{1w};\ \ldots;\ p_{1v}q_{11}\, p_{1v}q_{12} \cdots p_{1v}q_{1w} \\ \cdots p_{12}q_{2w};\ \ldots;\ p_{1v}q_{21}\, p_{1v}q_{22} \cdots p_{1v}q_{2w} \\ \\ \cdots p_{v2}q_{ww};\ \ldots;\ p_{vv}q_{w1}\, p_{vv}q_{w2} \cdots p_{vv}q_{ww} \end{matrix}\ \Bigg]. \tag{6.13}$$

The matrix of conditional distribution functions [16]

$$HC_{bl\,mn}(t) = P(\theta C_{bl\,mn} < t),\ t \in <0, \infty),$$
$$b = 1, 2, \ldots, v,\ l = 1, 2, \ldots, w,\ m = 1, 2, \ldots, v,\ n = 1, 2, \ldots, w, \tag{6.14}$$

of the critical infrastructure operation process related to climate-weather change $ZC(t)$ conditional sojourn times $\theta C_{bl\,mn}$, $b = 1, 2, \ldots, v$, $l = 1, 2, \ldots, w$, $m = 1, 2, \ldots, v$, $n = 1, 2, \ldots, w$, at the operation and climate-weather state $zc_{bl}$, $b = 1, 2, \ldots, v$, $l = 1, 2, \ldots, w$, when the next operation and climate-weather state is $zc_{mn}$, $m = 1, 2, \ldots, v$, $n = 1, 2, \ldots, w$, takes the following form [16]

$$[HC_{bl\ mn}(t)]_{vwxvw} = \begin{bmatrix} HC_{11\ 11}(t)HC_{11\ 12}(t)\ldots HC_{11\ 1w}(t);\ HC_{11\ 21}(t) & HC_{11\ 22}(t)\ldots HC_{112w}(t);\ldots;\ HC_{11\ v1}(t)HC_{11\ v2}(t)\ldots HC_{11\ vw}(t) \\ HC_{12\ 11}(t)HC_{12\ 12}(t)\ldots HC_{12\ 1w}(t);\ HC_{12\ 21}(t) & HC_{12\ 22}(t)\ldots HC_{12\ 2w}(t);\ldots;\ HC_{12\ v1}(t)HC_{12\ v2}(t)HC_{12\ vw}(t) \\ \ldots & \\ HC_{vw\ 11}(t)HC_{vw\ 12}(t)\ldots HC_{vw\ 1w}(t);\ HC_{vw\ 21}(t) & HC_{vw\ 22}(t)\ldots HC_{vw\ 2w}(t);\ldots;\ HC_{vw\ v1}(t)HC_{vw\ v2}(t)\ldots HC_{vw\ vw}(t) \end{bmatrix} \quad (6.15)$$

and the matrix of their corresponding conditional density functions

$$hc_{bl\ mn}(t) = \frac{d}{dt}[HC_{bl\ mn}(t)] \text{ for } t \in <0,\infty)$$
$$b = 1,2,\ldots,v,\ l = 1,2,\ldots,w,\ m = 1,2,\ldots,v,\ n = 1,2,\ldots,w, \quad (6.16)$$

has the form [16]

$$[hc_{bl\ mn}(t)]_{vwxvw} = \begin{bmatrix} hc_{11\ 11}(t)hc_{11\ 12}(t)\ldots hc_{11\ 1w}(t);\ hc_{11\ 21}(t)hc_{11\ 22}(t) & \ldots hc_{11\ 2w}(t);\ldots;\ hc_{11\ v1}(t)hc_{11\ v2}(t)\ldots hc_{11\ vw}(t) \\ hc_{12\ 11}(t)hc_{12\ 12}(t)\ldots hc_{12\ 1w}(t);\ hc_{1221}(t)hc_{12\ 22}(t) & \ldots hc_{12\ 2w}(t);\ldots;\ hc_{12\ v1}(t)hc_{12\ v2}(t)hc_{12\ vw}(t) \\ \ldots & \\ hc_{vw\ 11}(t)hc_{vw\ 12}(t)\ldots hc_{vw\ 1w}(t);\ hc_{vw21}(t)hc_{vw\ 22}(t) & \ldots hc_{vw\ 2w}(t);\ldots;\ hc_{vw\ v1}(t)hc_{vw\ v2}(t)\ldots hc_{vw\ vw}(t) \end{bmatrix} \quad (6.17)$$

From the assumption that the critical infrastructure operation process $Z(t)$ and climate-weather change process $C(t)$ are independent, it follows that [16]

$$HC_{bl\ mn}(t) = P(\theta C_{bl\ mn} < t) = P((\theta_{bm} < t) \cap (C_{ln} < t)) = H_{bm}(t)C_{ln}(t), \quad (6.18)$$

$$t \in <0,\infty),\ b = 1,2,\ldots,v,\ l = 1,2,\ldots,w,\ m = 1,2,\ldots,v,\ n = 1,2,\ldots,w,$$

and

$$hc_{bl\ mn}(t) = \frac{d}{dt}[HC_{bl\ mn}(t)] = \frac{d}{dt}[H_{bm}(t)C_{ln}(t)] = h_{bm}(t)C_{ln}(t) + H_{bm}(t)c_{ln}(t), \quad (6.19)$$

$$t \in <0,\infty),\ b = 1,2,\ldots,v,\ l = 1,2,\ldots,w,\ m = 1,2,\ldots,v,\ n = 1,2,\ldots,w,$$

where

$$H_{bm}(t),\ b = 1,2,\ldots,v,\ m = 1,2,\ldots,v, \text{ and}$$
$$C_{ln}(t),\ l = 1,2,\ldots,w\ n = 1,2,\ldots,w, \quad (6.20)$$

are conditional distribution functions of the critical infrastructure operation process lifetimes at the operation states and conditional distribution functions of the climate-weather change process lifetimes at climate-weather states and

$$h_{bm}(\mathrm{t}), b = 1, 2, \ldots, v, m = 1, 2, \ldots, v, \text{ and}$$
$$c_{ln}(\mathrm{t}) l = 1, 2, \ldots, w\ n = 1, 2, \ldots, w, \tag{6.21}$$

are conditional density functions correspond to them, respectively defined in [10].

Hence, the matrix of the conditional distribution functions and the matrix of the conditional density functions of the critical infrastructure operation process related to climate-weather change $ZC(t)$ conditional sojourn times defined by (6.15) respectively take the following forms [16]

$$[HC_{bl\ mn}(t)]_{vwxvw} = [H_{bm}(t)C_{ln}(t)]_{vwxvw}$$
$$= \begin{bmatrix} H_{11}(t)C_{11}(t)H_{11}(t)C_{12}(t)\ldots H_{11}(t)C_{1w}(t);\ldots;H_{1v}(t)C_{11}(t)H_{1v}(t)C_{12}(t)\ldots H_{1v}(t)C_{1w}(t) \\ H_{11}(t)C_{12}(t)H_{11}(t)C_{22}(t)\ldots H_{11}(t)C_{2w}(t);\ldots;H_{1v}(t)C_{21}(t)H_{1v}(t)C_{22}(t)\ldots H_{1v}(t)C_{1w}(t) \\ \ldots \\ H_{v1}(t)C_{w1}(t)H_{v1}(t)C_{w2}(t)\ldots H_{v1}(t)C_{ww}(t);\ldots;H_{vv}(t)C_{w1}(t)H_{vv}(t)C_{w2}(t)\ldots H_{vv}(t)C_{ww}(t) \end{bmatrix} \tag{6.22}$$

and

$$[hc_{blmn(t)}]_{vwxvw} = [h_{bm}(t)C_{ln}(t) + H_{bm}(t)c_{ln}(t)]_{vwxvw} \begin{bmatrix} h_{11}(t)C_{11}(t) + H_{11}(t)c_{11}(t)\ldots \\ h_{11}(t)C_{21}(t) + H_{11}(t)c_{21}(t)\ldots \\ \ldots \\ h_{v1}(t)C_{w1}(t) + H_{v1}(t)c_{w1}(t)\ldots \end{bmatrix}$$
$$\begin{matrix} h_{11}(t)C_{1w}(t) + H_{11}(t)c_{1w}(t);\ldots;h_{1v}(t)C_{11}(t) + H_{1v}(t)c_{11}(t)\ldots h_{1v}(t)C_{1w}(t) + H_{1v}(t)c_{1w}(t) \\ h_{11}(t)C_{1w}(t) + H_{11}(t)c_{2w}(t);\ldots;h_{1v}(t)C_{21}(t) + H_{1v}(t)c_{21}(t)\ldots h_{1v}(t)C_{2w}(t) + H_{1v}(t)c_{2w}(t) \\ \\ h_{v1}(t)C_{ww}(t) + H_{v1}(t)c_{ww}(t);\ldots;h_{vv}(t)C_{w1}(t) + H_{vv}(t)c_{w1}(t)\ldots h_{vv}(t)C_{ww}(t) + H_{vv}(t)c_{ww}(t) \end{matrix}\Bigg]. \tag{6.23}$$

We assume that the suitable and typical distributions to describe the critical infrastructure operation process related to climate-weather change $ZC(t)$ conditional sojourn times $\theta C_{bl\ mn}$, $b = 1, 2, \ldots, v$, $l = 1, 2, \ldots, w$, $m = 1, 2, \ldots, v$, $n = 1, 2, \ldots, w$, at the particular states are of the same kind as those distinguished in [10] for the critical infrastructure operation process $Z(t)$ conditional sojourn times $\theta_{bl}$, $b = 1, 2, \ldots, v$, $l = 1, 2, \ldots, w$.

### 6.1.2 Joint Model of Critical Infrastructure Dependent Operation Process and Climate-Weather Change Process

Under the assumption that the critical infrastructure operation process $Z(t)$, $t \in < 0, \infty)$, and the climate-weather change process $C(t)$ at its operating area are dependent, we introduce the joint process of critical infrastructure operation process and climate-weather change process called the critical infrastructure operation process related to climate-weather change marked by [16]

$$ZC(t), t \in< 0, \infty), \tag{6.24}$$

and we assume that it can take $vw$, $v$, $w \in N$, different operation and climate-weather states

$$zc_{11}, zc_{12}, \ldots, zc_{vw} b = 1, 2, \ldots, v, l = 1, 2, \ldots, w. \tag{6.25}$$

We assume that the critical infrastructure operation process related to climate-weather change $ZC(t)$, at the moment $t \in< 0, \infty)$, is at the operation and climate-weather state $zc_{bl}$, $b = 1, 2, \ldots, v$, $l = 1, 2, \ldots, w$, if and only if at that moment $t$, the operation process $Z(t)$ is at the operation states $z_b$, $b = 1, 2, \ldots, v$, and the climate-weather change process $C(t)$ is at the climate-weather state $c_l$, $l = 1, 2, \ldots, w$, what we express as follows [16]:

$$(ZC(t) = zc_{bl}) \leftrightarrow ((Z(t) = z_b) \cap (C(t) = c_l)),$$
$$t \in< 0, \infty), b = 1, 2, \ldots, v, l = 1, 2, \ldots, w. \tag{6.26}$$

Further, we define the initial probabilities

$$pq_{bl}(0) = P(ZC(0) = zc_{bl}), b = 1, 2, \ldots, v, l = 1, 2, \ldots, w, \tag{6.27}$$

of the critical infrastructure operation process related to climate-weather change $ZC(t)$, at the initial moment $t = 0$ at the operation and climate-weather state $zc_{bl}$, $b = 1, 2, \ldots, v \in N$, $l = 1, 2, \ldots, w$, and this way we have the vector [16]

$$[pq_{bl}(0)]_{1xvw} = [pq_{11}(0), pq_{12}(0), \ldots pq_{1w}(0); pq_{21}(0), pq_{22}(0), \ldots pq_{2w}(0);$$
$$\ldots; pq_{v1}(0), pq_{v2}(0), \ldots, pq_{vw}(0),] \tag{6.28}$$

of the initial probabilities of the critical infrastructure operation process related to climate-weather change $ZC(t)$ staying at the particular operation and climate-weather state at the initial moment $t = 0$.

In the case when the processess $Z(t)$ and $C(t)$ are dependent the initial probabilities existing in (6.28) can be express either by [16]

$$\begin{aligned} pq_{bl}(0) &= P(ZC(0) = zc_{bl}) = P((Z(0) = z_b) \cap (C(0) = c_l)) \\ &= P(Z(0) = z_b) \cdot P(C(0) = c_l | Z(0) = z_b) \\ &= p_b(0) \cdot q_{l/b}(0), b = 1, 2, \ldots, v, l = 1, 2, \ldots, w, \end{aligned} \tag{6.29}$$

where

$$p_b(0) = P(Z(0) = z_b), b = 1, 2, \ldots, v, \tag{6.30}$$

are the initial probabilities of the operation process $Z(t)$ at the operation state $z_b$ defined in Chap. 4 and

$$q_{l/b}(0) = P(C(0) = c_l | Z(0) = z_b), b = 1, 2, \ldots, v, l = 1, 2, \ldots, w, \quad (6.31)$$

are conditional initial probabilities of the climate-weather change process $C(t)$ at the climate-weather state corresponding to those defined in Chap. 5 in case they are not conditional or by [16]

$$\begin{aligned} q_{bl}(0) &= P(ZC(0) = zc_{bl}) = P((Z(0) = z_b) \cap (C(0) = c_l)) \\ &= P(C(0) = c_l) \cdot P(Z(0) = z_b | C(0) = c_l) \\ &= q_l(0) \cdot p_{b/l}(0), b = 1, 2, \ldots, v, l = 1, 2, \ldots, w, \end{aligned} \quad (6.32)$$

where

$$q_l(0) = P(C(0) = c_l, l = 1, 2, \ldots, w, \quad (6.33)$$

are initial probabilities of the climate-weather change process $C(t)$ at the climate-weather state $c_l$ defined in Chap. 5 and

$$p_{b/l}(0) = P(Z(0) = z_b | C(0) = c_l), b = 1, 2, \ldots, v, l = 1, 2, \ldots, w, \quad (6.34)$$

are conditional initial probabilities of the operation process $Z(t)$ at the operation state corresponding to those defined in Chap. 4 in case they are not conditional.

Further, we introduce the probabilities

$$pq_{bl\ mn}, b = 1, 2, \ldots, v, l = 1, 2, \ldots, w, m = 1, 2, \ldots, v, n = 1, 2, \ldots, w, \quad (6.35)$$

of the transitions of the critical infrastructure operation process related to climate-weather change $ZC(t)$ between the operation and climate-weather states

$$zc_{bl} \text{ and } zc_{mn}, b = 1, 2, \ldots, v, l = 1, 2, \ldots, w, m = 1, 2, \ldots, v, n = 1, 2, \ldots, w, \quad (6.36)$$

and get their following matrix form [16]

$$[pq_{bl\ mn}]_{vwxvw} = \begin{bmatrix} pq_{11\ 11}\, pq_{11\ 12} \cdots pq_{11\ 1w};\ pq_{11\ 12}\, pq_{11\ 22} \cdots pq_{11\ 2w}; \cdots;\ pq_{11\ v1}\, pq_{11\ v2} \cdots pq_{11\ vw} \\ pq_{12\ 11}\, pq_{12\ 12} \cdots pq_{12\ 1w};\ pq_{12\ 21}\, pq_{12\ 12} \cdots pq_{12\ 2w}; \cdots;\ pq_{12\ v1}\, pq_{12\ v2} \cdots pq_{12\ vw} \\ \cdots \\ pq_{vw\ 11}\, pq_{vw\ 12} \cdots pq_{vw\ 1w};\ pq_{vw\ 21}\, pq_{vw\ 22} \cdots pq_{vw\ 2w}; \cdots;\ pq_{vw\ v1}\, pq_{vw\ v2} \cdots pq_{vw\ vw} \end{bmatrix}. \quad (6.37)$$

In the case when the processess $Z(t)$ and $C(t)$ are dependent the probabilities of transitions between the operation and climate-weather states existing in (6.37) can be express either by [16]

$$pq_{bl\,mn} = p_{bm} \cdot q_{ln/bm},\, b = 1, 2, \ldots, v,$$
$$l = 1, 2, \ldots, w,\, m = 1, 2, \ldots, v,\, n = 1, 2, \ldots, w, \tag{6.38}$$

where

$$p_{bm},\, b = 1, 2, \ldots, v,\, m = 1, 2, \ldots, v, \tag{6.39}$$

are transient probabilities between the operation states of the operation process $Z(t)$ defined in Chap. 4 and

$$q_{ln/bm},\, b = 1, 2, \ldots, v,\, l = 1, 2, \ldots, w,\, m = 1, 2, \ldots, v,\, n = 1, 2, \ldots, w, \tag{6.40}$$

are conditional transient probabilities between the climate-weather states of the climate-weather change process $C(t)$ coressponding to those defined in Chap. 5 in case they are not conditional or by [16]

$$pq_{bl\,mn} = q_{ln} \cdot p_{bm/ln},\, b = 1, 2, \ldots, v,$$
$$l = 1, 2, \ldots, w,\, m = 1, 2, \ldots, v,\, n = 1, 2, \ldots, w, \tag{6.41}$$

where

$$q_{ln},\, l = 1, 2, \ldots, w,\, n = 1, 2, \ldots, w, \tag{6.42}$$

are transient probabilities beetwen the climate-weather states of the climate-weather change process $C(t)$ defined in Chap. 5 and

$$p_{bm/ln},\, p_{bm/ln} b = 1, 2, \ldots, v,\, l = 1, 2, \ldots, w,\, m = 1, 2, \ldots, v,\, n = 1, 2, \ldots, w, \tag{6.43}$$

are conditional transient probabilities between the operation states of the operation process $Z(t)$ coressponding to those defined in Chap. 4 in case they are not conditional.

The matrix of conditional distribution functions

$$HC_{bl\,mn}(t) = P(\theta C_{bl\,mn} < t),\, t \in < 0, \infty),\, b = 1, 2, \ldots, v,$$
$$l = 1, 2, \ldots, w,\, m = 1, 2, \ldots, v,\, n = 1, 2, \ldots, w, \tag{6.44}$$

of the critical infrastructure operation process related to climate-weather change $ZC(t)$ conditional sojourn times $\theta C_{bl\,mn}$, $b = 1, 2, \ldots, v$, $l = 1, 2, \ldots, w$, $m = 1, 2, \ldots, v$, $n = 1, 2, \ldots, w$, at the operation and climate-weather state $zc_{bl}$, $b = 1, 2, \ldots, v$,

$l = 1, 2, \ldots, w$, when the next operation and climate-weather state is $zc_{mn}$ $m = 1, 2, \ldots, v$, $n = 1, 2, \ldots, w$, takes the following form [16]

$$[HC_{bl\,mn}(t)]_{vwxvw} = \begin{bmatrix} HC_{11\,11}(t) HC_{11\,12}(t) \ldots HC_{11\,1w}(t);\ HC_{11\,21}(t) \\ HC_{12\,11}(t) HC_{12\,12}(t) \ldots HC_{12\,1w}(t);\ HC_{12\,21}(t) \\ \ldots \\ HC_{vw\,11}(t) HC_{vw\,12}(t) \ldots HC_{vw\,1w}(t);\ HC_{vw\,21}(t) \end{bmatrix}$$

$$= \begin{matrix} HC_{11\,22}(t) \ldots HC_{11\,2w}(t); \ldots;\ HC_{11\,v1}(t) HC_{11\,v2}(t) \ldots HC_{11\,vw}(t) \\ HC_{12\,22}(t) \ldots HC_{12\,2w}(t); \ldots;\ HC_{12\,v1}(t) HC_{12\,v2}(t) HC_{12\,vw}(t) \\ \\ HC_{vw\,22}(t) \ldots HC_{vw\,2w}(t); \ldots;\ HC_{vw\,v1}(t) HC_{vw\,v2}(t) \ldots HC_{vw\,vw}(t) \end{matrix} \Bigg] \qquad (6.45)$$

and the matrix of their corresponding conditional density functions

$$hc_{bl\,mn}(t) = \frac{d}{dt}[HC_{bl\,mn}(t)] \text{ for } t \in\!> 0, \infty),\ b = 1, 2, \ldots, v,$$
$$l = 1, 2, \ldots, w,\ m = 1, 2, \ldots, v,\ n = 1, 2, \ldots, w \qquad (6.46)$$

has form [16]

$$[hc_{bl\,mn}(t)]_{vwxvw} = \begin{bmatrix} hc_{11\,11}(t) hc_{11\,12}(t) \ldots hc_{11\,1w}(t);\ hc_{11\,21}(t) hc_{11\,22}(t) \\ hc_{12\,11}(t) hc_{12\,12}(t) \ldots hc_{12\,1w}(t);\ hc_{12\,21}(t) hc_{12\,22}(t) \\ \ldots \\ hc_{vw\,11}(t) hc_{vw\,12}(t) \ldots hc_{vw\,1w}(t);\ hc_{vw\,21}(t) hc_{vw\,22}(t) \end{bmatrix}$$

$$\begin{matrix} \ldots hc_{11\,2w}(t); \ldots;\ hc_{11\,v1}(t) hc_{11\,v2}(t) \ldots hc_{11\,vw}(t) \\ \ldots hc_{12\,2w}(t); \ldots;\ hc_{12\,v1}(t) hc_{12\,v2}(t) hc_{12\,vw}(t) \\ \\ \ldots hc_{vw\,2w}(t); \ldots;\ hc_{vw\,v1}(t) hc_{vw\,v2}(t) \ldots hc_{vw\,vw}(t) \end{matrix} \Bigg] \qquad (6.47)$$

In the case when the critical infrastructure operation process $Z(t)$ and the climate-weather change process $C(t)$ at its operating area are dependent, the distribution functions existing in the matrix (6.45) can be expressed either by [16]

$$HC_{bl\,mn}(t) = P(\theta C_{bl\,mn} < t) = P((\theta_{bm} < t) \cap (C_{ln} < t)) = H_{bm}(t) C_{ln}(t), \qquad (6.48)$$

$$t \in\!< 0, \infty),\ b = 1, 2, \ldots, v,\ l = 1, 2, \ldots, w,\ m = 1, 2, \ldots, v,\ n = 1, 2, \ldots, w,$$

where

$$H_{bm}(t),\ b = 1, 2, \ldots, v,\ m = 1, 2, \ldots, v, \qquad (6.49)$$

are distribution functions of the sojourn lifetimes at the operation states of the critical infrastructure operation process $Z(t)$ defined in Chap. 4 and

$$C_{ln/bm}(t) = P(C_{ln} < t|\theta_{bm} < t), b = 1, 2, \ldots, v,$$
$$l = 1, 2, \ldots, w, m = 1, 2, \ldots, v, n = 1, 2, \ldots, w, \quad (6.50)$$

are conditional distributions of the sojourn lifetimes at the climate-weather states of the climate-weather change process $C(t)$ at the critical infrastructure operating area coressponding to those defined in Chap. 5 in case they are not conditional or by [16]

$$HC_{bl\,mn}(t) = P(\theta C_{bl\,mn} < t) = P(\theta_{bm} < t \cap C_{ln} < t)$$
$$= C_{ln}(t) H_{bm/ln}(t), t \in < 0, \infty), \quad (6.51)$$

$$b = 1, 2, \ldots, v, l = 1, 2, \ldots, w, m = 1, 2, \ldots, v, n = 1, 2, \ldots, w,$$

where

$$C_{ln}(t), l = 1, 2, \ldots, w, n = 1, 2, \ldots, w, \quad (6.52)$$

are distribution functions of the sojourn lifetimes at the climate-weather states of the climate-weather change process $C(t)$ at the critical infrastructure operating area defined in Chap. 5 and

$$H_{bm/ln}(t) = P(\theta_{bm} < t|C_{ln} < t), b = 1, 2, \ldots, v, l = 1, 2, \ldots,$$
$$w, m = 1, 2, \ldots, v, n = 1, 2, \ldots, w, \quad (6.53)$$

are conditional distributions of the sojourn lifetimes at the operation states of the critical infrastructure operation process $Z(t)$ coressponding to those defined in Chap. 4 in case they are not conditional.

Hence, the density functions existing in the matrix (6.47) can be expressed either by [16]

$$hc_{bl\,mn}(t) = \frac{d}{dt}[HC_{bl\,mn}(t)] = \frac{d}{dt}\left[H_{bm}(t)C_{ln/bm}(t)\right]$$
$$= h_{bm}(t)C_{ln/bm}(t) + H_{bm}(t)c_{ln/bm}(t), \quad (6.54)$$

$$t \in < 0, \infty), b = 1, 2, \ldots, v, l = 1, 2, \ldots, w, m = 1, 2, \ldots, v, n = 1, 2, \ldots, w,$$

where

$$H_{bm}(t), b = 1, 2, \ldots, v, m = 1, 2, \ldots, v, \quad (6.55)$$

are conditional distribution functions given by (6.49) and

$$C_{ln/bm}(t), l = 1, 2, \ldots, w, n = 1, 2, \ldots, w, \quad (6.56)$$

are conditional distribution functions given by (6.50) and

$$h_{bm}(t)b = 1, 2, \ldots, v, m = 1, 2, \ldots, v, \text{ and}$$
$$c_{ln/bm}(t), l = 1, 2, \ldots, w, n = 1, 2, \ldots, w,$$

are their derivatives or by [16]

$$\begin{aligned} hc_{bl\,mn}(t) &= \frac{d}{dt}[HC_{bl\,mn}(t)] = \frac{d}{dt}\left[C_{ln}(t)H_{bm/ln}(t)\right] \\ &= c_{ln}(t)H_{bm/ln}(t) + C_{ln}(t)h_{bm/ln}(t), \end{aligned} \tag{6.57}$$

$t \in < 0, \infty), b = 1, 2, \ldots, v, l = 1, 2, \ldots, w, m = 1, 2, \ldots, v, n = 1, 2, \ldots, w,$

where

$$C_{ln}(t), l = 1, 2, \ldots, w, n = 1, 2, \ldots, w, \tag{6.58}$$

are distribution functions given by (6.52) and

$$H_{bm/ln}(t), b = 1, 2, \ldots, v, m = 1, 2, \ldots, v, \tag{6.59}$$

are conditional distribution functions given by (6.53) and

$$\begin{aligned} &c_{ln}(t), l = 1, 2, \ldots, w, n = 1, 2, \ldots, w, \text{ and} \\ &h_{bm/ln}(t), b = 1, 2, \ldots, v, m = 1, 2, \ldots, v, \end{aligned} \tag{6.60}$$

are their derivatives.

We assume that the suitable and typical distributions to describe the critical infrastructure operation process related to climate-weather change $ZC(t)$ conditional sojourn times $\theta C_{blmn}$, $b = 1, 2, \ldots, v$, $l = 1, 2, \ldots, w$, $m = 1, 2, \ldots, v$, $n = 1, 2, \ldots, w$, at the particular operation and climate-weather states are of the same kind as those distinguished in [10] for the critical infrastructure operation process $Z(t)$ conditional sojourn times $\theta_{bl}$, $b = 1, 2, \ldots, v$, $l = 1, 2, \ldots, w$.

## 6.2 Critical Infrastructure Operation Process Related to Climate-Weather Change Prediction

Assuming that we have identified the unknown parameters of the critical infrastructure operation process related to climate-weather change $ZC(t)$, $t \in < 0, \infty)$ that can take $vw$, $v, w \in N$, different operation and climate-weather states

$zc_{11}, zc_{12}, \ldots, zc_{vw}$, $b = 1, 2, \ldots, v$, $l = 1, 2, \ldots, w$, defined in Sects. 6.1.1 and 6.1.2 and described by [16]:

- the vector $[pq_{bl}(0)]_{1xvw}$ of initial probabilities of the critical infrastructure operation process related to climate-weather change $ZC(t)$ staying at the initial moment $t = 0$ at the operation and climate-weather states $zc_{bl}$, $b = 1, 2, \ldots, v$, $l = 1, 2, \ldots, w$;
- the matrix $[pq_{blmn}]_{vwxvw}$ of the probabilities of transitions of the critical infrastructure operation process related to climate-weather change $ZC(t)$ between the operation and climate-weather states $zc_{bl}$ and $zc_{mn}$, $b = 1, 2, \ldots, v$, $l = 1, 2, \ldots, w$, $m = 1, 2, \ldots, v$, $n = 1, 2, \ldots, w$,
- the matrix $[HC_{bl\,mn}(t)]_{vwxvw}$of conditional distribution functions of the critical infrastructure operation process related to climate-weather change $ZC(t)$ conditional sojourn times $\theta C_{bl\,mn}$, $b = 1, 2, \ldots, v$, $l = 1, 2, \ldots, w$, $m = 1, 2, \ldots, v$, $n = 1, 2, \ldots, w$, at the operation and climate-weather state $zc_{bl}$, $b = 1, 2, \ldots, v$, $l = 1, 2, \ldots, w$, when the next operation and climate-weather state is $zc_{mn} m = 1, 2, \ldots, v$, $n = 1, 2, \ldots, w$,

we can predict this process basic characteristics.

### 6.2.1 Critical Infrastructure Operation Process Related to Climate-Weather Change Characteristics in Case of Independent Critical Infrastructure Operation Process and Climate-Weather Change Process

The mean values of the conditional sojourn times $\theta C_{bl\,mn}$, $b = 1, 2, \ldots, v$, $l = 1, 2, \ldots, w$, $m = 1, 2, \ldots, v$, $n = 1, 2, \ldots, w$,

at the operation and climate-weather state $zc_{bl}$, $b = 1, 2, \ldots, v$, $l = 1, 2, \ldots, w$, when the next operation and climate-weather state is $zc_{mn}$, $m = 1, 2, \ldots, v$, $n = 1, 2, \ldots, w$, are defined by [10, 16, 22]

$$MN_{blmn} = E[\theta C_{bl\,mn}] = \int_0^\infty t dHC_{bl\,mn} dt = \int_0^\infty t hc_{bl\,mn}(t) dt, \tag{6.61}$$

$$b = 1, 2, \ldots, v, \quad l = 1, 2, \ldots, w, \quad m = 1, 2, \ldots, v, \quad n = 1, 2, \ldots, w,$$

where $HC_{bl\,mn}(t)$ and $hc_{bl\,mn}(t)$ are respectively defined by (6.15) and (6.16).

In the case when the processess $Z(t)$ and $C(t)$ are independent, according to (6.19), the expessions (6.61) takes the form [16]

$$MN_{bl\ mn} = E[\theta C_{bl\ mn}] = \int_0^\infty t[h_{bm}(t)C_{ln}(t) + H_{bm}(t)c_{ln}(t)]dt, \tag{6.62}$$

$$b = 1, 2, \ldots, v, \quad l = 1, 2, \ldots, w, \quad m = 1, 2, \ldots, v, \quad n = 1, 2, \ldots, w,$$

where the distribution functions $H_{bm}(t)$ and $C_{ln}(t)$ are defined by (6.20) and the density functions $h_{bm}(t)$, $c_{ln}(t)$ are defined by (6.21).

From the formula for total probability, it follows that the unconditional distribution functions of the conditional sojourn times $\theta C_{bl}$, $b = 1, 2, \ldots, v$, $l = 1, 2, \ldots, w$, of the critical infrastructure operation process related to climate-weather change $ZC(t)$ at the operation and climate-weather states $zc_{bl}$, $b = 1, 2, \ldots, v$, $l = 1, 2, \ldots, w$, are given by [16]

$$HC_{bl}(t) = \sum_{m=1}^{v}\sum_{n=1}^{w} pq_{bl\ mn} HC_{bl\ mn}(t), \quad t \in < 0, \infty,$$
$$b = 1, 2, \ldots, v, \quad l = 1, 2, \ldots, w, \tag{6.63}$$

where the probabilities of transitions between operation and climate-weather states are given by (6.8), (6.9) and the distributions $HC_{bl\ mn}(t)$ are defined by (6.14).

In the case when the processes $Z(t)$ and $C(t)$ are independent, according to (6.11), (6.12) and (6.18) the expessions (6.63) takes the form [16]

$$HC_{bl}(t) = \sum_{m=1}^{w}\sum_{n=1}^{w} p_{bm} q_{ln} H_{bm}(t) C_{ln}(t), t \in < 0, \infty),$$
$$b = 1, 2, \ldots, v, \quad l = 1, 2, \ldots, w. \tag{6.64}$$

where the probabilities of transitions $p_{bm}$, $q_{ln}$ are defined by (6.12) and the distribution functions $H_{bm}(t)$ and $C_{ln}(t)$ are defined by (6.20).

From (6.63) it follows that the mean values $E[\theta C_{bl}]$, of the unconditional distribution functions of the conditional sojourn times $\theta C_{bl}$, of the critical infrastructure operation process related to climate-weather change $ZC(t)$ at the operation and climate-weather states $zc_{bl}$, $b = 1, 2, \ldots, v$, $l = 1, 2, \ldots, w$, are given by [16]

$$MN_{bl} = E[\theta C_{bl}] = \sum_{m=1}^{v}\sum_{n=1}^{w} pq_{bl\ mn} MN_{bl\ mn}, \quad b = 1, 2, \ldots, v, \quad l = 1, 2, \ldots, w, \tag{6.65}$$

where $MN_{bl\ mn}$, are given by the formula (6.61).

In the case when the processess $Z(t)$ and $C(t)$ are independent, considering (6.11) and (6.64) the expession (6.65) takes the form [16]

$$MN_{bl} = E[\theta C_{bl}] = \sum_{m=1}^{v}\sum_{n=1}^{w} p_{bm} q_{ln} MN_{bl\ mn}, \quad b = 1, 2, \ldots, v, \quad l = 1, 2, \ldots, w, \tag{6.66}$$

where the probabilities of transitions $p_{bm}$, $q_{ln}$ are defined by (6.12) and the mean values $MN_{bl\ mn}$, are given by the formula (6.62).

The transient probabilities of the critical infrastructure operation process related to climate-weather change $ZC(t)$ at the operation and climate-weather states $zc_{bl}$, $b = 1, 2, \ldots, v$, $l = 1, 2, \ldots, w$, can be defined by [16]

$$pq_{bl}(t) = P(ZC(t) = zc_{bl}), t \in< 0, \infty), \quad b = 1, 2, \ldots, v, \quad l = 1, 2, \ldots, w. \tag{6.67}$$

In the case when the processes $Z(t)$ and $C(t)$ are independent the expression (6.67) for the transient probabilities can be expressed in the following way [16]

$$\begin{aligned} pq_{bl}(t) &= P(ZC(t) = zc_{bl}) = P((Z(t) = z_b) \cap (C(t) = c_l)) \\ &= P(Z(t) = z_b) \cdot P(C(t) = c_l) = p_b(t) \cdot q_l(t), \\ & t \in<, \infty, \quad b = 1, 2, \ldots, v, \quad l = 1, 2, \ldots, w, \end{aligned} \tag{6.68}$$

where

$$p_b(t) = P(Z(t) = z_b), \quad t \in< 0, \infty, \quad b = 1, 2, \ldots, v, \tag{6.69}$$

are the transient probabilities at the operation states of the operation process $Z(t)$ defined in Chap. 4 and

$$q_l(t) = P(C(t) = c_l), \quad t \in< 0, \infty, \quad l = 1, 2, \ldots, w, \tag{6.70}$$

are the transient probabilities at the climate-weather states of the climate-weather change process $C(t)$ defined in Chap. 5.

The limit values of transient probabilities $P(ZC(t) = zc_{bl})$, $b = 1, 2, \ldots, v$, $l = 1, 2, \ldots, w$, of the critical infrastructure operation process related to climate-weather change $ZC(t)$ at the operation and climate-weather states $zc_{bl}$, $b = 1, 2, \ldots, v$, $l = 1, 2, \ldots, w$, can be found from the formula [10, 16]

$$pq_{bl} = \lim_{t\to\infty} \frac{\pi_{bl} MN_{bl}}{\sum_{b=1}^{v}\sum_{l=1}^{w} \pi_{bl} MN_{bl}}, \quad b = 1, 2, \ldots, v, \quad l = 1, 2, \ldots, w, \tag{6.71}$$

where $MN_{bl}$, $b = 1, 2, \ldots, v$, $l = 1, 2, \ldots, w$, are given by (6.66), while the steady probabilities $\pi_{bl}$, $b = 1, 2, \ldots, v$, $l = 1, 2, \ldots, w$, of the vector $[\pi_{bl}]_{1xvw}$ satisfy the system of equations

$$\begin{cases} [\pi_{bl}][pq_{blmn}] = [\pi_{bl}] \\ \sum_{b=1}^{v} \sum_{l=1}^{w} \pi_{bl} = 1, \end{cases} \tag{6.72}$$

where $pq_{bl\,mn}$, $b = 1, 2, \ldots, v$, $l = 1, 2, \ldots, w$, $m = 1, 2, \ldots, v$, $n = 1, 2, \ldots, w$, are given by (6.8), (6.9).

In the case of a periodic system operation process, the limit transient probabilities $pq_{bl}$, $b = 1, 2, \ldots, v$, $l = 1, 2, \ldots, w$, at the operation and climate-weather states given by (6.71), are the long term proportions of the critical infrastructure operation process related to climate-weather change $ZC_{bl}(t)$ sojourn times at the particular operation and climate-weather states $zc_{bl}$, $b = 1, 2, \ldots, v$, $l = 1, 2, \ldots, w$, [16].

Other interesting characteristics of the critical infrastructure operation process related to climate-weather change $ZC_{bl}(t)$ possible to obtain are its total sojourn times $\hat{\theta}\widehat{C_{bl}}$ $b = 1, 2, \ldots, v$, $l = 1, 2, \ldots, w$, at the particular operation and climate-weather states $zc_{bl}$,$b = 1, 2, \ldots, v$, $l = 1, 2, \ldots, w$, during the fixed system operation time. It is well known [10] that the system operation process related to climate-weather change total sojourn times $\hat{\theta}\widehat{C_{bl}}$ at the particular operation and climate-weather states $zc_{bl}$, for sufficiently large operation time $\theta$, have approximately normal distributions with the mean values given by [16]

$$\widehat{M}\widehat{N_{bl}} = E\left[\hat{\theta}\widehat{C_{bl}}\right] = pq_{bl}\theta, \quad b = 1, 2, \ldots, v, \quad l = 1, 2, \ldots, w, \tag{6.73}$$

where $pq_{bl}$, $b = 1, 2, \ldots, v$, $l = 1, 2, \ldots, w$, are given by (6.71).

From (6.68) it follows that in case of the independent processes $Z(t)$ and $C(t)$ the formula (6.71) for limit values of transient probabilities takes simpler following form

$$pq_{bl} = p_b \cdot q_l, \quad b = 1, 2, \ldots, v, \quad l = 1, 2, \ldots, w, \tag{6.74}$$

where $p_b$, $b = 1, 2, \ldots, v$, are the limit values of the transient probabilities $P(Z(t) = z_b)$ $b = 1, 2, \ldots, v$, of the operation process $Z(t)$ defined in Chap. 4 by (4.8) and $q_l$, $l = 1, 2, \ldots, w$, are the limit values of the transient probabilities $P(C(t) = q_l)$ $l = 1, 2, \ldots, w$, of the climate-weather change process $C(t)$ defined in Chap. 5 by (5.4) and consequently the formula for total sojourn times $\hat{\theta}\widehat{C_{bl}}$, $b = 1, 2, \ldots, v$, $l = 1, 2, \ldots, w$, at the particular operation and climate-weather states $zc_{bl}$, $b = 1, 2, \ldots, v$, $l = 1, 2, \ldots, w$, for sufficiently large operation time $\theta$, takes simplified following form

$$\widehat{M}\widehat{N_{bl}} = E\left[\hat{\theta}\widehat{C_{bl}}\right] = p_b \cdot q_l \cdot \theta, \quad b = 1, 2, \ldots, v, \quad l = 1, 2, \ldots, w, \tag{6.75}$$

where $p_b$, $b = 1, 2, \ldots, v$, $q_l$, $l = 1, 2, \ldots, w$, are respectively given by (4.8) in Chap. 4 and (5.4) in Chap. 5.

### 6.2.2 Critical Infrastructure Operation Process Related to Climate-Weather Change Characteristics in Case of Dependent Critical Infrastructure Operation Process and Climate-Weather Change Process

The mean values of the conditional sojourn times $\theta C_{bl\,mn}$, $b = 1, 2, \ldots, v$, $l = 1, 2, \ldots, w$, $m = 1, 2, \ldots, v$, $n = 1, 2, \ldots, w$,

at the operation and climate-weather state $zc_{bl}$, $b = 1, 2, \ldots, v$, $l = 1, 2, \ldots, w$, when the next operation and climate-weather state is $zc_{mn}$, $m = 1, 2, \ldots, v$, $n = 1, 2, \ldots, w$, are defined by [10, 16, 22]

$$MN_{bl\,mn} = E[\theta C_{bl\,mn}] = \int_0^\infty t dHC_{blmn}(t)dt = \int_0^\infty t hc_{bl\,mn}(t)dt, \quad (6.76)$$

$$b = 1, 2, \ldots, v, \; l = 1, 2, \ldots, w, \; m = 1, 2, \ldots, v, \; n = 1, 2, \ldots, w,$$

where $HC_{bl\,mn}(t)$ and $hc_{bl\,mn}(t)$ are respectively defined by (6.15) and (6.16).

Since from the formula for total probability, it follows that the unconditional distribution functions of the conditional sojourn times $\theta C_{bl}$, $b = 1, 2, \ldots, v$, $l = 1, 2, \ldots, w$, of the critical infrastructure operation process related to climate-weather change $ZC(t)$ at the operation and climate-weather state $zc_{bl}$, $b = 1, 2, \ldots, v$, $l = 1, 2, \ldots, w$, are given by [16]

$$C_{bl}(t) = \sum_{m=1}^{v}\sum_{n=1}^{w} pq_{bl\,mn} HC_{bl\,mn}(t), \quad t \in < 0, \infty,$$
$$b = 1, 2, \ldots, v, \quad l = 1, 2, \ldots, w. \quad (6.77)$$

Then, the mean values $E[\theta C_{bl}]$ of the unconditional distribution functions of the conditional sojourn times $\theta C_{bl}$, $b = 1, 2, \ldots, v$, $l = 1, 2, \ldots, w$, of the critical infrastructure operation process related to climate-weather change $ZC(t)$ at the operation and climate-weather states $zc_{bl}$, $b = 1, 2, \ldots, v$, $l = 1, 2, \ldots, w$, are given by [16]

$$MN_{bl} = E[\theta C_{bl}] = \sum_{m=1}^{v}\sum_{n=1}^{w} p_{bl\,mn} MN_{bl\,mn}, \quad b = 1, 2, \ldots, v, \quad l = 1, 2, \ldots, w, \quad (6.78)$$

where probabilities of transitions between the operation and climate-weather states are defined by (6.38) and the mean values $MN_{bl\,mn}$ are defined by the formula (6.76).

The transient probabilities of the critical infrastructure operation process related to climate-weather change $ZC(t)$ at the operation and climate-weather states $zc_{bl}$, $b = 1, 2, \ldots, v$, $l = 1, 2, \ldots, w$, can be defined by [16]

$$pq_{bl}(t) = P(ZC(t) = zc_{bl}, t \in <0, \infty), \quad b = 1, 2, \ldots, v, \quad l = 1, 2, \ldots, w. \tag{6.79}$$

In the case when the processess $Z(t)$ and $C(t)$ are dependent the transient probabilities can be expresed either by [16]

$$\begin{aligned} pq_{bl}(t) &= P(ZC(t) = zc_{bl} = P((Z(t) = z_b) \cap (C(t) = c_l)) \\ &= P(Z(t) = z_b) \cdot P(C(t) = c_l | Z(t) = z_b) \\ &= p_b(t) \cdot q_{l/b}(t), t \in <0, \infty), \quad b = 1, 2, \ldots, v, \quad l = 1, 2, \ldots, w, \end{aligned} \tag{6.80}$$

where

$$p_b(t) = P(Z(t) = z_b, t \in <0, \infty), \quad b = 1, 2, \ldots, v, \tag{6.81}$$

are transient probabilities at the operation state of the operation process $Z(t)$ defined in Chap. 4 and

$$q_{l/b}(t) = P(C(t) = c_l | Z(t) = z_b), t \in <0, \infty), \quad b = 1, 2, \ldots, \quad v, l = 1, 2, \ldots, w, \tag{6.82}$$

are conditional transient probabilities at the climate-weather states of the climate-weather change process $C(t)$ corresponding to those defined in Chap. 5 in case they are not conditional or by [16]

$$\begin{aligned} pq_{bl}(t) &= P(ZC(t) = zc_{bl} = P((Z(t) = z_b) \cap (C(t) = c_l)) \\ &= P(C(t) = c_l) \cdot P(Z(t) = z_b | C(t) = c_l) \\ &= q_l(t) \cdot p_{b/l}(t), \quad t \in <0, \infty, \quad b = 1, 2, \ldots, v, \quad l = 1, 2, \ldots, w, \end{aligned} \tag{6.83}$$

where

$$q_l(t) = P(C(t) = c_l), t \in <0, \infty, \quad l = 1, 2, \ldots, w, \tag{6.84}$$

are transient probabilities at the climate-weather state of the climate-weather change process $C(t)$ defined in Chap. 5 and

$$p_{b/l}(t) = P(Z(t) = z_b | C(t) = c_l), t \in <0, \infty, \quad b = 1, 2, \ldots, v, \quad l = 1, 2, \ldots, w, \tag{6.85}$$

are conditional transient probabilities at operation state of the operation proces process $Z(t)$ corresponding to those defined in Chap. 4 in case they are not conditional.

The limit values of the critical infrastructure operation process related to climate-weather change $ZC(t)$ at the operation and climate-weather state $zc_{bl}$, $b = 1, 2, \ldots, v$, $l = 1, 2, \ldots, w$, can be found from [10, 16, 22]

$$pq_{bl} = \lim_{t\to\infty} pq_{bl}(t) = \lim_{t\to\infty} \frac{\pi_{bl} MN_{bl}}{\sum_{b=1}^{v} \sum_{l=1}^{w} \pi_{bl} MN_{bl}}, \quad b = 1, 2, \ldots, v, \quad l = 1, 2, \ldots, w, \quad (6.86)$$

where $MN_{bl}$, $b = 1, 2, \ldots, v$, $l = 1, 2, \ldots, w$, are given by (6.78), while the steady probabilities $\pi_{bl}$, $b = 1, 2, \ldots, v$, $l = 1, 2, \ldots, w$, of the vector $[\pi_{bl}]_{1xvw}$ satisfy the system of equations

$$\begin{cases} [\pi_{bl}][pq_{blmn}] = [\pi_{bl}] \\ \sum_{b=1}^{v} \sum_{l=1}^{w} \pi_{bl} = 1. \end{cases}$$

where $pq_{bl\,mn}$, $b = 1, 2, \ldots, v$, $l = 1, 2, \ldots, w$, $m = 1, 2, \ldots, v$, $n = 1, 2, \ldots, w$, are given by (6.8), (6.9).

In the case of a periodic system operation process, the limit transient probabilities $pq_{bl}$, $b = 1, 2, \ldots, v$, $l = 1, 2, \ldots, w$, at the operation and climate-weather states given by (6.86), are the long term proportions of the critical infrastructure operation process related to climate-weather change $ZC_{bl}(t)$ sojourn times at the particular operation and climate-weather states $zc_{bl}$, $b = 1, 2, \ldots, v$, $l = 1, 2, \ldots, w$, [16].

Other interesting characteristics of the critical infrastructure operation process related to climate-weather change $ZC_{bl}(t)$ possible to obtain are its total sojourn times $\widehat{\hat{\theta} C_{bl}}$, $b = 1, 2, \ldots, v$, $l = 1, 2, \ldots, w$, at the particular operation and climate-weather states $zc_{bl}$, $b = 1, 2, \ldots, v$, $l = 1, 2, \ldots, w$, during the fixed system operation time. It is well known [10] that the system operation process related to climate-weather change total sojourn times $\widehat{\hat{\theta} C_{bl}}$ $b = 1, 2, \ldots, v$, $l = 1, 2, \ldots, w$, at the particular operation and climate-weather states $zc_{bl}$, $b = 1, 2, \ldots, v$, $l = 1, 2, \ldots, w$, for sufficiently large operation time $\theta$, have approximately normal distributions with the expected value given by [16]

$$\widehat{\hat{M} N_{bl}} = E\left[\widehat{\hat{\theta} C_{bl}}\right] = pq_{bl}\theta, \quad b = 1, 2, \ldots, v, \quad l = 1, 2, \ldots, w, \quad (6.87)$$

where $pq_{bl}$, $b = 1, 2, \ldots, v$, $l = 1, 2, \ldots, w$, are given by (6.86).

## 6.3 Safety and Resilience Indicators of Critical Infrastructure Impacted by Operation Process Related to Climate-Weather Change

We denote by $[T^3(u)]^{bl}$, $u = 1, 2, \ldots, z$, $b = 1, 2, \ldots, v$, $l = 1, 2, \ldots, w$, the critical infrastructure conditional lifetime in the safety state subset $\{u, u + 1, \ldots, z\}$, $u = 1, 2, \ldots, z$, while its operation process related to the climate-weather change $ZC(t)$,

$t \in < 0, \infty)$, is at the operation and climate-weather state $zc_{bl}$, $b = 1, 2, \ldots, v$, $l = 1, 2, \ldots, w$, and we introduce the conditional safety function of the critical infrastructure impacted by the operation process related to the climate-weather change $ZC(t)$, $t \in < 0, \infty)$, defined by the vector [10, 22]

$$[\boldsymbol{S}^3(t, \cdot)]^{(bl)} = [1, [\boldsymbol{S}^3(t, 1)]^{bl}, \ldots, [\boldsymbol{S}^3(t, z)]^{(bl)}],$$
$$t \in < 0, \infty), \quad b = 1, 2, \ldots, v, \quad l = 1, 2, \ldots, w \tag{6.88}$$

with the coordinates

$$[\boldsymbol{S}^3(t, u)]^{(bl)} = P([T^3(u)]^{(bl)} > t | ZC(t) = zc_{bl})$$
$$\text{for } t \in < 0, \infty),\ u = 1, 2, \ldots, z, \quad b = 1, 2, \ldots, v, \quad l = 1, 2, \ldots, w. \tag{6.89}$$

The safety function $[\boldsymbol{S}^3(t, u)]^{(bl)}$, $t \in < 0, \infty$, $u = 1, 2, \ldots, z$, $b = 1, 2, \ldots, v$, $l = 1, 2, \ldots, w$, is the conditional probability that the critical infrastructure impacted by the operation process related to the climate-weather change $ZC(t)$, $t \in < 0, \infty)$, lifetime $[T^3(u)]^{(bl)}$, $u = 1, 2, \ldots, z$, $b = 1, 2, \ldots, v$, $l = 1, 2, \ldots, w$, in the safety state subset $\{u, u + 1, \ldots, z\}$, $u = 1, 2, \ldots, z$ is greater than $t$, while the operation process related to the climate-weather change $ZC(t)$, $t \in < 0, \infty)$, is at the operation and climate-weather state $zc_{bl}$, $b = 1, 2, \ldots, v$, $l = 1, 2, \ldots, w$.

Next, we denote by $T^3(u)$, $u = 1, 2, \ldots, z$, the critical infrastructure impacted by the operation process related to the climate-weather change $ZC(t)$, $t \in < 0, \infty$, unconditional lifetime in the safety state subset $\{u, u + 1, \ldots, z\}$ $u = 1, 2, \ldots, z$, and the unconditional safety function (SafI1) of the critical infrastructure impacted by the operation process related to the climate-weather change $ZC(t)$, $t \in < 0, \infty)$, by the vector

$$\boldsymbol{S}^3(t, \cdot) = \left[1, \boldsymbol{S}^3(t, 1), \ldots, \boldsymbol{S}^3(t, z)\right], \tag{6.90}$$

with the coordinates defined by

$$\boldsymbol{S}^3(t, u) = P(T^3(u) > t) \quad \text{for } t \in < 0, \infty, \quad u = 1, 2, \ldots, z. \tag{6.91}$$

In the case when the critical infrastructure operation time $\theta$ is large enough, the coordinates of the unconditional safety function of the critical infrastructure impacted by the operation process related to the climate-weather change $ZC(t)$, $t \in < 0, \infty$, defined by (6.91), are given by [10]

$$\boldsymbol{S}^3(t, u) \cong \sum_{b=1}^{v} \sum_{l=1}^{w} pq_{bl} [\boldsymbol{S}^3(t, u)]^{(bl)} \quad \text{for } t \in < 0, \infty, \quad u = 1, 2, \ldots, z, \tag{6.92}$$

where $S^3(t, u)]^{(bl)}$, $u = 1, 2, \ldots, z$, $b = 1, 2, \ldots, v$, $l = 1, 2, \ldots, w$, are the coordinates of the critical infrastructure impacted by the operation process related to the climate-weather change $ZC(t)$, $t \in <0, \infty$, conditional safety functions defined by (6.88), (6.89) and $pq_{bl}$, $b = 1, 2, \ldots, v$, $l = 1, 2, \ldots, w$, are the limit transient probabilities of the operation process related to climate-weather change $ZC(t)$, $t \in <0, \infty$, at the operation and climate-weather states $zc_{bl}$,, $b = 1, 2, \ldots, v$, $l = 1, 2, \ldots, w$, given either by (6.71) or (6.86).

If $r$ is the critical safety state, then the second safety indicator of the critical infrastructure impacted by the operation process related to the climate-weather change $ZC(t)$, $t \in <0, \infty$, the risk function (SafI2)

$$\boldsymbol{r}^3(t) = P(s(t) < r | s(0) = z) = P(T^3(r) \leq t), \quad t \in <0, \infty, \tag{6.93}$$

is defined as a probability that the critical infrastructure impacted by the operation process related to the climate-weather change $ZC(t)$, $t \in <0, \infty)$, is in the subset of safety states worse than the critical safety state $r$, $r \in \{1, \ldots, z\}$ while it was in the best safety state $z$ at the moment $t = 0$ and given by [10, 19, 22]

$$\boldsymbol{r}^3(t) = 1 - \boldsymbol{S}^3(t, r), \quad t \in <0, \infty, \tag{6.94}$$

where $\boldsymbol{S}^3(t, r)$ is the coordinate of the critical infrastructure impacted by the operation process related to climate-weather change $ZC(t)$, $t \in <0, \infty)$, unconditional safety function given by (6.92) for $u = r$.

The graph of the critical infrastructure risk function $\boldsymbol{r}^3(t)$, $t \in <0, \infty)$, defined by (6.94), is the safety indicator called the fragility curve (SafI3) of the critical infrastructure impacted by the operation process related to the climate-weather change $ZC(t)$, $t \in <0, \infty)$.

Other practically useful safety indicators of the critical infrastructure impacted by the operation process related to the climate-weather change $ZC(t)$, $t \in <0, \infty)$, are:

– the mean value of the critical infrastructure unconditional lifetime $T^3(r)$ up to exceeding critical safety state $r$ (SafI4) given by

$$\mu^3(r) = \int_0^\infty [\boldsymbol{S}^3(t, r)]dt \cong \sum_{b=1}^{v} \sum_{l=1}^{w} pq_{bl}[\boldsymbol{\mu}^3(r)]^{(bl)}, \tag{6.95}$$

where $[\boldsymbol{\mu}^3(r)]^{(bl)}$, $b = 1, 2, \ldots, v$, $l = 1, 2, \ldots, w$, are the mean values of the critical infrastructure conditional lifetimes $[T^3(r)]^{(bl)}$, $b = 1, 2, \ldots, v$, $l = 1, 2, \ldots, w$, in the safety state subset $\{r, r + 1, \ldots, z\}$at the operation and climate-weather state $zc_{bl}$, $b = 1, 2, \ldots, v$, $l = 1, 2, \ldots, w$, of the operation process related to the climate-weather change $ZC(t)$, $t \in <0, \infty)$, given by

$$[\boldsymbol{\mu}^3(r)]^{(bl)} = \int_0^\infty [\boldsymbol{S}^3(t,r)]^{(bl)}dt, \quad b = 1, 2, \ldots, v, \quad l = 1, 2, \ldots, w, \tag{6.96}$$

and $[\boldsymbol{S}^3(t,r)]^{(bl)}$, $b = 1, 2, \ldots, v$, $l = 1, 2, \ldots, w$, are defined by (6.88), (6.89) and $pq_{bl}$, $b = 1, 2, \ldots, v$, $l = 1, 2, \ldots, w$, are given either by (6.71) or by (6.86);

- the standard deviation of the critical infrastructure lifetime $T^3(r)$ up to the exceeding the critical safety state $r$ (SafI5) given by

$$\boldsymbol{\sigma}^3(r) = \sqrt{\boldsymbol{n}^3(r) - [\boldsymbol{\mu}^3(r)]^2}, \tag{6.97}$$

where

$$\boldsymbol{n}^3(r) = 2\int_0^\infty t\boldsymbol{S}^3(t,r)dt, \tag{6.98}$$

and $\boldsymbol{S}^3(t,r)$ is defined by (6.91) for $u = r$, and $\boldsymbol{\mu}^3(r)$ is given by (6.95);

- the moment $\boldsymbol{\tau}^3$, of exceeding acceptable value of critical infrastructure risk function level $\delta$ (SafI6) given by

$$\boldsymbol{\tau}^3 = \boldsymbol{r}^{3^{-1}}(\delta), \tag{6.99}$$

where $\boldsymbol{r}^{3^{-1}}(\delta)$ is the inverse function of the risk function $\boldsymbol{r}^3(t)$ given by (6.93);

- the mean lifetime of the critical infrastructure in the safety state subsets $\{u, u + 1, \ldots, z\}$ $u = 1, 2, \ldots, z$, (SafI7), given by

$$\boldsymbol{\mu}^3(u) = \int_0^\infty [\boldsymbol{S}^3(t,u)]dt \cong \sum_{b=1}^{v} \sum_{l=1}^{w} pq_{bl}[\boldsymbol{\mu}^3(u)]^{(bl)}, \quad u = 1, 2, \ldots, z, \tag{6.100}$$

where $[\boldsymbol{\mu}^3(u)]^{(bl)}$, $u = 1, 2, \ldots, z$, $b = 1, 2, \ldots, v$, $l = 1, 2, \ldots, w$, are the mean values of the critical infrastructure conditional lifetimes $[\boldsymbol{T}^3(u)]^{(bl)}$, $u = 1, 2, \ldots, z$, $b = 1, 2, \ldots, v$, $l = 1, 2, \ldots, w$, in the safety state subset $\{u, u + 1, \ldots, z\}$ at the critical infrastructure operation process related to the climate-weather change state $zc_{bl}$, $b = 1, 2, \ldots, v$, $l = 1, 2, \ldots, w$, given by

$$[\boldsymbol{\mu}^3(u)]^{(bl)} = \int_0^\infty [\boldsymbol{S}^3(t,u)]^{(bl)} dt, \quad u = 1, 2, \ldots, z, \quad b = 1, 2, \ldots, v, \quad l = 1, 2, \ldots, w, \qquad (6.101)$$

and $[\boldsymbol{S}^3(t,u)]^{(bl)}$, $u = 1, 2, \ldots, z$, $b = 1, 2, \ldots, v$, $l = 1, 2, \ldots, w$, are defined by (6.88), (6.89) and $pq_{bl}$, $b = 1, 2, \ldots, v$, $l = 1, 2, \ldots, w$, are given by (6.71) or by (6.86);

- the standard deviations of the critical infrastructure lifetimes in the safety state subsets $\{u, u+1, \ldots, z\}$, $u = 1, 2, \ldots, z$, (SafI8), given by

$$\sigma^3(u) = \sqrt{\boldsymbol{n}^3(u) - [\boldsymbol{\mu}^3(u)]^2}, \quad u = 1, 2, \ldots, z, \qquad (6.102)$$

where

$$\boldsymbol{n}^3(u) = 2\int_0^\infty t\boldsymbol{S}^3(t,u)dt, \quad u = 1, 2, \ldots, z; \qquad (6.103)$$

- the mean lifetimes $\overline{\boldsymbol{\mu}}^3(u)$, $u = 1, 2, \ldots, z$, of the critical infrastructure in the particular safety states (SafI9)

$$\overline{\boldsymbol{\mu}}^3(u) = \boldsymbol{\mu}^3(u) - \boldsymbol{\mu}^3(u+1), \quad u = 1, 2, \ldots, z-1, \overline{\boldsymbol{\mu}}^3(z) = \boldsymbol{\mu}^3(z); \qquad (6.104)$$

- the intensities of degradation of the critical infrastructure / the intensities of critical infrastructure departure from the safety state subset $\{u, u+1, \ldots, z\}$, $u = 1, 2, \ldots, z$, (SafI10), i.e. the coordinates of the vector

$$\boldsymbol{\lambda}^3(t,\cdot) = \left[0, \boldsymbol{\lambda}^3(t,1), \ldots \boldsymbol{\lambda}^3(t,z)\right], \quad t \in <0, \infty, \qquad (6.105)$$

where

$$\boldsymbol{\lambda}^3(t,u) = \frac{-\frac{d\boldsymbol{S}^3(t,\mathrm{u})}{dt}}{\boldsymbol{S}^3(t,\mathrm{u})}, \quad t \in <0, \infty), \quad u = 1, 2, \ldots, z; \qquad (6.106)$$

- the coefficients of the operation process related to the climate-weather change impact on the critical infrastructure intensities of degradation/the coefficients of the operation process related to the climate-weather change impact on critical infrastructure intensities of departure from the safety state subset $\{u, u+1, \ldots, z\}$, (ResI1), i.e. the coordinates of the vector

$$\boldsymbol{\rho}^3(t,\cdot) = \left[0, \boldsymbol{\rho}^3(t,1), \ldots, \boldsymbol{\rho}^3(t,z)\right], \quad t \in <0, \infty, \tag{6.107}$$

where

$$\boldsymbol{\lambda}^3(t,u) = \boldsymbol{\rho}^3(t,u) \cdot \boldsymbol{\lambda}^0(t,u), \quad t \in <0, \infty, \quad u = 1, 2, \ldots, z, \tag{6.108}$$

i.e.

$$\boldsymbol{\rho}^3(t,u) = \frac{\boldsymbol{\lambda}^3(t,u)}{\boldsymbol{\lambda}^0(t,u)}, \quad t \in <0, \infty), \quad u = 1, 2, \ldots, z, \tag{6.109}$$

and $\boldsymbol{\lambda}^0(t,u)$, $t \in <0, \infty)$, $u = 1, 2, \ldots, z$, defined by (3.14), are the intensities of degradation of the critical infrastructure without of the operation process related to the climate-weather change impact, i.e. the coordinate of the vector

$$\boldsymbol{\lambda}^0(t,\cdot) = \left[0, \boldsymbol{\lambda}^0(t,1), \ldots, \boldsymbol{\lambda}^0(t,z)\right], \tag{6.110}$$

and $\boldsymbol{\lambda}^3(t,u)$, , $t \in <0, \infty$, $u = 1, 2, \ldots, z$, defined by (6.106), are the intensities of degradation of the critical infrastructure with of the operation process related to the climate-weather change impact, i.e. the coordinate of the vector

$$\boldsymbol{\lambda}^3(t,\cdot) = \left[0, \boldsymbol{\lambda}^3(t,1), \ldots, \boldsymbol{\lambda}^3(t,z)\right], \quad t \in <0, \infty; \tag{6.111}$$

– the indicator of critical infrastructure resilience to operation process related to climate-weather change impact (ResI2) defined by

$$\boldsymbol{RI}^3(t,r) = \frac{1}{\boldsymbol{\rho}^3(t,r)}, \quad t \in <0, \infty, \tag{6.112}$$

where $\boldsymbol{\rho}^3(t,r)$, $t \in <0, \infty)$ is the coefficients of operation process related to the climate-weather change impact on the critical infrastructure intensities of degradation given by (6.109) for $u = r$.

In the case, the critical infrastructure have the picewise exponential safety functions, i.e.

$$\boldsymbol{S}^3(t,\cdot) = \left[1, \boldsymbol{S}^3(t,1), \ldots, \boldsymbol{S}^3(t,z)\right], \quad t \in <0, \infty, \tag{6.113}$$

where

$$\boldsymbol{S}^3(t,u) = \exp\left[-\boldsymbol{\lambda}^3(u)t\right], \quad t \in <0, \infty), \boldsymbol{\lambda}^3(u) \geq 0, \quad u = 1, 2, \ldots, z, \tag{6.114}$$

the critical infrastructure safety and resilience indicators defined by (6.105)–(6.112) take forms:

– the intensities of degradation of the critical infrastructure related to the operation process related to climate-weather change impact, i.e. the coordinates of the vector

$$\boldsymbol{\lambda}^3(\cdot) = \left[0, \boldsymbol{\lambda}^3(1), \ldots, \boldsymbol{\lambda}^3(z)\right], \tag{6.115}$$

are constant and

$$\boldsymbol{\lambda}^3(u) = \frac{1}{\boldsymbol{\mu}^3(u)}, \quad u = 1, 2, \ldots, z, \tag{6.116}$$

where $\boldsymbol{\mu}^3(u)$ are given by (6.100);

– the coefficients of the operation process related to the climate-weather change impact on the critical infrastructure intensities of degradation / the coefficients of the operation process related to the climate-weather change impact on critical infrastructure intensities of departure from the safety state subset $\{u, u+1, \ldots, z\}$, (ResII) i.e. the coordinates of the vector

$$\boldsymbol{\rho}^3(\cdot) = \left[0, \boldsymbol{\rho}^3(1), \ldots, \boldsymbol{\rho}^3(z)\right], \tag{6.117}$$

where

$$\boldsymbol{\rho}^3(u) = \frac{\boldsymbol{\lambda}^3(u)}{\boldsymbol{\lambda}^0(u)} = \frac{\boldsymbol{\mu}^0(u)}{\boldsymbol{\mu}^3(u)}, \quad u = 1, 2, \ldots, z, \tag{6.118}$$

and $\boldsymbol{\lambda}^0(u)$, $u = 1, 2, \ldots, z$, defined by (3.18) are the intensities of degradation of the critical infrastructure without of the operation process related to the climate-weather change impact, i.e. the coordinate of the vector

$$\boldsymbol{\lambda}^3(u) = \left[0, \boldsymbol{\lambda}^0(1), \ldots, \boldsymbol{\lambda}^0(z)\right], \tag{6.119}$$

and $\boldsymbol{\lambda}^3(u)$, $u = 1, 2, \ldots, z$, defined by (6.116) are the intensities of degradation of the critical infrastructure related to the operation process and the climate-weather change process impact, i.e. the coordinates of the vector

$$\boldsymbol{\lambda}^3(\cdot) = \left[0, \boldsymbol{\lambda}^3(1), \ldots, \boldsymbol{\lambda}^3(z)\right]; \tag{6.120}$$

– the indicator of critical infrastructure resilience to the operation process related to the climate-weather change impact (ResI2) defined by

$$\boldsymbol{RI}^3(r) = \frac{1}{\boldsymbol{\rho}^3(r)}, \quad t \in <0, +\infty), \tag{6.121}$$

where $\boldsymbol{\rho}^3(r)$ is the coefficient of the operation process related to the climate-weather-change impact on the critical infrastructure intensities of degradation given by (6.118) for $u = r$.

## 6.4 Critical Infrastructure Assets Safety Parameters

We assume that the changes of the critical infrastructure operating process related to the climate-weather change $ZC(t)$, $t \in <0, \infty$, states at the system operating area have an influence on the critical infrastructure asset $A_i$, $I = 1, 2, \ldots, n$, safety.

We denote by $[T_i^3(u)]^{(bl)}$, $b = 1, 2, \ldots, v$, $l = 1, 2, \ldots, w$, $i = 1, 2, \ldots, n$, the critical infrastructure asset $A_i$, $i = 1, 2, \ldots, n$, conditional lifetime in the safety state subset $\{u, u + 1, \ldots, z\}$, while the operation process related to the climate-weather change $ZC(t)$, $t \in <0, \infty$, is at the operation and climate-weather state $zc_{bl}$, $b = 1, 2, \ldots, v$, $l = 1, 2, \ldots, w$, and its conditional safety function (SafI1) by the vector [14, 19, 22]

$$[S_i^3(t,\cdot)]^{(bl)} = [1, [S_i^3(t,1)]^{bl}, \ldots, [S_i^3(t,1)]^{bl}], \quad t \in <0, +\infty,$$
$$b = 1, 2, \ldots, v, \quad l = 1, 2, \ldots, w, \quad i = 1, 2, \ldots, n, \tag{6.122}$$

with the coordinates defined by

$$[S_i^3(t,u)]^{(bl)} = P([T_i^3(u)]^{(bl)} > t | ZC(t) = zc_{bl}) \tag{6.123}$$

for $t \in <0, +\infty$, $u = 1, 2, \ldots, z$, $b = 1, 2, \ldots, v$, $l = 1, 2, \ldots, w$, $i = 1, 2, \ldots, n$.

The safety function $[S_i^3(t,u)]^{(bl)}$, $t \in <0, \infty$, $u = 1, 2, \ldots, z$, $b = 1, 2, \ldots, v$, $l = 1, 2, \ldots, w$, $i = 1, 2, \ldots, n$, is the conditional probability that the asset $A_i$ lifetime $[T_i^3(u)]^{(bl)}$, $t \in <0, \infty$, $u = 1, 2, \ldots, z$, $b = 1, 2, \ldots, v$, $l = 1, 2, \ldots, w$, $i = 1, 2, \ldots, n$, in the safety state subset $\{u, u + 1, \ldots, z\}$, is greater than $t$, while the operation process related to the climate-weather change $ZC(t)$, $t \in <0, \infty)$, is at the operation and climate-weather state $zc_{bl}$, $b = 1, 2, \ldots, v$, $l = 1, 2, \ldots, w$.

The conditional safety functions $[S_i^3(t,u)]^{(bl)}$, $t \in <0, \infty$, $u = 1, 2, \ldots, z$, $b = 1, 2, \ldots, v$, $l = 1, 2, \ldots, w$, $i = 1, 2, \ldots, n$, defined by (6.123) are called the coordinates of the asset $A_i$, $i = 1, 2, \ldots, n$, conditional safety function $[S_i^3(t,\cdot)]^{(bl)}$, $t \in <0, \infty)$, $b = 1, 2, \ldots, v$, $l = 1, 2, \ldots, w$, $i = 1, 2, \ldots, n$, while the operation process related to the climate-weather change $ZC(t)$, $t \in <0, \infty)$, is at the operation and climate-weather state $zc_{bl}$, $b = 1, 2, \ldots, v$, $l = 1, 2, \ldots, w$, given by (6.122). Thus, the relationship between the conditional distribution function $[F_i^3(t,u)]^{(bl)}$, $t \in <0, \infty)$, $u = 1, 2, \ldots, z$, $b = 1, 2, \ldots, v$, $l = 1, 2, \ldots, w$, $i = 1, 2, \ldots, n$, of the asset $A_i$, $i = 1, 2, \ldots, n$, lifetime $[T_i^3(u)]^{(bl)}$ $u = 1, 2, \ldots, z$, $b = 1, 2, \ldots, v$, $l = 1, 2, \ldots, w$, $i = 1, 2, \ldots, n$, in the safety state subset $\{u, u + 1, \ldots, z\}$, $u = 1, 2, \ldots, z$, and the coordinate $[S_i^3(t,u)]^{(bl)}$,

$t \in < 0, \infty)$, $u = 1, 2, \ldots, z$, $b = 1, 2, \ldots, v$, $l = 1, 2, \ldots, w$, $i = 1, 2, \ldots, n$, of its conditional safety function is given by

$$[F_i^3(t,u)]^{(bl)} = P([T_i^3(u)]^{(bl)} \le t) = 1 - P([T_i^3(u)]^{(bl)} > t)$$
$$= 1 - [S_i^3(t,u)]^{(bl)}, \quad t \in < 0, \infty, \tag{6.124}$$

$$u = 1, 2, \ldots, z, \quad b = 1, 2, \ldots, v, \quad l = 1, 2, \ldots, w, \quad i = 1, 2, \ldots, n.$$

The function

$$[r_i^3(t)]^{(bl)} = 1 - [S_i^3(t,r)]^{(bl)}, \quad t \in < 0, \infty),$$
$$b = 1, 2, \ldots, v, \quad l = 1, 2, \ldots, w, \quad i = 1, 2, \ldots, n, \tag{6.125}$$

is the asset $A_i$, $i = 1, 2, \ldots, n$, the conditional risk function (SafI2) and its graph is the asset $A_i$, $i = 1, 2, \ldots, n$, fragility curve (SafI3) while the operation process related to the climate-weather change is at the operation and climate-weather state $zc_{bl}$, $b = 1, 2, \ldots, v$, $l = 1, 2, \ldots, w$.

Moreover, the conditional mean lifetimes of the asset $A_i$ in the safety state subsets $\{u, u + 1, \ldots, z\}$ $u = 1, 2, \ldots, z$, while the operation process related to the climate-weather change is at the operation and climate-weather state $zc_{bl}$, $b = 1, 2, \ldots, v$, $l = 1, 2, \ldots, w$, are given by

$$[\mu_i^3(u)]^{(bl)} = \int_0^\infty [S_i^3(t,u)]^{(bl)} dt, \quad u = 1, 2, \ldots, z,$$
$$b = 1, 2, \ldots, v, \quad l = 1, 2, \ldots, w, \quad i = 1, 2, \ldots, n, \tag{6.126}$$

and the conditional standard deviations of the asset $A_i$ lifetime in the safety state subsets $\{u, u + 1, \ldots, z\}$, $u = 1, 2, \ldots, z$, while the operation process related to the climate-weather change is at the operation and climate-weather state $zc_{bl}$, $b = 1, 2, \ldots, v$, $l = 1, 2, \ldots, w$, are given by

$$[\sigma_i^3(u)]^{(bl)} = \sqrt{[n_i^3(u)]^{bl} - [[\mu_i^3(u)]^{(bl)}]^2}, \quad u = 1, 2, \ldots, z,$$
$$b = 1, 2, \ldots, v, \quad l = 1, 2, \ldots, w, \quad i = 1, 2, \ldots, n, \tag{6.127}$$

where

$$[n_i^3(u)]^{(bl)} = 2\int_0^\infty t[S_i^3(t,u)]^{(bl)} dt, \quad u = 1, 2, \ldots, z,$$
$$b = 1, 2, \ldots, v, \quad l = 1, 2, \ldots, w, \quad i = 1, 2, \ldots, n. \tag{6.128}$$

Next, it can be justified that the conditional mean lifetimes $[\overline{\mu_i^3(u)}]^{(bl)}$, $u = 1, 2, \ldots, z$, $b = 1, 2, \ldots, v$, $l = 1, 2, \ldots, w$, $i = 1, 2, \ldots, n$, of the asset $A_i$, $1, 2, \ldots, n$, in the particular safety states, while the operation process related to the climate-weather change is at the operation and climate-weather state $zc_{bl}$, $b = 1, 2, \ldots, v$, $l = 1, 2, \ldots, w$, are

$$[\overline{\mu_i^3(u)}]^{(bl)} = [\mu_i^3(u)]^{(bl)} - [\mu_i^3(u+1)]^{(bl)},$$
$$u = 1, 2, \ldots, z, \quad [\overline{\mu_i^3(z)}]^{(bl)} = [\mu_i^3(z)]^{(bl)} \tag{6.129}$$

$$u = 1, 2, \ldots, z, \quad b = 1, 2, \ldots, v, \quad l = 1, 2, \ldots, w, \quad i = 1, 2, \ldots, n.$$

In the case, when the critical infrastructure assets $A_i$, $i = 1, 2, \ldots, n$, lifetimes at the operation process related to the climate-weather change states $zc_{bl}$, $b = 1, 2, \ldots, v$, $l = 1, 2, \ldots, w$, have the exponential safety functions, the coordinates (6.123) of the vector (6.122) are given by

$$\left[S_i^3(t,u)\right]^{(bl)} = P\left(\left[T_i^3(u)\right]^{(bl)} > t | ZC(t) = zc_{bl}\right)$$
$$= \exp\left[-\left[\lambda_i^3(u)\right]^{(bl)} t\right], \quad t \in <0, \infty \tag{6.130}$$

$$u = 1, 2, \ldots, z, \quad b = 1, 2, \ldots, v, \quad l = 1, 2, \ldots, w, \quad i = 1, 2, \ldots, n.$$

Existing in (6.130) the intensities of degradation of the critical infrastructure asset $A_i$, $i = 1, 2, \ldots, n$, with the operation process related to the climate-weather change impact at the operation and climate-weather state $zc_{bl}$, $b = 1, 2, \ldots, v$, $l = 1, 2, \ldots, w$, (SafI10), i.e. the coordinates of the vector

$$\left[\lambda_i^3(\cdot)\right]^{(bl)} = \left[0, \left[\lambda_i^3(1)\right]^{(bl)}, \ldots, \left[\lambda_i^3(z)\right]^{(bl)}\right],$$
$$b = 1, 2, \ldots, v, \quad l = 1, 2, \ldots, w, \quad i = 1, 2, \ldots, n. \tag{6.131}$$

are constant and given by

$$\left[\lambda_i^3(u)\right]^{(bl)} = \frac{1}{\left[\overline{\mu_i^3(u)}\right]^{(bl)}}, \quad u = 1, 2, \ldots, z,$$
$$b = 1, 2, \ldots, v, \quad l = 1, 2, \ldots, w, \quad i = 1, 2, \ldots, n, \tag{6.132}$$

and moreover

$$\left[\lambda_i^3(u)\right]^{(bl)} = \left[\rho_i^3(u)\right]^{(bl)} \cdot \lambda_i^0(u), \quad u = 1, 2, \ldots, z,$$
$$b = 1, 2, \ldots, v, \quad l = 1, 2, \ldots, w, \quad i = 1, 2, \ldots, n, \tag{6.133}$$

where $\lambda_i^0(u)$, $u = 1, 2, \ldots, z$, $i = 1, 2, \ldots, n$, defined by (3.31) are the intensities of degradation of the critical infrastructure asset $A_i$, $i = 1, 2, \ldots, n$, without the operation process related to the climate-weather change impact (SafI10), i.e. the coordinate of the vector

$$\lambda_i^0(\cdot) = \left[0, \lambda_i^0(1), \ldots, \lambda_i^0(z)\right], \quad i = 1, 2, \ldots, n, \tag{6.134}$$

and $\left[\rho_i^3(u)\right]^{(bl)}$, $u = 1, 2, \ldots, z$, $b = 1, 2, \ldots, v$, $l = 1, 2, \ldots, w$, $i = 1, 2, \ldots, n$, are the coefficients of the operation process related to the climate-weather change impact on the critical infrastructure asset $A_i$, $i = 1, 2, \ldots, n$, intensities of degradation at the operation and climate-weather state $zc_{bl}$, $b = 1, 2, \ldots, v$, $l = 1, 2, \ldots, w$, (ResI1), i.e. the coordinate of the vector

$$\begin{aligned} &\left[\rho_i^3(u)\right]^{(bl)} = \left[\rho_i^3(u)\right]^{(bl)} \cdot \lambda_i^0(u), \quad u = 1, 2, \ldots, z, \\ &b = 1, 2, \ldots, v, \quad l = 1, 2, \ldots, \quad w, i = 1, 2, \ldots, n, \left[\rho_i^3(\cdot)\right]^{(bl)} \\ &= \left[0, \left[\rho_i^3(1)\right]^{(bl)}, \ldots, \left[\rho_i^3(z)\right]^{(bl)}\right], \quad b = 1, 2, \ldots, v, \\ &l = 1, 2, \ldots, w, \quad i = 1, 2, \ldots, n, \end{aligned} \tag{6.135}$$

where by (6.132) and (3.31)

$$\begin{aligned} &\left[\rho_i^3(u)\right]^{(bl)} = \frac{\left[\lambda_i^3(u)\right]^{(bl)}}{\lambda_i^0(u)} = \frac{\mu_i^0(u)}{\left[\mu_i^3(u)\right]^{(bl)}}, \quad u = 1, 2, \ldots, z, \\ &b = 1, 2, \ldots, v, \quad l = 1, 2, \ldots, w, \quad i = 1, 2, \ldots, n. \end{aligned} \tag{6.136}$$

The conditional safety function (6.122), (6.123), the conditional risk function (6.125), the conditional intensities' of degradation (6.132), (6.133) and the coefficients of the operation process related to the climate-weather change impact on the critical infrastructure assets intensities of degradation (6.136) are main critical infrastructure asset safety and resilience indicators impacted by the operation process related to the climate-weather change.

The indicators of critical infrastructure asset $A_i$, $i = 1, 2, \ldots, n$, resilience to the operation process related to the climate-weather change impact (ResI2) at the operation and climate-weather state $zc_{bl}$, $b = 1, 2, \ldots, v$, $l = 1, 2, \ldots, w$, can be defined by

$$\begin{aligned} &\left[RI_i^3(r)\right]^{(bl)} = \frac{1}{\left[\rho_i^3(r)\right]^{(bl)}}, \quad t \in \langle 0, \infty), \\ &b = 1, 2, \ldots, v, \quad l = 1, 2, \ldots, w, \quad i = 1, 2, \ldots, n, \end{aligned} \tag{6.137}$$

where $\left[\rho_i^3(r)\right]^{(bl)}$, $b = 1, 2, \ldots, v$, $l = 1, 2, \ldots, w$, $i = 1, 2, \ldots, n$, are the coefficients of the operation process related to the climate-weather change at the critical infrastructure operating area impact on the critical infrastructure assets intensities of degradation at the operation and climate-weather state $zc_{bl}$, $b = 1, 2, \ldots, v$, $l = 1, 2, \ldots, w$, given by (6.136) for $u = r$.

## 6.5 Application

We consider the port oil terminal critical infrastructure impacted by its operation process related to the climate-weather change process at its operating area. The considered port oil terminal placed at the Baltic seaside is described and its safety indicators without considering outside impacts are determined in Chap. 3. This port oil terminal assets are presented in Chaps. 1 and 3. In this section this port oil terminal characteristics of the operation process related to the climate-weather change at its operating area are evaluated. Next, the safety analysis of the port oil terminal critical infrastructure impacted by the operation process related to the climate-weather change is performed and its safety indicators are evaluated.

### *6.5.1 Parameters and Characteristics of Port Oil Terminal Critical Infrastructure Operation Process Related to the Climate-Weather Change*

The port oil terminal critical infrastructure operation process $Z(t), t \in< 0, \infty)$, was analysed in Sect. 4.4 and its seven operation states $z_b$, $b = 1, 2, \ldots, 7$, were fixed in Sect. 4.4.1

The port oil terminal critical infrastructure was analyzed in Chap. 5 and the climate-weather change process $C(t), t \in< 0, \infty)$, at its operating area was defined in Sect. 5.4 and its 96 climate-weather states $c_l$, $l = 1, 2, \ldots, 96$, were fixed in Sect. 5.4.4.

Considering these findings, we assume that the port oil terminal operation process $Z(t), t \in< 0, \infty)$, and the climate-weather change process $C(t), t \in< 0, \infty)$, are independent and we introduce the port oil terminal critical infrastructure operation process related to climate-weather change $ZC(t), t \in< 0, \infty)$, at its operating area and we fix its operation and climate-weather states $zc_{bl}$, $b = 1, 2, \ldots, 7$, $l = 1, 2, \ldots, 96$.

To simplify the calculations, we omit the climate-weather states that did not occur, and we consider only these weather states $c_l$, $l = 1, 2, 6, 7, 10, 11, 17, 18, 22, 23, 26, 27, 33, 34, 38, 39, 42, 43, 65, 66, 70, 71, 74, 75, 81, 82, 86, 87, 90, 91$, that occurred with a non-zero probability.

Thus, we consider the following port oil terminal critical infrastructure operation process related to climate-weather change $ZC(t), t \in< 0, \infty)$, operation and climate-weather states $zc_{bl}$, namely:

- the state $zc_{11}$—the operation process $Z(t)$ is at the state $z_1$, and the climate-weather change process $C(t)$, is at the weather state $c_1$,
- the state $zc_{12}$—the operation process $Z(t)$, is at the state $z_1$, and the climate-weather change process $C(t)$, is at the weather state $c_2$,
- the state $zc_{16}$—the operation process $Z(t)$, is at the state $z_1$, and the climate-weather change process $C(t)$, is at the weather state $c_6$,
- the state $zc_{17}$—the operation process $Z(t)$, is at the state $z_1$, and the climate-weather change process $C(t)$, is at the weather state $c_7$,
- the state $zc_{1\,10}$- the operation process $Z(t)$, is at the state $z_1$, and the climate-weather change process $C(t)$, is at the weather state $c_{10}$,
- the state $zc_{1\,11}$—the operation process $Z(t)$, is at the state $z_1$, and the climate-weather change process $C(t)$, is at the weather state $c_{11}$,
- the state $zc_{1\,17}$—the operation process $Z(t)$, is at the state $z_1$, and the climate-weather change process $C(t)$, is at the weather state $c_{17}$,
- the state $zc_{1\,18}$—the operation process $Z(t)$, is at the state $z_1$, and the climate-weather change process $C(t)$, is at the weather state $c_{18}$,
- the state $zc_{1\,22}$—the operation process $Z(t)$, is at the state $z_1$, and the climate-weather change process $C(t)$, is at the weather state $c_{22}$,
- the state $zc_{1\,23}$—the operation process $Z(t)$, is at the state $z_1$, and the climate-weather change process $C(t)$, is at the weather state $c_{23}$,
- the state $zc_{1\,26}$—the operation process $Z(t)$, is at the state $z_1$, and the climate-weather change process $C(t)$, is at the weather state $c_{26}$,
- the state $zc_{1\,27}$- the operation process $Z(t)$, is at the state $z_1$, and the climate-weather change process $C(t)$, is at the weather state $c_{27}$,
- the state $zc_{1\,33}$—the operation process $Z(t)$, is at the state $z_1$, and the climate-weather change process $C(t)$, is at the weather state $c_{33}$,
- the state $zc_{1\,34}$—the operation process $Z(t)$, is at the state $z_1$, and the climate-weather change process $C(t)$, is at the weather state $c_{34}$,
- the state $zc_{1\,38}$—the operation process $Z(t)$, is at the state $z_1$, and the climate-weather change process $C(t)$, is at the weather state $c_{38}$,
- the state $zc_{1\,39}$—the operation process $Z(t)$, is at the state $z_1$, and the climate-weather change process $C(t)$, is at the weather state $c_{39}$,
- the state $zc_{1\,42}$—the operation process $Z(t)$, is at the state $z_1$, and the climate-weather change process $C(t)$, is at the weather state $c_{42}$,
- the state $zc_{1\,43}$—the operation process $Z(t)$, is at the state $z_1$, and the climate-weather change process $C(t)$, is at the weather state $c_{43}$,
- the state $zc_{1\,65}$—the operation process $Z(t)$, is at the state $z_1$, and the climate-weather change process $C(t)$, is at the weather state $c_{65}$,
- the state $zc_{1\,66}$—the operation process $Z(t)$, is at the state $z_1$, and the climate-weather change process $C(t)$, is at the weather state $c_{66}$,
- the state $zc_{1\,70}$—the operation process $Z(t)$, is at the state $z_1$, and the climate-weather change process $C(t)$, is at the weather state $c_{70}$,
- the state $zc_{1\,71}$—the operation process $Z(t)$, is at the state $z_1$, and the climate-weather change process $C(t)$, is at the weather state $c_{71}$,

- the state $zc_{1\,74}$—the operation process $Z(t)$, is at the state $z_1$, and the climate-weather change process $C(t)$, is at the weather state $c_{74}$,
- the state $zc_{1\,75}$- the operation process $Z(t)$, is at the state $z_1$, and the climate-weather change process $C(t)$, is at the weather state $c_{75}$,
- the state $zc_{1\,81}$—the operation process $Z(t)$, is at the state $z_1$, and the climate-weather change process $C(t)$, is at the weather state $c_{81}$,
- the state $zc_{1\,82}$—the operation process $Z(t)$, is at the state $z_1$, and the climate-weather change process $C(t)$, is at the weather state $c_{82}$,
- the state $zc_{1\,86}$—the operation process $Z(t)$, is at the state $z_1$, and the climate-weather change process $C(t)$, is at the weather state $c_{86}$,
- the state $zc_{1\,87}$—the operation process $Z(t)$, is at the state $z_1$, and the climate-weather change process $C(t)$, is at the weather state $c_{87}$,
- the state $zc_{1\,90}$—the operation process $Z(t)$, is at the state $z_1$, and the climate-weather change process $C(t)$, is at the weather state $c_{90}$,
- the state $zc_{1\,91}$—the operation process $Z(t)$, is at the state $z_1$, and the climate-weather change process $C(t)$, is at the weather state $c_{91}$,
- the state $zc_{21}$—the operation process $Z(t)$, is at the state $z_2$, and the climate-weather change process $C(t)$, is at the weather state $c_1$,
- the state $zc_{22}$—the operation process $Z(t)$, is at the state $z_2$, and the climate-weather change process $C(t)$, is at the weather state $c_2$,
- the state $zc_{26}$—the operation process $Z(t)$, is at the state $z_2$, and the climate-weather change process $C(t)$, is at the weather state $c_6$,
- the state $zc_{27}$—the operation process $Z(t)$, is at the state $z_2$, and the climate-weather change process $C(t)$, is at the weather state $c_7$,
- the state $zc_{2\,10}$—the operation process $Z(t)$, is at the state $z_2$, and the climate-weather change process $C(t)$, is at the weather state $c_{10}$,
- the state $zc_{2\,11}$—the operation process $Z(t)$, is at the state $z_2$, and the climate-weather change process $C(t)$, is at the weather state $c_{11}$,
- the state $zc_{2\,17}$—the operation process $Z(t)$, is at the state $z_2$, and the climate-weather change process $C(t)$, is at the weather state $c_{17}$,
- the state $zc_{2\,18}$—the operation process $Z(t)$, is at the state $z_2$, and the climate-weather change process $C(t)$, is at the weather state $c_{18}$,
- the state $zc_{2\,22}$—the operation process $Z(t)$, is at the state $z_2$, and the climate-weather change process $C(t)$, is at the weather state $c_{22}$,
- the state $zc_{2\,23}$—the operation process $Z(t)$, is at the state $z_2$, and the climate-weather change process $C(t)$, is at the weather state $c_{23}$,
- the state $zc_{2\,26}$—the operation process $Z(t)$, is at the state $z_2$, and the climate-weather change process $C(t)$, is at the weather state $c_{26}$,
- the state $zc_{2\,27}$—the operation process $Z(t)$, is at the state $z_2$, and the climate-weather change process $C(t)$, is at the weather state $c_{27}$,
- the state $zc_{2\,33}$—the operation process $Z(t)$, is at the state $z_2$, and the climate-weather change process $C(t)$, is at the weather state $c_{33}$,
- the state $zc_{2\,34}$—the operation process $Z(t)$, is at the state $z_2$, and the climate-weather change process $C(t)$, is at the weather state $c_{34}$,

- the state $zc_{2\,38}$—the operation process $Z(t)$, is at the state $z_2$, and the climate-weather change process $C(t)$, is at the weather state $c_{38}$,
- the state $zc_{2\,39}$—the operation process $Z(t)$, is at the state $z_2$, and the climate-weather change process $C(t)$, is at the weather state $c_{39}$,
- the state $zc_{2\,42}$—the operation process $Z(t)$, is at the state $z_2$, and the climate-weather change process $C(t)$, is at the weather state $c_{42}$,
- the state $zc_{2\,43}$—the operation process $Z(t)$, is at the state $z_2$, and the climate-weather change process $C(t)$, is at the weather state $c_{43}$,
- the state $zc_{2\,65}$—the operation process $Z(t)$, is at the state $z_2$, and the climate-weather change process $C(t)$, is at the weather state $c_{65}$;
- the state $zc_{2\,66}$—the operation process $Z(t)$, is at the state $z_2$, and the climate-weather change process $C(t)$, is at the weather state $c_{66}$,
- the state $zc_{2\,70}$—the operation process $Z(t)$, is at the state $z_2$, and the climate-weather change process $C(t)$, is at the weather state $c_{70}$,
- the state $zc_{2\,71}$—the operation process $Z(t)$, is at the state $z_2$, and the climate-weather change process $C(t)$, is at the weather state $c_{71}$,
- the state $zc_{2\,74}$—the operation process $Z(t)$, is at the state $z_2$, and the climate-weather change process $C(t)$, is at the weather state $c_{74}$,
- the state $zc_{2\,75}$—the operation process $Z(t)$, is at the state $z_2$, and the climate-weather change process $C(t)$, is at the weather state $c_{75}$,
- the state $zc_{2\,81}$—the operation process $Z(t)$, is at the state $z_2$, and the climate-weather change process $C(t)$, is at the weather state $c_{81}$,
- the state $zc_{2\,82}$—the operation process $Z(t)$, is at the state $z_2$, and the climate-weather change process $C(t)$, is at the weather state $c_{82}$,
- the state $zc_{2\,86}$—the operation process $Z(t)$, is at the state $z_2$, and the climate-weather change process $C(t)$, is at the weather state $c_{86}$,
- the state $zc_{2\,87}$—the operation process $Z(t)$, is at the state $z_2$, and the climate-weather change process $C(t)$, is at the weather state $c_{87}$,
- the state $zc_{2\,90}$—the operation process $Z(t)$, is at the state $z_2$, and the climate-weather change process $C(t)$, is at the weather state $c_{90}$,
- the state $zc_{2\,91}$—the operation process $Z(t)$, is at the state $z_2$, and the climate-weather change process $C(t)$, is at the weather state $c_{91}$,
- the state $zc_{31}$—the operation process $Z(t)$, is at the state $z_3$, and the climate-weather change process $C(t)$, is at the weather state $c_1$,
- the state $zc_{32}$—the operation process $Z(t)$, is at the state $z_3$, and the climate-weather change process $C(t)$, is at the weather state $c_2$,
- the state $zc_{36}$—the operation process $Z(t)$, is at the state $z_3$, and the climate-weather change process $C(t)$, is at the weather state $c_6$,
- the state $zc_{37}$—the operation process $Z(t)$, is at the state $z_3$, and the climate-weather change process $C(t)$, is at the weather state $c_7$,
- the state $zc_{3\,10}$—the operation process $Z(t)$, is at the state $z_3$, and the climate-weather change process $C(t)$, is at the weather state $c_{10}$,
- the state $zc_{3\,11}$—the operation process $Z(t)$, is at the state $z_3$, and the climate-weather change process $C(t)$, is at the weather state $c_{11}$,

- the state $zc_{3\,17}$—the operation process $Z(t)$, is at the state $z_3$, and the climate-weather change process $C(t)$, is at the weather state $c_{17}$,
- the state $zc_{3\,18}$—the operation process $Z(t)$, is at the state $z_3$, and the climate-weather change process $C(t)$, is at the weather state $c_{18}$,
- the state $zc_{3\,22}$—the operation process $Z(t)$, is at the state $z_3$, and the climate-weather change process $C(t)$, is at the weather state $c_{22}$,
- the state $zc_{3\,23}$—the operation process $Z(t)$, is at the state $z_3$, and the climate-weather change process $C(t)$, is at the weather state $c_{23}$,
- the state $zc_{3\,26}$—the operation process $Z(t)$, is at the state $z_3$, and the climate-weather change process $C(t)$, is at the weather state $c_{26}$,
- the state $zc_{3\,27}$—the operation process $Z(t)$, is at the state $z_3$, and the climate-weather change process $C(t)$, is at the weather state $c_{27}$,
- the state $zc_{3\,33}$—the operation process $Z(t)$, is at the state $z_3$, and the climate-weather change process $C(t)$, is at the weather state $c_{33}$,
- the state $zc_{3\,34}$—the operation process $Z(t)$, is at the state $z_3$, and the climate-weather change process $C(t)$, is at the weather state $c_{34}$,
- the state $zc_{3\,38}$—the operation process $Z(t)$, is at the state $z_3$, and the climate-weather change process $C(t)$, is at the weather state $c_{38}$,
- the state $zc_{3\,39}$—the operation process $Z(t)$, is at the state $z_3$, and the climate-weather change process $C(t)$, is at the weather state $c_{39}$,
- the state $zc_{3\,42}$—the operation process $Z(t)$, is at the state $z_3$, and the climate-weather change process $C(t)$, is at the weather state $c_{42}$,
- the state $zc_{3\,43}$—the operation process $Z(t)$, is at the state $z_3$, and the climate-weather change process $C(t)$, is at the weather state $c_{43}$,
- the state $zc_{3\,65}$- the operation process $Z(t)$, is at the state $z_3$, and the climate-weather change process $C(t)$, is at the weather state $c_{65}$,
- the state $zc_{3\,66}$—the operation process $Z(t)$, is at the state $z_3$, and the climate-weather change process $C(t)$, is at the weather state $c_{66}$,
- the state $zc_{3\,70}$—the operation process $Z(t)$, is at the state $z_3$, and the climate-weather change process $C(t)$, is at the weather state $c_{70}$,
- the state $zc_{3\,71}$—the operation process $Z(t)$, is at the state $z_3$, and the climate-weather change process $C(t)$, is at the weather state $c_{71}$,
- the state $zc_{3\,74}$—the operation process $Z(t)$, is at the state $z_3$, and the climate-weather change process $C(t)$, is at the weather state $c_{74}$,
- the state $zc_{3\,75}$—the operation process $Z(t)$, is at the state $z_3$, and the climate-weather change process $C(t)$, is at the weather state $c_{75}$,
- the state $zc_{3\,81}$—the operation process $Z(t)$, is at the state $z_3$, and the climate-weather change process $C(t)$, is at the weather state $c_{81}$,
- the state $zc_{3\,82}$—the operation process $Z(t)$, is at the state $z_3$, and the climate-weather change process $C(t)$, is at the weather state $c_{82}$,
- the state $zc_{3\,86}$—the operation process $Z(t)$, is at the state $z_3$, and the climate-weather change process $C(t)$, is at the weather state $c_{86}$,
- the state $zc_{3\,87}$—the operation process $Z(t)$, is at the state $z_3$, and the climate-weather change process $C(t)$, is at the weather state $c_{87}$,

- the state $zc_{3\,90}$—the operation process $Z(t)$, is at the state $z_3$, and the climate-weather change process $C(t)$, is at the weather state $c_{90}$,
- the state $zc_{3\,91}$—the operation process $Z(t)$, is at the state $z_3$, and the climate-weather change process $C(t)$, is at the weather state $c_{91}$,
- the state $zc_{41}$—the operation process $Z(t)$, is at the state $z_4$, and the climate-weather change process $C(t)$, is at the weather state $c_1$,
- the state $zc_{42}$—the operation process $Z(t)$, is at the state $z_4$, and the climate-weather change process $C(t)$, is at the weather state $c_2$,
- the state $zc_{46}$—the operation process $Z(t)$, is at the state $z_4$, and the climate-weather change process $C(t)$, is at the weather state $c_6$,
- the state $zc_{47}$—the operation process $Z(t)$, is at the state $z_4$, and the climate-weather change process $C(t)$, is at the weather state $c_7$,
- the state $zc_{4\,10}$—the operation process $Z(t)$, is at the state $z_4$, and the climate-weather change process $C(t)$, is at the weather state $c_{10}$,
- the state $zc_{4\,11}$—the operation process $Z(t)$, is at the state $z_4$, and the climate-weather change process $C(t)$, is at the weather state $c_{11}$,
- the state $zc_{4\,17}$—the operation process $Z(t)$, is at the state $z_4$, and the climate-weather change process $C(t)$, is at the weather state $c_{17}$,
- the state $zc_{4\,18}$—the operation process $Z(t)$, is at the state $z_4$, and the climate-weather change process $C(t)$, is at the weather state $c_{18}$,
- the state $zc_{4\,22}$—the operation process $Z(t)$, is at the state $z_4$, and the climate-weather change process $C(t)$, is at the weather state $c_{22}$,
- the state $zc_{4\,23}$—the operation process $Z(t)$, is at the state $z_4$, and the climate-weather change process $C(t)$, is at the weather state $c_{23}$,
- the state $zc_{4\,26}$—the operation process $Z(t)$, is at the state $z_4$, and the climate-weather change process $C(t)$, is at the weather state $c_{26}$,
- the state $zc_{4\,27}$—the operation process $Z(t)$, is at the state $z_4$, and the climate-weather change process $C(t)$, is at the weather state $c_{27}$,
- the state $zc_{4\,33}$—the operation process $Z(t)$, is at the state $z_4$, and the climate-weather change process $C(t)$, is at the weather state $c_{33}$,
- the state $zc_{4\,34}$—the operation process $Z(t)$, is at the state $z_4$, and the climate-weather change process $C(t)$, is at the weather state $c_{34}$,
- the state $zc_{4\,38}$—the operation process $Z(t)$, is at the state $z_4$, and the climate-weather change process $C(t)$, is at the weather state $c_{38}$,
- the state $zc_{4\,39}$—the operation process $Z(t)$, is at the state $z_4$, and the climate-weather change process $C(t)$, is at the weather state $c_{39}$,
- the state $zc_{4\,42}$—the operation process $Z(t)$, is at the state $z_4$, and the climate-weather change process $C(t)$, is at the weather state $c_{42}$,
- the state $zc_{4\,43}$—the operation process $Z(t)$, is at the state $z_4$, and the climate-weather change process $C(t)$, is at the weather state $c_{43}$,
- the state $zc_{4\,65}$—the operation process $Z(t)$, is at the state $z_4$, and the climate-weather change process $C(t)$, is at the weather state $c_{65}$,
- the state $zc_{4\,66}$—the operation process $Z(t)$, is at the state $z_4$, and the climate-weather change process $C(t)$, is at the weather state $c_{65}$,

- the state $zc_{4\,70}$—the operation process $Z(t)$, is at the state $z_4$, and the climate-weather change process $C(t)$, is at the weather state $c_{70}$,
- the state $zc_{4\,71}$—the operation process $Z(t)$, is at the state $z_4$, and the climate-weather change process $C(t)$, is at the weather state $c_{71}$,
- the state $zc_{4\,74}$—the operation process $Z(t)$, is at the state $z_4$, and the climate-weather change process $C(t)$, is at the weather state $c_{74}$,
- the state $zc_{4\,75}$—the operation process $Z(t)$, is at the state $z_4$, and the climate-weather change process $C(t)$, is at the weather state $c_{75}$,
- the state $zc_{4\,81}$—the operation process $Z(t)$, is at the state $z_4$, and the climate-weather change process $C(t)$, is at the weather state $c_{81}$,
- the state $zc_{4\,82}$—the operation process $Z(t)$, is at the state $z_4$, and the climate-weather change process $C(t)$, is at the weather state $c_{82}$,
- the state $zc_{4\,86}$—the operation process $Z(t)$, is at the state $z_4$, and the climate-weather change process $C(t)$, is at the weather state $c_{86}$,
- the state $zc_{4\,87}$—the operation process $Z(t)$, is at the state $z_4$, and the climate-weather change process $C(t)$, is at the weather state $c_{87}$,
- the state $zc_{4\,90}$—the operation process $Z(t)$, is at the state $z_4$, and the climate-weather change process $C(t)$, is at the weather state $c_{90}$,
- the state $zc_{4\,91}$—the operation process $Z(t)$, is at the state $z_4$, and the climate-weather change process $C(t)$, is at the weather state $c_{91}$,
- the state $zc_{51}$—the operation process $Z(t)$, is at the state $z_5$, and the climate-weather change process $C(t)$, is at the weather state $c_1$,
- the state $zc_{52}$—the operation process $Z(t)$, is at the state $z_5$, and the climate-weather change process $C(t)$, is at the weather state $c_2$,
- the state $zc_{56}$—the operation process $Z(t)$, is at the state $z_5$, and the climate-weather change process $C(t)$, is at the weather state $c_6$,
- the state $zc_{57}$—the operation process $Z(t)$, is at the state $z_5$, and the climate-weather change process $C(t)$, is at the weather state $c_7$,
- the state $zc_{5\,10}$—the operation process $Z(t)$, is at the state $z_5$, and the climate-weather change process $C(t)$, is at the weather state $c_{10}$,
- the state $zc_{5\,11}$—the operation process $Z(t)$, is at the state $z_5$, and the climate-weather change process $C(t)$, is at the weather state $c_{11}$,
- the state $zc_{5\,17}$—the operation process $Z(t)$, is at the state $z_5$, and the climate-weather change process $C(t)$, is at the weather state $c_{17}$,
- the state $zc_{5\,18}$—the operation process $Z(t)$, is at the state $z_5$, and the climate-weather change process $C(t)$, is at the weather state $c_{18}$,
- the state $zc_{5\,22}$—the operation process $Z(t)$, is at the state $z_5$, and the climate-weather change process $C(t)$, is at the weather state $c_{22}$,
- the state $zc_{5\,23}$—the operation process $Z(t)$, is at the state $z_5$, and the climate-weather change process $C(t)$, is at the weather state $c_{23}$,
- the state $zc_{5\,26}$—the operation process $Z(t)$, is at the state $z_5$, and the climate-weather change process $C(t)$, is at the weather state $c_{26}$,
- the state $zc_{5\,27}$—the operation process $Z(t)$, is at the state $z_5$, and the climate-weather change process $C(t)$, is at the weather state $c_{27}$,

- the state $zc_{5\,33}$—the operation process $Z(t)$, is at the state $z_5$, and the climate-weather change process $C(t)$, is at the weather state $c_{33}$,
- the state $zc_{5\,34}$—the operation process $Z(t)$, is at the state $z_5$, and the climate-weather change process $C(t)$, is at the weather state $c_{34}$,
- the state $zc_{5\,38}$—the operation process $Z(t)$, is at the state $z_5$, and the climate-weather change process $C(t)$, is at the weather state $c_{38}$,
- the state $zc_{5\,39}$—the operation process $Z(t)$, is at the state $z_5$, and the climate-weather change process $C(t)$, is at the weather state $c_{39}$,
- the state $zc_{5\,42}$—the operation process $Z(t)$, is at the state $z_5$, and the climate-weather change process $C(t)$, is at the weather state $c_{42}$,
- the state $zc_{5\,43}$—the operation process $Z(t)$, is at the state $z_5$, and the climate-weather change process $C(t)$, is at the weather state $c_{43}$,
- the state $zc_{5\,65}$—the operation process $Z(t)$, is at the state $z_5$, and the climate-weather change process $C(t)$, is at the weather state $c_{65}$,
- the state $zc_{5\,66}$—the operation process $Z(t)$, is at the state $z_5$, and the climate-weather change process $C(t)$, is at the weather state $c_{66}$,
- the state $zc_{5\,70}$—the operation process $Z(t)$, is at the state $z_5$, and the climate-weather change process $C(t)$, is at the weather state $c_{70}$,
- the state $zc_{5\,71}$—the operation process $Z(t)$, is at the state $z_5$, and the climate-weather change process $C(t)$, is at the weather state $c_{71}$,
- the state $zc_{5\,74}$—the operation process $Z(t)$, is at the state $z_5$, and the climate-weather change process $C(t)$, is at the weather state $c_{74}$,
- the state $zc_{5\,75}$—the operation process $Z(t)$, is at the state $z_5$, and the climate-weather change process $C(t)$, is at the weather state $c_{75}$,
- the state $zc_{5\,81}$—the operation process $Z(t)$, is at the state $z_5$, and the climate-weather change process $C(t)$, is at the weather state $c_{81}$,
- the state $zc_{5\,82}$—the operation process $Z(t)$, is at the state $z_s$, and the climate-weather change process $C(t)$, is at the weather state $c_{82}$,
- the state $zc_{5\,86}$—the operation process $Z(t)$, is at the state $z_5$, and the climate-weather change process $C(t)$, is at the weather state $c_{86}$,
- the state $zc_{5\,87}$—the operation process $Z(t)$, is at the state $z_5$, and the climate-weather change process $C(t)$, is at the weather state $c_{87}$,
- the state $zc_{5\,90}$—the operation process $Z(t)$, is at the state $z_5$, and the climate-weather change process $C(t)$, is at the weather state $c_{90}$,
- the state $zc_{5\,91}$—the operation process $Z(t)$, is at the state $z_5$, and the climate-weather change process $C(t)$, is at the weather state $c_{91}$,
- the state $zc_{61}$—the operation process $Z(t)$, is at the state $z_6$, and the climate-weather change process $C(t)$, is at the weather state $c_1$,
- the state $zc_{62}$—the operation process $Z(t)$, is at the state $z_6$, and the climate-weather change process $C(t)$, is at the weather state $c_2$,
- the state $zc_{66}$—the operation process $Z(t)$, is at the state $z_6$, and the climate-weather change process $C(t)$, is at the weather state $c_6$,
- the state $zc_{67}$—the operation process $Z(t)$, is at the state $z_6$, and the climate-weather change process $C(t)$, is at the weather state $c_7$,

- the state $zc_{6\,10}$—the operation process $Z(t)$, is at the state $z_6$, and the climate-weather change process $C(t)$, is at the weather state $c_{10}$,
- the state $zc_{6\,11}$—the operation process $Z(t)$, is at the state $z_6$, and the climate-weather change process $C(t)$, is at the weather state $c_{11}$,
- the state $zc_{6\,17}$—the operation process $Z(t)$, is at the state $z_6$, and the climate-weather change process $C(t)$, is at the weather state $c_{17}$,
- the state $zc_{6\,18}$—the operation process $Z(t)$, is at the state $z_6$, and the climate-weather change process $C(t)$, is at the weather state $c_{18}$,
- the state $zc_{6\,22}$—the operation process $Z(t)$, is at the state $z_6$, and the climate-weather change process $C(t)$, is at the weather state $c_{22}$,
- the state $zc_{6\,23}$—the operation process $Z(t)$, is at the state $z_6$, and the climate-weather change process $C(t)$, is at the weather state $c_{23}$,
- the state $zc_{6\,26}$—the operation process $Z(t)$, is at the state $z_6$, and the climate-weather change process $C(t)$, is at the weather state $c_{26}$,
- the state $zc_{6\,27}$—the operation process $Z(t)$, is at the state $z_6$, and the climate-weather change process $C(t)$, is at the weather state $c_{27}$,
- the state $zc_{6\,33}$—the operation process $Z(t)$, is at the state $z_6$, and the climate-weather change process $C(t)$, is at the weather state $c_{33}$,
- the state $zc_{6\,34}$—the operation process $Z(t)$, is at the state $z_6$, and the climate-weather change process $C(t)$, is at the weather state $c_{34}$,
- the state $zc_{6\,38}$—the operation process $Z(t)$, is at the state $z_6$, and the climate-weather change process $C(t)$, is at the weather state $c_{38}$,
- the state $zc_{6\,39}$—the operation process $Z(t)$, is at the state $z_6$, and the climate-weather change process $C(t)$, is at the weather state $c_{39}$,
- the state $zc_{6\,42}$—the operation process $Z(t)$, is at the state $z_6$, and the climate-weather change process $C(t)$, is at the weather state $c_{42}$,
- the state $zc_{6\,43}$—the operation process $Z(t)$, is at the state $z_6$, and the climate-weather change process $C(t)$, is at the weather state $c_{43}$,
- the state $zc_{6\,65}$—the operation process $Z(t)$, is at the state $z_6$, and the climate-weather change process $C(t)$, is at the weather state $c_{65}$,
- the state $zc_{6\,66}$—the operation process $Z(t)$, is at the state $z_6$, and the climate-weather change process $C(t)$, is at the weather state $c_{66}$,
- the state $zc_{6\,70}$—the operation process $Z(t)$, is at the state $z_6$, and the climate-weather change process $C(t)$, is at the weather state $c_{70}$,
- the state $zc_{6\,71}$—the operation process $Z(t)$, is at the state $z_6$, and the climate-weather change process $C(t)$, is at the weather state $c_{71}$,
- the state $zc_{6\,74}$—the operation process $Z(t)$, is at the state $z_6$, and the climate-weather change process $C(t)$, is at the weather state $c_{74}$,
- the state $zc_{6\,75}$—the operation process $Z(t)$, is at the state $z_6$, and the climate-weather change process $C(t)$, is at the weather state $c_{75}$,
- the state $zc_{6\,81}$—the operation process $Z(t)$, is at the state $z_6$, and the climate-weather change process $C(t)$, is at the weather state $c_{81}$,
- the state $zc_{6\,82}$—the operation process $Z(t)$, is at the state $z_6$, and the climate-weather change process $C(t)$, is at the weather state $c_{82}$,

- the state $zc_{6\,86}$—the operation process $Z(t)$, is at the state $z_6$, and the climate-weather change process $C(t)$, is at the weather state $c_{86}$,
- the state $zc_{6\,87}$—the operation process $Z(t)$, is at the state $z_6$, and the climate-weather change process $C(t)$, is at the weather state $c_{87}$,
- the state $zc_{6\,90}$—the operation process $Z(t)$, is at the state $z_6$, and the climate-weather change process $C(t)$, is at the weather state $c_{90}$,
- the state $zc_{6\,91}$—the operation process $Z(t)$, is at the state $z_6$, and the climate-weather change process $C(t)$, is at the weather state $c_{91}$,
- the state $zc_{71}$—the operation process $Z(t)$, is at the state $z_7$, and the climate-weather change process $C(t)$, is at the weather state $c_1$,
- the state $zc_{72}$—the operation process $Z(t)$, is at the state $z_7$, and the climate-weather change process $C(t)$, is at the weather state $c_2$,
- the state $zc_{76}$—the operation process $Z(t)$, is at the state $z_7$, and the climate-weather change process $C(t)$, is at the weather state $c_6$,
- the state $zc_{77}$—the operation process $Z(t)$, is at the state $z_7$, and the climate-weather change process $C(t)$, is at the weather state $c_7$,
- the state $zc_{7\,10}$—the operation process $Z(t)$, is at the state $z_7$, and the climate-weather change process $C(t)$, is at the weather state $c_{10}$,
- the state $zc_{7\,11}$—the operation process $Z(t)$, is at the state $z_7$, and the climate-weather change process $C(t)$, is at the weather state $c_{11}$,
- the state $zc_{7\,17}$—the operation process $Z(t)$, is at the state $z_7$, and the climate-weather change process $C(t)$, is at the weather state $c_{17}$,
- the state $zc_{7\,18}$—the operation process $Z(t)$, is at the state $z_7$, and the climate-weather change process $C(t)$, is at the weather state $c_{18}$,
- the state $zc_{7\,22}$—the operation process $Z(t)$, is at the state $z_7$, and the climate-weather change process $C(t)$, is at the weather state $c_{22}$,
- the state $zc_{7\,23}$—the operation process $Z(t)$, is at the state $z_7$, and the climate-weather change process $C(t)$, is at the weather state $c_{23}$,
- the state $zc_{7\,26}$—the operation process $Z(t)$, is at the state $z_7$, and the climate-weather change process $C(t)$, is at the weather state $c_{26}$,
- the state $zc_{7\,27}$—the operation process $Z(t)$, is at the state $z_7$, and the climate-weather change process $C(t)$, is at the weather state $c_{27}$,
- the state $zc_{7\,33}$—the operation process $Z(t)$, is at the state $z_7$, and the climate-weather change process $C(t)$, is at the weather state $c_{33}$,
- the state $zc_{7\,34}$—the operation process $Z(t)$, is at the state $z_7$, and the climate-weather change process $C(t)$, is at the weather state $c_{34}$,
- the state $zc_{7\,38}$—the operation process $Z(t)$, is at the state $z_7$, and the climate-weather change process $C(t)$, is at the weather state $c_{38}$,
- the state $zc_{7\,39}$—the operation process $Z(t)$, is at the state $z_7$, and the climate-weather change process $C(t)$, is at the weather state $c_{39}$,
- the state $zc_{7\,42}$—the operation process $Z(t)$, is at the state $z_7$, and the climate-weather change process $C(t)$, is at the weather state $c_{42}$,
- the state $zc_{7\,43}$—the operation process $Z(t)$, is at the state $z_7$, and the climate-weather change process $C(t)$, is at the weather state $c_{43}$,

- the state $zc_{7\,65}$—the operation process $Z(t)$, is at the state $z_7$, and the climate-weather change process $C(t)$, is at the weather state $c_{65}$,
- the state $zc_{7\,66}$—the operation process $Z(t)$, is at the state $z_7$, and the climate-weather change process $C(t)$, is at the weather state $c_{66}$,
- the state $zc_{7\,70}$—the operation process $Z(t)$, is at the state $z_7$, and the climate-weather change process $C(t)$, is at the weather state $c_{70}$,
- the state $zc_{7\,71}$—the operation process $Z(t)$, is at the state $z_7$, and the climate-weather change process $C(t)$, is at the weather state $c_{71}$,
- the state $zc_{7\,74}$—the operation process $Z(t)$, is at the state $z_7$, and the climate-weather change process $C(t)$, is at the weather state $c_{74}$,
- the state $zc_{7\,75}$—the operation process $Z(t)$, is at the state $z_7$, and the climate-weather change process $C(t)$, is at the weather state $c_{75}$,
- the state $zc_{7\,81}$—the operation process $Z(t)$, is at the state $z_7$, and the climate-weather change process $C(t)$, is at the weather state $c_{81}$,
- the state $zc_{7\,82}$—the operation process $Z(t)$, is at the state $z_7$, and the climate-weather change process $C(t)$, is at the weather state $c_{82}$,
- the state $zc_{7\,86}$—the operation process $Z(t)$, is at the state $z_7$, and the climate-weather change process $C(t)$, is at the weather state $c_{86}$,
- the state $zc_{7\,87}$—the operation process $Z(t)$, is at the state $z_7$, and the climate-weather change process $C(t)$, is at the weather state $c_{87}$,
- the state $zc_{7\,90}$—the operation process $Z(t)$, is at the state $z_7$, and the climate-weather change process $C(t)$, is at the weather state $c_{90}$,
- the state $zc_{7\,91}$—the operation process $Z(t)$, is at the state $z_7$, and the climate-weather change process $C(t)$, is at the weather state $c_{91}$.

Next considering the assumption that the processes $Z(t)$ and $C(t)$ are independent the transient probabilities of the port oil terminal critical infrastructure operation process related to climate-weather change $ZC(t)$, $t \in <0, \infty)$, can be expressed in the following way

$$
\begin{aligned}
pq_{bl}(t) &= P(ZC(t) = zc_{bl}) = P((Z(t) = z_b) \cap (C(t) = c_l) \\
&= P(Z(t) = z_b) \cdot P(C(t) = c_l) = p_b(t) \cdot q_l(t), \\
t \in &< 0, \infty), \quad b = 1, 2, \ldots, 7, \quad l = 1, 2, 6, 7, 10, 11, 17, 18, 22, 23, 26, 27, 33, \\
&34, 38, 39, 42, 43, 65, 66, 70, 71, 74, 75, 81, 82, 86, 87, 90, 91, \qquad (6.138)
\end{aligned}
$$

where $p_b(t)$, $b = 1, 2, \ldots, 7$, are transient probabilities of the operation process $Z(t)$ at the operation states $z_b$, and $q_l(t)$, $l = 1, 2, 6, 7, 10, 11, 17, 18, 22, 23, 26, 27, 33, 34, 38, 39, 42, 43, 65, 66, 70, 71, 74, 75, 81, 82, 86, 87, 90, 91$, are the transient probabilities of the climate-weather change process $C(t)$ at the climate-weather states $c_l$. Hence the limit values of the transient probabilities of the port oil terminal critical infrastructure operation process related to climate-weather change $ZC(t)$, at the operation and climate-weather state $zc_{bl}$, $b = 1, 2, \ldots, 7$, $l = 1, 2, 6, 7, 10, 11, 17, 18, 22, 23, 26, 27, 33, 34, 38, 39, 42, 43, 65, 66, 70, 71, 74, 75, 81, 82, 86, 87, 90, 91$,

$$pq_{bl} = \lim_{t\to\infty} pq_{bl}(t), \quad b = 1, 2, \ldots, 7, \quad l = 1, 2, 6, 7, 10, 11, 17, 18, 22, 23, 26,$$
$$27, 33, 34, 38, 39, 42, 43, 65, 66, 70, 71, 74, 75, 81, 82, 86, 87, 90, 91. \quad (6.139)$$

According to (6.74) can be found from the formula [21]

$$pq_{bl} = p_b q_l, \quad b = 1, 2, \ldots, 7, l = 1, 2, 6, 7, 10, 11, 17, 18, 22, 23, 26, 27,$$
$$33, 34, 38, 39, 42, 43, 65, 66, 70, 71, 74, 75, 81, 82, 86, 87, 90, 91, \quad (6.140)$$

where $p_b, b = 1, 2, \ldots, 7$, are the limit transient probabilities of the port oil terminal operation process $Z(t)$ at the particular operation states $z_b$, $b = 1, 2, \ldots, 7$, and $q_l$, $l$ = 1, 2, 6, 7, 10, 11, 17, 18, 22, 23, 26, 27, 33, 34, 38, 39, 42, 43, 65, 66, 70, 71, 74, 75, 81, 82, 86, 87, 90, 91, are the limit transient probabilities of the port oil terminal climate-weather change process $C(t)$ at its operating area, at the particular climate-weather states $c_l$, $l = 1, 2, 6, 7, 10, 11, 17, 18, 22, 23, 26, 27, 33, 34, 38, 39, 42, 43, 65, 66, 70, 71, 74, 75, 81, 82, 86, 87, 90, 91$.

Next applying (6.140) to (4.66) and (5.69)–(5.74), we determine the limit transient probabilities of the port oil terminal critical infrastructure operation process related to climate-weather change $ZC(t)$, $t \in< 0, +\infty)$ at its operation and climate-weather states $zc_{bl}$ and they are as follows:

$$pq_{11} = p_1 \cdot q_1 = 0.395 \cdot 0.000926 = 0.00036577,$$
$$pq_{12} = p_1 \cdot q_2 = 0.395 \cdot 0.107416 = 0.04242932,$$
$$pq_{16} = p_1 \cdot q_6 = 0.395 \cdot 0.718576 = 0.28383752,$$
$$pq_{17} = p_1 \cdot q_7 = 0.395 \cdot 0.019446 = 0.00768117,$$
$$pq_{1\,10} = p_1 \cdot q_{10} = 0.395 \cdot 0.012964 = 0.00512078,$$
$$pq_{1\,11} = p_1 \cdot q_{11} = 0.395 \cdot 0.066672 = 0.02633544,$$
$$pq_{1\,17} = p_1 \cdot q_{17} = 0.395 \cdot 0.000064 = 0.00002528,$$
$$pq_{1\,18} = p_1 \cdot q_{18} = 0.395 \cdot 0.007424 = 0.00293248,$$
$$pq_{1\,22} = p_1 \cdot q_{22} = 0.395 \cdot 0.049664 = 0.01961728,$$
$$pq_{1\,23} = p_1 \cdot q_{23} = 0.395 \cdot 0.001344 = 0.00053088,$$
$$pq_{1\,26} = p_1 \cdot q_{26} = 0.395 \cdot 0.000896 = 0.00035392,$$
$$pq_{1\,27} = p_1 \cdot q_{27} = 0.395 \cdot 0.004608 = 0.00182016,$$
$$pq_{1\,33} = p_1 \cdot q_{33} = 0.395 \cdot 0.000001 = 0.000000395,$$
$$pq_{1\,34} = p_1 \cdot q_{34} = 0.395 \cdot 0.000116 = 0.00004582,$$
$$pq_{1\,38} = p_1 \cdot q_{38} = 0.395 \cdot 0.000776 = 0.00030652,$$
$$pq_{1\,39} = p_1 \cdot q_{39} = 0.395 \cdot 0.000021 = 0.000008295,$$
$$pq_{1\,42} = p_1 \cdot q_{42} = 0.395 \cdot 0.000014 = 0.00000553,$$
$$pq_{1\,43} = p_1 \cdot q_{43} = 0.395 \cdot 0.000072 = 0.00002844,$$
$$pq_{1\,65} = p_1 \cdot q_{65} = 0.395 \cdot 0.000006 = 0.00000237,$$

$$
\begin{aligned}
pq_{1\,66} &= p_1 \cdot q_{66} = 0.395 \cdot 0.000696 = 0.00027492,\\
pq_{1\,70} &= p_1 \cdot q_{70} = 0.395 \cdot 0.004656 = 0.00183912,\\
pq_{1\,71} &= p_1 \cdot q_{71} = 0.395 \cdot 0.000126 = 0.00004977,\\
pq_{1\,74} &= p_1 \cdot q_{74} = 0.395 \cdot 0.000084 = 0.00003318,\\
pq_{1\,75} &= p_1 \cdot q_{75} = 0.395 \cdot 0.000432 = 0.00017064,\\
pq_{1\,81} &= p_1 \cdot q_{81} = 0.395 \cdot 0.000003 = 0.000001185,\\
pq_{1\,82} &= p_1 \cdot q_{82} = 0.395 \cdot 0.000348 = 0.00013746,\\
pq_{1\,86} &= p_1 \cdot q_{86} = 0.395 \cdot 0.002328 = 0.00091956,\\
pq_{1\,87} &= p_1 \cdot q_{87} = 0.395 \cdot 0.000063 = 0.000024885,\\
pq_{1\,90} &= p_1 \cdot q_{90} = 0.395 \cdot 0.000042 = 0.00001659,\\
pq_{1\,91} &= p_1 \cdot q_{91} = 0.395 \cdot 0.000216 = 0.00008532;
\end{aligned}
\tag{6.141}
$$

$$
\begin{aligned}
pq_{21} &= p_2 \cdot q_1 = 0.060 \cdot 0.000926 = 0.00005556,\\
pq_{22} &= p_2 \cdot q_2 = 0.060 \cdot 0.107416 = 0.00644496,\\
pq_{26} &= p_2 \cdot q_6 = 0.060 \cdot 0.718576 = 0.04311456,\\
pq_{27} &= p_2 \cdot q_7 = 0.060 \cdot 0.019446 = 0.00116676,\\
pq_{2\,10} &= p_2 \cdot q_{10} = 0.060 \cdot 0.012964 = 0.00077784,\\
pq_{2\,11} &= p_2 \cdot q_{11} = 0.060 \cdot 0.066672 = 0.00400032,\\
pq_{2\,17} &= p_2 \cdot q_{17} = 0.060 \cdot 0.000064 = 0.00000384,\\
pq_{2\,18} &= p_2 \cdot q_{18} = 0.060 \cdot 0.007424 = 0.00044544,\\
pq_{2\,22} &= p_2 \cdot q_{22} = 0.060 \cdot 0.049664 = 0.00297984,\\
pq_{2\,23} &= p_2 \cdot q_{23} = 0.060 \cdot 0.001344 = 0.00008064,\\
pq_{2\,26} &= p_2 \cdot q_{26} = 0.060 \cdot 0.000896 = 0.00005376,\\
pq_{2\,27} &= p_2 \cdot q_{27} = 0.060 \cdot 0.004608 = 0.00027648,\\
pq_{2\,33} &= p_2 \cdot q_{33} = 0.060 \cdot 0.000001 = 0.00000006,\\
pq_{2\,34} &= p_2 \cdot q_{34} = 0.060 \cdot 0.000116 = 0.00000696,\\
pq_{2\,38} &= p_2 \cdot q_{38} = 0.060 \cdot 0.000776 = 0.00004656,\\
pq_{2\,39} &= p_2 \cdot q_{39} = 0.060 \cdot 0.000021 = 0.00000126,\\
pq_{2\,42} &= p_2 \cdot q_{42} = 0.060 \cdot 0.000014 = 0.00000084,\\
pq_{2\,43} &= p_2 \cdot q_{43} = 0.060 \cdot 0.000072 = 0.00000432,\\
pq_{2\,65} &= p_2 \cdot q_{65} = 0.060 \cdot 0.000006 = 0.00000036,\\
pq_{2\,66} &= p_2 \cdot q_{66} = 0.060 \cdot 0.000696 = 0.00004176,\\
pq_{2\,70} &= p_2 \cdot q_{70} = 0.060 \cdot 0.004656 = 0.00027936,\\
pq_{2\,71} &= p_2 \cdot q_{71} = 0.060 \cdot 0.000126 = 0.00000756,\\
pq_{2\,74} &= p_2 \cdot q_{74} = 0.060 \cdot 0.000084 = 0.00000504,\\
pq_{2\,75} &= p_2 \cdot q_{75} = 0.060 \cdot 0.000432 = 0.00002592,
\end{aligned}
$$

$$
\begin{aligned}
pq_{2\,81} &= p_2 \cdot q_{81} = 0.060 \cdot 0.000003 = 0.00000018,\\
pq_{2\,82} &= p_2 \cdot q_{82} = 0.060 \cdot 0.000348 = 0.00002088,\\
pq_{2\,86} &= p_2 \cdot q_{86} = 0.060 \cdot 0.002328 = 0.00013968,\\
pq_{2\,87} &= p_2 \cdot q_{87} = 0.060 \cdot 0.000063 = 0.00000378,\\
pq_{2\,90} &= p_2 \cdot q_{90} = 0.060 \cdot 0.000042 = 0.00000252,\\
pq_{2\,91} &= p_2 \cdot q_{91} = 0.060 \cdot 0.000216 = 0.00001296;
\end{aligned}
\tag{6.142}
$$

$$
\begin{aligned}
pq_{31} &= p_3 \cdot q_1 = 0.003 \cdot 0.000926 = 0.000002778,\\
pq_{32} &= p_3 \cdot q_2 = 0.003 \cdot 0.107416 = 0.000322248,\\
pq_{36} &= p_3 \cdot q_6 = 0.003 \cdot 0.718576 = 0.002155728,\\
pq_{37} &= p_3 \cdot q_7 = 0.003 \cdot 0.019446 = 0.000058338,\\
pq_{3\,10} &= p_3 \cdot q_{10} = 0.003 \cdot 0.012964 = 0.000038892,\\
pq_{3\,11} &= p_3 \cdot q_{11} = 0.003 \cdot 0.066672 = 0.000200016,\\
pq_{3\,17} &= p_3 \cdot q_{17} = 0.003 \cdot 0.000064 = 0.000000192,\\
pq_{3\,18} &= p_3 \cdot q_{18} = 0.003 \cdot 0.007424 = 0.0000022272,\\
pq_{3\,22} &= p_3 \cdot q_{22} = 0.003 \cdot 0.049664 = 0.000148992,\\
pq_{3\,23} &= p_3 \cdot q_{23} = 0.003 \cdot 0.001344 = 0.000004032,\\
pq_{3\,26} &= p_3 \cdot q_{26} = 0.003 \cdot 0.000896 = 0.000002688,\\
pq_{3\,27} &= p_3 \cdot q_{27} = 0.003 \cdot 0.004608 = 0.000013824,\\
pq_{3\,33} &= p_3 \cdot q_{33} = 0.003 \cdot 0.000001 = 0.000000003,\\
pq_{3\,34} &= p_3 \cdot q_{34} = 0.003 \cdot 0.000116 = 0.000000348,\\
pq_{3\,38} &= p_3 \cdot q_{38} = 0.003 \cdot 0.000776 = 0.000002328,\\
pq_{3\,39} &= p_3 \cdot q_{39} = 0.003 \cdot 0.000021 = 0.000000063,\\
pq_{3\,42} &= p_3 \cdot q_{42} = 0.003 \cdot 0.000014 = 0.000000042,\\
pq_{3\,43} &= p_3 \cdot q_{43} = 0.003 \cdot 0.000072 = 0.000000216,\\
pq_{3\,65} &= p_3 \cdot q_{65} = 0.003 \cdot 0.000006 = 0.000000018,\\
pq_{3\,66} &= p_3 \cdot q_{66} = 0.003 \cdot 0.000696 = 0.000002088,\\
pq_{3\,70} &= p_3 \cdot q_{70} = 0.003 \cdot 0.004656 = 0.000013968,\\
pq_{3\,71} &= p_3 \cdot q_{71} = 0.003 \cdot 0.000126 = 0.000000378,\\
pq_{3\,74} &= p_3 \cdot q_{74} = 0.003 \cdot 0.000084 = 0.000000252,\\
pq_{3\,75} &= p_3 \cdot q_{75} = 0.003 \cdot 0.000432 = 0.000001296,\\
pq_{3\,81} &= p_3 \cdot q_{81} = 0.003 \cdot 0.000003 = 0.000000009,\\
pq_{3\,82} &= p_3 \cdot q_{82} = 0.003 \cdot 0.000348 = 0.000001044,\\
pq_{3\,86} &= p_3 \cdot q_{86} = 0.003 \cdot 0.002328 = 0.000006984,\\
pq_{3\,87} &= p_3 \cdot q_{87} = 0.003 \cdot 0.000063 = 0.000000189,\\
pq_{3\,90} &= p_3 \cdot q_{90} = 0.003 \cdot 0.000042 = 0.000000126,
\end{aligned}
$$

$$pq_{3\,91} = p_3 \cdot q_{91} = 0.003 \cdot 0.000216 = 0.000000648; \tag{6.143}$$

$$\begin{aligned}
pq_{41} &= p_4 \cdot q_1 = 0.002 \cdot 0.000926 = 0.000001852,\\
pq_{42} &= p_4 \cdot q_2 = 0.002 \cdot 0.107416 = 0.000214832,\\
pq_{46} &= p_4 \cdot q_6 = 0.002 \cdot 0.718576 = 0.001437152,\\
pq_{47} &= p_4 \cdot q_7 = 0.002 \cdot 0.019446 = 0.000038892,\\
pq_{4\,10} &= p_4 \cdot q_{10} = 0.002 \cdot 0.012964 = 0.000025928,\\
pq_{4\,11} &= p_4 \cdot q_{11} = 0.002 \cdot 0.066672 = 0.000133344,\\
pq_{4\,17} &= p_4 \cdot q_{17} = 0.002 \cdot 0.000064 = 0.000000128,\\
pq_{4\,18} &= p_4 \cdot q_{18} = 0.002 \cdot 0.007424 = 0.000014848,\\
pq_{4\,22} &= p_4 \cdot q_{22} = 0.002 \cdot 0.049664 = 0.000099328,\\
pq_{4\,23} &= p_4 \cdot q_{23} = 0.002 \cdot 0.001344 = 0.000002688,\\
pq_{4\,26} &= p_4 \cdot q_{26} = 0.002 \cdot 0.000896 = 0.000001792,\\
pq_{4\,27} &= p_4 \cdot q_{27} = 0.002 \cdot 0.004608 = 0.000009216,\\
pq_{4\,33} &= p_4 \cdot q_{33} = 0.002 \cdot 0.000001 = 0.000000002,\\
pq_{4\,34} &= p_4 \cdot q_{34} = 0.002 \cdot 0.000116 = 0.000000232,\\
pq_{4\,38} &= p_4 \cdot q_{38} = 0.002 \cdot 0.000776 = 0.000001552,\\
pq_{4\,39} &= p_4 \cdot q_{39} = 0.002 \cdot 0.000021 = 0.000000042,\\
pq_{4\,42} &= p_4 \cdot q_{42} = 0.002 \cdot 0.000014 = 0.000000028,\\
pq_{4\,43} &= p_4 \cdot q_{43} = 0.002 \cdot 0.000072 = 0.000000144,\\
pq_{4\,65} &= p_4 \cdot q_{65} = 0.002 \cdot 0.000006 = 0.000000012,\\
pq_{4\,66} &= p_4 \cdot q_{66} = 0.002 \cdot 0.000696 = 0.000001392,\\
pq_{4\,70} &= p_4 \cdot q_{70} = 0.002 \cdot 0.004656 = 0.000009312,\\
pq_{4\,71} &= p_4 \cdot q_{71} = 0.002 \cdot 0.000126 = 0.000000252,\\
pq_{4\,74} &= p_4 \cdot q_{74} = 0.002 \cdot 0.000084 = 0.000000168,\\
pq_{4\,75} &= p_4 \cdot q_{75} = 0.002 \cdot 0.000432 = 0.000000864,\\
pq_{4\,81} &= p_4 \cdot q_{81} = 0.002 \cdot 0.000003 = 0.000000006,\\
pq_{4\,82} &= p_4 \cdot q_{82} = 0.002 \cdot 0.000348 = 0.000000696,\\
pq_{4\,86} &= p_4 \cdot q_{86} = 0.002 \cdot 0.002328 = 0.000004656,\\
pq_{4\,87} &= p_4 \cdot q_{87} = 0.002 \cdot 0.000063 = 0.000000126,\\
pq_{4\,90} &= p_4 \cdot q_{90} = 0.002 \cdot 0.000042 = 0.000000084,\\
pq_{4\,91} &= p_4 \cdot q_{91} = 0.002 \cdot 0.000216 = 0.000000432;
\end{aligned} \tag{6.144}$$

$$
\begin{aligned}
pq_{51} &= p_5 \cdot q_1 = 0.2 \cdot 0.000926 = 0.0001852,\\
pq_{52} &= p_5 \cdot q_2 = 0.2 \cdot 0.107416 = 0.0214832,\\
pq_{56} &= p_5 \cdot q_6 = 0.2 \cdot 0.718576 = 0.1437152,\\
pq_{57} &= p_5 \cdot q_7 = 0.2 \cdot 0.019446 = 0.0038892,\\
pq_{5\,10} &= p_5 \cdot q_{10} = 0.2 \cdot 0.012964 = 0.0025928,\\
pq_{5\,11} &= p_5 \cdot q_{11} = 0.2 \cdot 0.066672 = 0.0133344,\\
pq_{5\,17} &= p_5 \cdot q_{17} = 0.2 \cdot 0.000064 = 0.0000128,\\
pq_{5\,18} &= p_5 \cdot q_{18} = 0.2 \cdot 0.007424 = 0.0014848,\\
pq_{5\,22} &= p_5 \cdot q_{22} = 0.2 \cdot 0.049664 = 0.0099328,\\
pq_{5\,23} &= p_5 \cdot q_{23} = 0.2 \cdot 0.001344 = 0.0002688,\\
pq_{5\,26} &= p_5 \cdot q_{26} = 0.2 \cdot 0.000896 = 0.0001792,\\
pq_{5\,27} &= p_5 \cdot q_{27} = 0.2 \cdot 0.004608 = 0.0009216,\\
pq_{5\,33} &= p_5 \cdot q_{33} = 0.2 \cdot 0.000001 = 0.0000002,\\
pq_{5\,34} &= p_5 \cdot q_{34} = 0.2 \cdot 0.000116 = 0.0000232,\\
pq_{5\,38} &= p_5 \cdot q_{38} = 0.2 \cdot 0.000776 = 0.0001552,\\
pq_{5\,39} &= p_5 \cdot q_{39} = 0.2 \cdot 0.000021 = 0.0000042,\\
pq_{5\,42} &= p_5 \cdot q_{42} = 0.2 \cdot 0.000014 = 0.0000028,\\
pq_{5\,43} &= p_5 \cdot q_{43} = 0.2 \cdot 0.000072 = 0.0000144,\\
pq_{5\,65} &= p_5 \cdot q_{65} = 0.2 \cdot 0.000006 = 0.0000012,\\
pq_{5\,66} &= p_5 \cdot q_{66} = 0.2 \cdot 0.000696 = 0.0001392,\\
pq_{5\,70} &= p_5 \cdot q_{70} = 0.2 \cdot 0.004656 = 0.0009312,\\
pq_{5\,71} &= p_5 \cdot q_{71} = 0.2 \cdot 0.000126 = 0.0000252,\\
pq_{5\,74} &= p_5 \cdot q_{74} = 0.2 \cdot 0.000084 = 0.0000168,\\
pq_{5\,75} &= p_5 \cdot q_{75} = 0.2 \cdot 0.000432 = 0.0000864,\\
pq_{5\,81} &= p_5 \cdot q_{81} = 0.2 \cdot 0.000003 = 0.0000006,\\
pq_{5\,82} &= p_5 \cdot q_{82} = 0.2 \cdot 0.000348 = 0.0000696,\\
pq_{5\,86} &= p_5 \cdot q_{86} = 0.2 \cdot 0.002328 = 0.0004656,\\
pq_{5\,87} &= p_5 \cdot q_{87} = 0.2 \cdot 0.000063 = 0.0000126,\\
pq_{5\,90} &= p_5 \cdot q_{90} = 0.2 \cdot 0.000042 = 0.0000084,\\
pq_{5\,91} &= p_5 \cdot q_{91} = 0.2 \cdot 0.000216 = 0.0000432; \qquad (6.145)
\end{aligned}
$$

$$
\begin{aligned}
pq_{61} &= p_6 \cdot q_1 = 0.058 \cdot 0.000926 = 0.000053708,\\
pq_{62} &= p_6 \cdot q_2 = 0.058 \cdot 0.107416 = 0.006230128,\\
pq_{66} &= p_6 \cdot q_6 = 0.058 \cdot 0.718576 = 0.041677408,\\
pq_{67} &= p_6 \cdot q_7 = 0.058 \cdot 0.019446 = 0.001127868,\\
pq_{6\,10} &= p_6 \cdot q_{10} = 0.058 \cdot 0.012964 = 0.000751912,\\
pq_{6\,11} &= p_6 \cdot q_{11} = 0.058 \cdot 0.066672 = 0.003866976,\\
pq_{6\,17} &= p_6 \cdot q_{17} = 0.058 \cdot 0.000064 = 0.000003712,\\
pq_{6\,18} &= p_6 \cdot q_{18} = 0.058 \cdot 0.007424 = 0.000430592,\\
pq_{6\,22} &= p_6 \cdot q_{22} = 0.058 \cdot 0.049664 = 0.002880512,\\
pq_{6\,23} &= p_6 \cdot q_{23} = 0.058 \cdot 0.001344 = 0.000077952,\\
pq_{6\,26} &= p_6 \cdot q_{26} = 0.058 \cdot 0.000896 = 0.0000051968,\\
pq_{6\,27} &= p_6 \cdot q_{27} = 0.058 \cdot 0.004608 = 0.000267264,\\
pq_{6\,33} &= p_6 \cdot q_{33} = 0.058 \cdot 0.000001 = 0.000000058,\\
pq_{6\,34} &= p_6 \cdot q_{34} = 0.058 \cdot 0.000116 = 0.000006728,\\
pq_{6\,38} &= p_6 \cdot q_{38} = 0.058 \cdot 0.000776 = 0.000045008,\\
pq_{6\,39} &= p_6 \cdot q_{39} = 0.058 \cdot 0.000021 = 0.000001218,\\
pq_{6\,42} &= p_6 \cdot q_{42} = 0.058 \cdot 0.000014 = 0.000000812,\\
pq_{6\,43} &= p_6 \cdot q_{43} = 0.058 \cdot 0.000072 = 0.000004176,\\
pq_{6\,65} &= p_6 \cdot q_{65} = 0.058 \cdot 0.000006 = 0.000000348,\\
pq_{6\,66} &= p_6 \cdot q_{66} = 0.058 \cdot 0.000696 = 0.000040368,\\
pq_{6\,70} &= p_6 \cdot q_{70} = 0.058 \cdot 0.004656 = 0.000270048,\\
pq_{6\,71} &= p_6 \cdot q_{71} = 0.058 \cdot 0.000126 = 0.000007308,\\
pq_{6\,74} &= p_6 \cdot q_{74} = 0.058 \cdot 0.000084 = 0.000004872,\\
pq_{6\,75} &= p_6 \cdot q_{75} = 0.058 \cdot 0.000432 = 0.000025056,\\
pq_{6\,81} &= p_6 \cdot q_{81} = 0.058 \cdot 0.000003 = 0.000000174,\\
pq_{6\,82} &= p_6 \cdot q_{82} = 0.058 \cdot 0.000348 = 0.000020184,\\
pq_{6\,86} &= p_6 \cdot q_{86} = 0.058 \cdot 0.002328 = 0.000135024,\\
pq_{6\,87} &= p_6 \cdot q_{87} = 0.058 \cdot 0.000063 = 0.000003654,\\
pq_{6\,90} &= p_6 \cdot q_{90} = 0.058 \cdot 0.000042 = 0.000002436,\\
pq_{6\,91} &= p_6 \cdot q_{91} = 0.058 \cdot 0.000216 = 0.000012528;
\end{aligned}
\tag{6.146}
$$

$$
\begin{aligned}
pq_{71} &= p_7 \cdot q_1 = 0.282 \cdot 0.000926 = 0.000261132,\\
pq_{72} &= p_7 \cdot q_2 = 0.282 \cdot 0.107416 = 0.030291312,\\
pq_{76} &= p_7 \cdot q_6 = 0.282 \cdot 0.718576 = 0.202638432,\\
pq_{77} &= p_7 \cdot q_7 = 0.282 \cdot 0.019446 = 0.005483772,
\end{aligned}
$$

$$
\begin{aligned}
pq_{7\,10} &= p_7 \cdot q_{10} = 0.282 \cdot 0.012964 = 0.003655848, \\
pq_{7\,11} &= p_7 \cdot q_{11} = 0.282 \cdot 0.066672 = 0.018801504, \\
pq_{7\,17} &= p_7 \cdot q_{17} = 0.282 \cdot 0.000064 = 0.000018048, \\
pq_{7\,18} &= p_7 \cdot q_{18} = 0.282 \cdot 0.007424 = 0.002093568, \\
pq_{7\,22} &= p_7 \cdot q_{22} = 0.282 \cdot 0.049664 = 0.014005248, \\
pq_{7\,23} &= p_7 \cdot q_{23} = 0.282 \cdot 0.001344 = 0.000379008, \\
pq_{7\,26} &= p_7 \cdot q_{26} = 0.282 \cdot 0.000896 = 0.000252672, \\
pq_{7\,27} &= p_7 \cdot q_{27} = 0.282 \cdot 0.004608 = 0.001299456, \\
pq_{7\,33} &= p_7 \cdot q_{33} = 0.282 \cdot 0.000001 = 0.000000282, \\
pq_{7\,34} &= p_7 \cdot q_{34} = 0.282 \cdot 0.000116 = 0.000032712, \\
pq_{7\,38} &= p_7 \cdot q_{38} = 0.282 \cdot 0.000776 = 0.000218832, \\
pq_{7\,39} &= p_7 \cdot q_{39} = 0.282 \cdot 0.000021 = 0.000005922, \\
pq_{7\,42} &= p_7 \cdot q_{42} = 0.282 \cdot 0.000014 = 0.000003948, \\
pq_{7\,43} &= p_7 \cdot q_{43} = 0.282 \cdot 0.000072 = 0.000020304, \\
pq_{7\,65} &= p_7 \cdot q_{65} = 0.282 \cdot 0.000006 = 0.000001692, \\
pq_{7\,66} &= p_7 \cdot q_{66} = 0.282 \cdot 0.000696 = 0.000196272, \\
pq_{7\,70} &= p_7 \cdot q_{70} = 0.282 \cdot 0.004656 = 0.001312992, \\
pq_{7\,71} &= p_7 \cdot q_{71} = 0.282 \cdot 0.000126 = 0.000035532, \\
pq_{7\,74} &= p_7 \cdot q_{74} = 0.282 \cdot 0.000084 = 0.000023688, \\
pq_{7\,75} &= p_7 \cdot q_{75} = 0.282 \cdot 0.000432 = 0.000121824, \\
pq_{7\,81} &= p_7 \cdot q_{81} = 0.282 \cdot 0.000003 = 0.000000846, \\
pq_{7\,82} &= p_7 \cdot q_{82} = 0.282 \cdot 0.000348 = 0.000098136, \\
pq_{7\,86} &= p_7 \cdot q_{86} = 0.282 \cdot 0.002328 = 0.000656496, \\
pq_{7\,87} &= p_7 \cdot q_{87} = 0.282 \cdot 0.000063 = 0.000017766, \\
pq_{7\,90} &= p_7 \cdot q_{90} = 0.282 \cdot 0.000042 = 0.000011844, \\
pq_{7\,91} &= p_7 \cdot q_{91} = 0.282 \cdot 0.000216 = 0.000060912.
\end{aligned} \tag{6.147}
$$

### 6.5.2 *Port Oil Terminal Critical Infrastructure Safety Parameters Impacted by Operation Process Related to the Climate-Weather Change*

In this section, we accept assumptions and agreements such as in Sect. 3.5.2. Next, considering the assumption that the processes $Z(t)$ and $C(t)$ are independent, we arbitrarily assume that the coefficients of the operation process related to the climate-weather change $ZC(t)$ impact on the port oil terminal critical infrastructure intensities of ageing are given by

$$[\rho_i^3(u)]^{(bl)} = [\rho_i^1(u)]^{(b)} \cdot [\rho_i^2(u)]^{(l)}, \quad u = 1, 2, \ldots, z,$$
$$b = 1, 2, \ldots, v, \quad l = 1, 2, \ldots, w, \tag{6.148}$$

where $[\rho_i^1(u)]^{(b)}$, $b = 1, 2, \ldots, v$, are the coefficients of the operation process impact on the port oil terminal critical infrastructure intensities of ageing given in Chap. 4, and $[\rho_i^2(u)]^{(l)}$, $l = 1, 2, \ldots, w$, are the coefficients of the climate-weather change process impact on the port oil terminal critical infrastructure intensities of ageing given in Chap. 5.

Thus, the coefficients of the operation process related to the climate-weather change impact on the port oil terminal critical infrastructure intensities of ageing at the operation and climate-weather states $zc_{bl}$, $b = 1, 2, \ldots, 7$, $l = 1, 2, 6, 7, 10, 11, 17, 18, 22, 23, 26, 27, 33, 34, 38, 39, 42, 43, 65, 66, 70, 71, 74, 75, 81, 82, 86, 87, 90, 91$, according to (6.148), (4.69) and (5.75)–(5.78) are as follows [18, 19, 22]:

- for asset $A_1$

$$[\rho_1^3(1)]^{(bl)} = 1.10 \cdot 1.05 = 1.155, \quad [\rho_1^3(2)]^{(bl)} = 1.10 \cdot 1.05 = 1.155,$$
$$b = 1, 2, 7, \quad l = 1, 2, 17, 18, \tag{6.149}$$

$$[\rho_1^3(1)]^{(bl)} = 1.10 \cdot 1.08 = 1.188, \quad [\rho_1^3(2)]^{(bl)} = 1.10 \cdot 1.09 = 1.199,$$
$$b = 1, 2, 7, \quad l = 33, 34, 65, 66, 81, 82, \tag{6.150}$$

$$[\rho_1^3(1)]^{(bl)} = 1.10 \cdot 1.03 = 1.133, \quad [\rho_1^3(2)]^{(bl)} = 1.10 \cdot 1.04 = 1.144,$$
$$b = 1, 2, 7, \quad l = 38, 39, 42, 43, 70, 71, 74, 75, 86, 87, 90, 91, \tag{6.151}$$

$$[\rho_1^3(1)]^{(bl)} = 1.10 \cdot 1 = 1.10, \quad [\rho_1^3(2)]^{(bl)} = 1.10 \cdot 1 = 1.10,$$
$$b = 1, 2, 7, \quad l = 6, 7, 10, 11, 22, 23, 26, 27, \tag{6.152}$$

$$[\rho_1^3(1)]^{(bl)} = 1.20 \cdot 1.05 = 1.26, \quad [\rho_1^3(2)]^{(bl)} = 1.20 \cdot 1.05 = 1.26,$$
$$b = 3, 5, \quad l = 1, 2, 17, 18, \tag{6.153}$$

$$[\rho_1^3(1)]^{(bl)} = 1.20 \cdot 1.08 = 1.296, \quad [\rho_1^3(2)]^{(bl)} = 1.20 \cdot 1.09 = 1.308,$$
$$b = 3, 5, \quad l = 33, 34, 65, 66, 81, 82, \tag{6.154}$$

$$[\rho_1^3(1)]^{(bl)} = 1.20 \cdot 1.03 = 1.236, \quad [\rho_1^3(2)]^{(bl)} = 1.20 \cdot 1.04 = 1.248,$$
$$b = 3, 5, \quad l = 38, 39, 42, 43, 70, 71, 74, 75, 86, 87, 90, 91, \tag{6.155}$$

$$\left[\rho_1^3(1)\right]^{(bl)} = 1.20 \cdot 1 = 1.20, \quad \left[\rho_1^3(2)\right]^{(bl)} = 1.20 \cdot 1 = 1.20, \\ b = 3, 5, \quad l = 6, 7, 10, 11, 22, 23, 26, 27, \tag{6.156}$$

$$\left[\rho_1^3(1)\right]^{(bl)} = 1.30 \cdot 1.05 = 1.365, \quad \left[\rho_1^3(2)\right]^{(bl)} = 1.30 \cdot 1.05 = 1.365, \\ b = 4, 6, \quad l = 1, 2, 17, 18, \tag{6.157}$$

$$\left[\rho_1^3(1)\right]^{(bl)} = 1.30 \cdot 1.08 = 1.404, \quad \left[\rho_1^3(2)\right]^{(bl)} = 1.30 \cdot 1.09 = 1.417, \\ b = 4, 6, \quad l = 33, 34, 65, 66, 81, 82, \tag{6.158}$$

$$\left[\rho_1^3(1)\right]^{(bl)} = 1.30 \cdot 1.03 = 1.339, \quad \left[\rho_1^3(2)\right]^{(bl)} = 1.30 \cdot 1.04 = 1.352, \\ b = 4, 6, \quad l = 38, 39, 42, 43, 70, 71, 74, 75, 86, 87, 90, 91, \tag{6.159}$$

$$\left[\rho_1^3(1)\right]^{(bl)} = 1.30 \cdot 1 = 1.30, \quad \left[\rho_1^3(2)\right]^{(bl)} = 1.30 \cdot 1 = 1.30, \\ b = 4, 6, \quad l = 6, 7, 10, 11, 22, 23, 26, 27; \tag{6.160}$$

- for asset $A_2$

$$\left[\rho_2^3(1)\right]^{(bl)} = 1.10 \cdot 1 = 1.10, \quad \left[\rho_2^3(2)\right]^{(bl)} = 1.10 \cdot 1 = 1.10, b = 1, 2, 7, \\ l = 6, 7, 10, 11, 22, 23, 26, 27, 38, 39, 42, 43, 70, 71, 74, 75, 86, 87, 90, 91, \tag{6.161}$$

$$\left[\rho_2^3(1)\right]^{(bl)} = 1.10 \cdot 1.10 = 1.21, \quad \left[\rho_2^3(2)\right]^{(bl)} = 1.10 \cdot 1.10 = 1.21, \\ b = 1, 2, 7, \quad l = 1, 2, 17, 18, 33, 34, 65, 66, 81, 82, \tag{6.162}$$

$$\left[\rho_2^3(1)\right]^{(bl)} = 1.20 \cdot 1 = 1.20, \quad \left[\rho_2^3(2)\right]^{(bl)} = 1.20 \cdot 1 = 1.20, b = 3, 5, \\ l = 6, 7, 10, 11, 22, 23, 26, 27, 38, 39, 42, 43, 70, 71, 74, 75, 86, 87, 90, 91, \tag{6.163}$$

$$\left[\rho_2^3(1)\right]^{(bl)} = 1.20 \cdot 1.10 = 1.32, \quad \left[\rho_2^3(2)\right]^{(bl)} = 1.20 \cdot 1.10 = 1.32, \\ b = 3, 5, \quad l = 1, 2, 17, 18, 33, 34, 65, 66, 81, 82, \tag{6.164}$$

$$\left[\rho_2^3(1)\right]^{(bl)} = 1.30 \cdot 1 = 1.30, \quad \left[\rho_2^3(2)\right]^{(bl)} = 1.30 \cdot 1 = 1.30, b = 4, 6, \\ l = 6, 7, 10, 11, 22, 23, 26, 27, 38, 39, 42, 43, 70, 71, 74, 75, 86, 87, 90, 91, \tag{6.165}$$

$$\left[\rho_2^3(1)\right]^{(bl)} = 1.30 \cdot 1.10 = 1.43, \quad \left[\rho_2^3(2)\right]^{(bl)} = 1.30 \cdot 1.10 = 1.43, \\ b = 4, 6, \quad l = 1, 2, 17, 18, 33, 34, 65, 66, 81, 82; \tag{6.166}$$

- for asset $A_3$

$$\left[\rho_3^3(1)\right]^{(bl)} = 1 \cdot 1 = 1, \quad \left[\rho_3^3(2)\right]^{(bl)} = 1 \cdot 1 = 1, b = 1, 2, 3, 7, \\ l = 6, 7, 10, 11, 22, 23, 26, 27, 38, 39, 42, 43, 70, 71, 74, 75, 86, 87, 90, 91, \tag{6.167}$$

$$\left[\rho_3^3(1)\right]^{(bl)} = 1 \cdot 1.10 = 1.10, \quad \left[\rho_3^3(2)\right]^{(bl)} = 1 \cdot 1.10 = 1.10, \\ b = 1, 2, 3, 7, \quad l = 1, 2, 17, 18, 33, 34, 65, 66, 81, 82, \tag{6.168}$$

$$\left[\rho_3^3(1)\right]^{(bl)} = 1.20 \cdot 1 = 1.20, \quad \left[\rho_3^3(2)\right]^{(bl)} = 1.20 \cdot 1 = 1.20, b = 5, \\ l = 6, 7, 10, 11, 22, 23, 26, 27, 38, 39, 42, 43, 70, 71, 74, 75, 86, 87, 90, 91, \tag{6.169}$$

$$\left[\rho_3^3(1)\right]^{(bl)} = 1.20 \cdot 1.10 = 1.32, \quad \left[\rho_3^3(2)\right]^{(bl)} = 1.20 \cdot 1.10 = 1.32, \\ b = 5, \quad l = 1, 2, 17, 18, 33, 34, 65, 66, 81, 82, \tag{6.170}$$

$$\left[\rho_3^3(1)\right]^{(bl)} = 1.30 \cdot 1 = 1.30, \quad \left[\rho_3^3(2)\right]^{(bl)} = 1.30 \cdot 1 = 1.30, b = 4, 6, \\ l = 6, 7, 10, 11, 22, 23, 26, 27, 38, 39, 42, 43, 70, 71, 74, 75, 86, 87, 90, 91, \tag{6.171}$$

$$\left[\rho_3^3(1)\right]^{(bl)} = 1.30 \cdot 1.10 = 1.43, \quad \left[\rho_3^3(2)\right]^{(bl)} = 1.30 \cdot 1.10 = 1.43, \\ b = 4, 6, \quad l = 1, 2, 17, 18, 33, 34, 65, 66, 81, 82; \tag{6.172}$$

- for asset $A_4$

$$\left[\rho_4^3(1)\right]^{(bl)} = 1.10 \cdot 1 = 1.10, \quad \left[\rho_4^3(2)\right]^{(bl)} = 1.10 \cdot 1 = 1.10, b = 1, 2, 7, \\ l = 6, 7, 10, 11, 22, 23, 26, 27, 38, 39, 42, 43, 70, 71, 74, 75, 86, 87, 90, 91, \tag{6.173}$$

$$\begin{aligned}&\left[\rho_4^3(1)\right]^{(bl)} = 1.10 \cdot 1.10 = 1.21, \quad \left[\rho_4^3(2)\right]^{(bl)} = 1.10 \cdot 1.10 = 1.21,\\ &b = 1, 2, 7, \quad l = 1, 2, 17, 18, 33, 34, 65, 66, 81, 82,\end{aligned} \tag{6.174}$$

$$\begin{aligned}&\left[\rho_4^3(1)\right]^{(bl)} = 1.20 \cdot 1 = 1.20, \quad \left[\rho_4^3(2)\right]^{(bl)} = 1.20 \cdot 1 = 1.20,\\ &b = 3, 5, \quad l = 6, 7, 10, 11, 22, 23, 26, 27, 38, 39,\\ &42, 43, 70, 71, 74, 75, 86, 87, 90, 91,\end{aligned} \tag{6.175}$$

$$\begin{aligned}&\left[\rho_4^3(1)\right]^{(bl)} = 1.20 \cdot 1.10 = 1.32, \quad \left[\rho_4^3(2)\right]^{(bl)} = 1.20 \cdot 1.10 = 1.32,\\ &b = 3, 5, \quad l = 1, 2, 17, 18, 33, 34, 65, 66, 81, 82,\end{aligned} \tag{6.176}$$

$$\begin{aligned}&\left[\rho_4^3(1)\right]^{(bl)} = 1.30 \cdot 1 = 1.30, \quad \left[\rho_4^3(2)\right]^{(bl)} = 1.30 \cdot 1 = 1.30, b = 4, 6,\\ &l = 6, 7, 10, 11, 22, 23, 26, 27, 38, 39, 42, 43, 70, 71, 74, 75, 86, 87, 90, 91,\end{aligned} \tag{6.177}$$

$$\begin{aligned}&\left[\rho_4^3(1)\right]^{(bl)} = 1.30 \cdot 1.10 = 1.43, \quad \left[\rho_4^3(2)\right]^{(bl)} = 1.30 \cdot 1.10 = 1.43,\\ &b = 4, 6, \quad l = 1, 2, 17, 18, 33, 34, 65, 66, 81, 82;\end{aligned} \tag{6.178}$$

- for asset $A_5$

$$\begin{aligned}&\left[\rho_5^3(1)\right]^{(bl)} = 1 \cdot 1 = 1, \quad \left[\rho_5^3(2)\right]^{(bl)} = 1 \cdot 1 = 1, b = 1, 2, 7,\\ &l = 6, 7, 10, 11, 22, 23, 26, 27, 38, 39, 42, 43, 70, 71, 74, 75, 86, 87, 90, 91,\end{aligned} \tag{6.179}$$

$$\begin{aligned}&\left[\rho_5^3(1)\right]^{(bl)} = 1 \cdot 1.10 = 1.10, \quad \left[\rho_5^3(2)\right]^{(bl)} = 1 \cdot 1.10 = 1.10,\\ &b = 1, 2, 7, \quad l = 1, 2, 17, 18, 33, 34, 65, 66, 81, 82,\end{aligned} \tag{6.180}$$

$$\begin{aligned}&\left[\rho_5^3(1)\right]^{(bl)} = 1.20 \cdot 1 = 1.20, \quad \left[\rho_5^3(2)\right]^{(bl)} = 1.20 \cdot 1 = 1.20, b = 3, 5,\\ &l = 6, 7, 10, 11, 22, 23, 26, 27, 38, 39, 42, 43, 70, 71, 74, 75, 86, 87, 90, 91,\end{aligned} \tag{6.181}$$

$$\begin{aligned}&\left[\rho_5^3(1)\right]^{(bl)} = 1.20 \cdot 1.10 = 1.32, \quad \left[\rho_5^3(2)\right]^{(bl)} = 1.20 \cdot 1.10 = 1.32,\\ &b = 3, 5, \quad l = 1, 2, 17, 18, 33, 34, 65, 66, 81, 82,\end{aligned} \tag{6.182}$$

$$\begin{aligned}&\left[\rho_5^3(1)\right]^{(bl)} = 1.30 \cdot 1 = 1.30, \quad \left[\rho_5^3(2)\right]^{(bl)} = 1.30 \cdot 1 = 1.30, b = 4, 6,\\ &l = 6, 7, 10, 11, 22, 23, 26, 27, 38, 39, 42, 43, 70, 71, 74, 75, 86, 87, 90, 91,\end{aligned} \tag{6.183}$$

$$\left[\rho_5^3(1)\right]^{(bl)} = 1.30 \cdot 1.10 = 1.43, \quad \left[\rho_5^3(2)\right]^{(bl)} = 1.30 \cdot 1.10 = 1.43, \quad b = 4, 6, \quad l = 1, 2, 17, 18, 33, 34, 65, 66, 81, 82; \tag{6.184}$$

- for asset $A_6$

$$\left[\rho_6^3(1)\right]^{(bl)} = 1 \cdot 1 = 1, \quad \left[\rho_6^3(2)\right]^{(bl)} = 1 \cdot 1 = 1, b = 2, 5, \quad l = 6, 7, 10, 11, 22, 23, 26, 27, 38, 39, 42, 43, 70, 71, 74, 75, 86, 87, 90, 91, \tag{6.185}$$

$$\left[\rho_6^3(1)\right]^{(bl)} = 1 \cdot 1.10 = 1.10, \quad \left[\rho_6^3(2)\right]^{(bl)} = 1 \cdot 1.10 = 1.10, \quad b = 2, 5, \quad l = 1, 2, 17, 18, 33, 34, 65, 66, 81, 82, \tag{6.186}$$

$$\left[\rho_6^3(1)\right]^{(bl)} = 1.10 \cdot 1 = 1.10, \quad \left[\rho_6^3(2)\right]^{(bl)} = 1.10 \cdot 1 = 1.10, b = 1, 7, \quad l = 6, 7, 10, 11, 22, 23, 26, 27, 38, 39, 42, 43, 70, 71, 74, 75, 86, 87, 90, 91, \tag{6.187}$$

$$\left[\rho_6^3(1)\right]^{(bl)} = 1.10 \cdot 1.10 = 1.21, \quad \left[\rho_6^3(2)\right]^{(bl)} = 1.10 \cdot 1.10 = 1.21, \quad b = 1, 7, \quad l = 1, 2, 17, 18, 33, 34, 65, 66, 81, 82, \tag{6.188}$$

$$\left[\rho_6^3(1)\right]^{(bl)} = 1.20 \cdot 1 = 1.20, \quad \left[\rho_6^3(2)\right]^{(bl)} = 1.20 \cdot 1 = 1.20, b = 3, \quad l = 6, 7, 10, 11, 22, 23, 26, 27, 38, 39, 42, 43, 70, 71, 74, 75, 86, 87, 90, 91, \tag{6.189}$$

$$\left[\rho_6^3(1)\right]^{(bl)} = 1.20 \cdot 1.10 = 1.32, \quad \left[\rho_6^3(2)\right]^{(bl)} = 1.20 \cdot 1.10 = 1.32, \quad b = 3, \quad l = 1, 2, 17, 18, 33, 34, 65, 66, 81, 82, \tag{6.190}$$

$$\left[\rho_6^3(1)\right]^{(bl)} = 1.30 \cdot 1 = 1.30, \quad \left[\rho_6^3(2)\right]^{(bl)} = 1.30 \cdot 1 = 1.30, b = 4, 6, \quad l = 6, 7, 10, 11, 22, 23, 26, 27, 38, 39, 42, 43, 70, 71, 74, 75, 86, 87, 90, 91, \tag{6.191}$$

$$\left[\rho_6^3(1)\right]^{(bl)} = 1.30 \cdot 1.10 = 1.43, \quad \left[\rho_6^3(2)\right]^{(bl)} = 1.30 \cdot 1.10 = 1.43, \quad b = 4, 6, \quad l = 1, 2, 17, 18, 33, 34, 65, 66, 81, 82; \tag{6.192}$$

- for asset $A_7$

$$\left[\rho_7^3(1)\right]^{(bl)} = 1 \cdot 1 = 1, \quad \left[\rho_7^3(2)\right]^{(bl)} = 1 \cdot 1 = 1, \quad b = 2, 3, 5, \quad l = 6, 7, 10, 11, 22, 23, 26, 27, 38, 39, 42, 43, 70, 71, 74, 75, 86, 87, 90, 91, \tag{6.193}$$

$$\left[\rho_7^3(1)\right]^{(bl)} = 1 \cdot 1.10 = 1.10, \quad \left[\rho_7^3(2)\right]^{(bl)} = 1 \cdot 1.10 = 1.10,$$
$$b = 2, 3, 5, \quad l = 1, 2, 17, 18, 33, 34, 65, 66, 81, 82, \tag{6.194}$$

$$\left[\rho_7^3(1)\right]^{(bl)} = 1.10 \cdot 1 = 1.10, \quad \left[\rho_7^3(2)\right]^{(bl)} = 1.10 \cdot 1 = 1.10,$$
$$b = 1, 7, \quad l = 6, 7, 10, 11, 22, 23, 26, 27, 38, 39, \tag{6.195}$$
$$42, 43, 70, 71, 74, 75, 86, 87, 90, 91,$$

$$\left[\rho_7^3(1)\right]^{(bl)} = 1.10 \cdot 1.10 = 1.21, \quad \left[\rho_7^3(2)\right]^{(bl)} = 1.10 \cdot 1.10 = 1.21,$$
$$b = 1, 7, \quad l = 1, 2, 17, 18, 33, 34, 65, 66, 81, 82, \tag{6.196}$$

$$\left[\rho_7^3(1)\right]^{(bl)} = 1.30 \cdot 1 = 1.30, \quad \left[\rho_7^3(2)\right]^{(bl)} = 1.30 \cdot 1 = 1.30,$$
$$b = 4, 6, \quad l = 6, 7, 10, 11, 22, 23, 26, 27, 38, 39, 42, \tag{6.197}$$
$$43, 70, 71, 74, 75, 86, 87, 90, 91,$$

$$\left[\rho_7^3(1)\right]^{(bl)} = 1.30 \cdot 1.10 = 1.43, \quad \left[\rho_7^3(2)\right]^{(bl)} = 1.30 \cdot 1.10 = 1.43,$$
$$b = 4, 6, \quad l = 1, 2, 17, 18, 33, 34, 65, 66, 81, 82; \tag{6.198}$$

- for asset $A_8$

$$\left[\rho_8^3(1)\right]^{(bl)} = 1 \cdot 1 = 1, \quad \left[\rho_8^3(2)\right]^{(bl)} = 1 \cdot 1 = 1,$$
$$b = 1, 3, 4, 5, 6, \quad l = 6, 7, 10, 11, 22, 23, 26, 27, 38, 39, \tag{6.199}$$
$$42, 43, 70, 71, 74, 75, 86, 87, 90, 91,$$

$$\left[\rho_8^3(1)\right]^{(bl)} = 1 \cdot 1.10 = 1.10, \quad \left[\rho_8^3(2)\right]^{(bl)} = 1 \cdot 1.10 = 1.10,$$
$$b = 1, 3, 4, 5, 6, \quad l = 1, 2, 17, 18, 33, 34, 65, 66, 81, 82, \tag{6.200}$$

$$\left[\rho_8^3(1)\right]^{(bl)} = 1.10 \cdot 1 = 1.10, \quad \left[\rho_3^3(2)\right]^{(bl)} = 1.10 \cdot 1 = 1.10, b = 2, 7,$$
$$l = 6, 7, 10, 11, 22, 23, 26, 27, 38, 39, 42, 43, 70, 71, 74, 75, 86, 87, 90, 91, \tag{6.201}$$

$$\left[\rho_8^3(1)\right]^{(bl)} = 1.10 \cdot 1.10 = 1.21, \quad \left[\rho_8^3(2)\right]^{(bl)} = 1.10 \cdot 1.10 = 1.21,$$
$$b = 2, 7, \quad l = 1, 2, 17, 18, 33, 34, 65, 66, 81, 82; \tag{6.202}$$

- for asset $A_9$

$$\left[\rho_9^3(1)\right]^{(bl)} = 1.10 \cdot 1 = 1.10, \quad \left[\rho_9^3(2)\right]^{(bl)} = 1.10 \cdot 1 = 1.10,$$
$$b = 1, 2, 7, \quad l = 6, 7, 10, 11, 22, 23, 26, 27, 38, 39, \tag{6.203}$$
$$42, 43, 70, 71, 74, 75, 86, 87, 90, 91,$$

$$\left[\rho_9^3(1)\right]^{(bl)} = 1.10 \cdot 1.10 = 1.21, \quad \left[\rho_9^3(2)\right]^{(bl)} = 1.10 \cdot 1.10 = 1.21,$$
$$b = 1, 2, 7, \quad l = 1, 2, 17, 18, 33, 34, 65, 66, 81, 82, \tag{6.204}$$

$$\left[\rho_9^3(1)\right]^{(bl)} = 1.20 \cdot 1 = 1.20, \quad \left[\rho_9^3(2)\right]^{(bl)} = 1.20 \cdot 1 = 1.20,$$
$$b = 3, 5, \quad l = 6, 7, 10, 11, 22, 23, 26, 27, 38, 39,$$
$$42, 43, 70, 71, 74, 75, 86, 87, 90, 91, \tag{6.205}$$

$$\left[\rho_9^3(1)\right]^{(bl)} = 1.20 \cdot 1.10 = 1.32, \quad \left[\rho_9^3(2)\right]^{(bl)} = 1.20 \cdot 1.10 = 1.32,$$
$$b = 3, 5, \quad l = 1, 2, 17, 18, 33, 34, 65, 66, 81, 82, \tag{6.206}$$

$$\left[\rho_9^3(1)\right]^{(bl)} = 1.30 \cdot 1 = 1.30, \quad \left[\rho_9^3(2)\right]^{(bl)} = 1.30 \cdot 1 = 1.30,$$
$$b = 4, 6, \quad l = 6, 7, 10, 11, 22, 23, 26, 27, 38,$$
$$39, 42, 43, 70, 71, 74, 75, 86, 87, 90, 91, \tag{6.207}$$

$$\left[\rho_9^3(1)\right]^{(bl)} = 1.30 \cdot 1.10 = 1.43, \quad \left[\rho_9^3(2)\right]^{(bl)} = 1.30 \cdot 1.10 = 1.43,$$
$$b = 4, 6, \quad l = 1, 2, 17, 18, 33, 34, 65, 66, 81, 82. \tag{6.208}$$

Considering this section assumptions and agreements, similarly to Sect. 3.5.2, we assume that the port oil terminal critical infrastructure assets $A_i$, $i = 1, 2, \ldots, 9$, at the operation process related to climate-weather change $ZC(t)$, at the operation and climate-weather states $zc_{bl}$, $b = 1, 2, \ldots, 7$, $l = 1$, 2, 6, 7, 10, 11, 17, 18, 22, 23, 26, 27, 33, 34, 38, 39, 42, 43, 65, 66, 70, 71, 74, 75, 81, 82, 86, 87, 90, 91, conditional safety functions (6.122), (6.123) are exponential with the coordinates $[S_i^3(t, u)]^{(bl)}$, $t \in < 0, \infty)$, $u = 1, 2$, $b = 1, 2, \ldots, 7$, $l = 1$, 2, 6, 7, 10, 11, 17, 18, 22, 23, 26, 27, 33, 34, 38, 39, 42, 43, 65, 66, 70, 71, 74, 75, 81, 82, 86, 87, 90, 91, $i = 1, 2, \ldots, 9$, given by (6.130). Whereas, the intensities of ageing $[\lambda_i^3(u)]^{(bl)}$, $u = 1, 2$, $b = 1, 2, \ldots, 7$, $l = 1, 2, 6, 7, 10, 11, 17, 18, 22, 23, 26, 27, 33, 34, 38, 39$, 42, 43, 65, 66, 70, 71, 74, 75, 81, 82, 86, 87, 90, 91, $i = 1, 2, \ldots, 9$, for the port oil terminal critical infrastructure assets $A_i$, $i = 1, 2, \ldots, 9$, at the operation process related to climate-weather change $ZC(t)$, at the operation and climate-weather states $zc_{bl}$, $b = 1, 2, \ldots, 7$, $l = 1$, 2, 6, 7, 10, 11, 17, 18, 22, 23, 26, 27, 33, 34, 38, 39, 42, 43, 65, 66, 70, 71, 74, 75, 81, 82, 86, 87, 90, 91, existing in (6.130) are given by (6.133), where $\lambda_i^0(u)$, $u = 1, 2$, $i = 1, 2, \ldots, 9$, are the intensities of ageing of the port oil terminal critical infrastructure assets $A_i$, $i = 1, 2, \ldots, 9$, without considering operation process related to climate-weather change impact and $[\rho_i^3(u)]^{(bl)}$, $u = 1, 2$, $b = 1, 2, \ldots, 7$, $l = 1$, 2, 6, 7, 10, 11, 17, 18, 22, 23, 26, 27, 33, 34, 38, 39, 42, 43, 65, 66, 70, 71, 74, 75, 81, 82, 86, 87, 90, 91, $i = 1, 2, \ldots, 9$, are the coefficients of operation process related to climate-weather change impact on the port oil terminal critical infrastructure assets $A_i$, $i = 1, 2, \ldots, 9$, intensities of ageing at the operation and climate-weather states $zc_{bl}$,$b = 1, 2, \ldots, 7$, $l = 1$, 2, 6, 7, 10, 11, 17, 18, 22, 23, 26, 27, 33, 34, 38, 39, 42, 43, 65, 66, 70, 71, 74, 75, 81, 82, 86, 87, 90, 91.

After that, from (6.133), considering (3.96)–(3.99) and (6.149)–(6.208), it follows that the intensities of ageing of the port oil terminal assets departure from the safety states subset {1, 2}, {2}, with considering operation process related to climate-weather change impact on their safety are:

- for asset $A_1$

$$\begin{aligned} &[\lambda_1^3(1)]^{(bl)} = [\rho_1^3(1)]^{(bl)} \cdot \lambda_1^0(1) = 1.155 \cdot 0.015873 = 0.018333, \\ &[\lambda_1^3(2)]^{(bl)} = [\rho_1^3(2)]^{(bl)} \cdot \lambda_1^0(2) = 1.155 \cdot 0.021739 = 0.025109, \\ &b = 1, 2, 7, \quad l = 1, 2, 17, 18, \end{aligned} \tag{6.209}$$

$$\begin{aligned} &[\lambda_1^3(1)]^{(bl)} = [\rho_1^3(1)]^{(bl)} \cdot \lambda_1^0(1) = 1.188 \cdot 0.015873 = 0.018857, \\ &[\lambda_1^3(2)]^{(bl)} = [\rho_1^3(2)]^{(bl)} \cdot \lambda_1^0(2) = 1.199 \cdot 0.021739 = 0.026065, \\ &b = 1, 2, 7, \quad l = 33, 34, 65, 66, 81, 82, \end{aligned} \tag{6.210}$$

$$\begin{aligned} &[\lambda_1^3(1)]^{(bl)} = [\rho_1^3(1)]^{(bl)} \cdot \lambda_1^0(1) = 1.133 \cdot 0.015873 = 0.017984, \\ &[\lambda_1^3(2)]^{(bl)} = [\rho_1^3(2)]^{(bl)} \cdot \lambda_1^0(2) = 1.144 \cdot 0.021739 = 0.024869, \\ &b = 1, 2, 7, \quad l = 38, 39, 42, 43, 70, 71, 74, 75, 86, 87, 90, 91, \end{aligned} \tag{6.211}$$

$$\begin{aligned} &[\lambda_1^3(1)]^{(bl)} = [\rho_1^3(1)]^{(bl)} \cdot \lambda_1^0(1) = 1.10 \cdot 0.015873 = 0.017460, \\ &[\lambda_1^3(2)]^{(bl)} = [\rho_1^3(2)]^{(bl)} \cdot \lambda_1^0(2) = 1.10 \cdot 0.021739 = 0.023913, \\ &b = 1, 2, 7, \quad l = 6, 7, 10, 11, 22, 23, 26, 27, \end{aligned} \tag{6.212}$$

$$\begin{aligned} &[\lambda_1^3(1)]^{(bl)} = [\rho_1^3(1)]^{(bl)} \cdot \lambda_1^0(1) = 1.26 \cdot 0.015873 = 0.019999, \\ &[\lambda_1^3(2)]^{(bl)} = [\rho_1^3(2)]^{(bl)} \cdot \lambda_1^0(2) = 1.26 \cdot 0.021739 = 0.027391, \\ &b = 3, 5, \quad l = 1, 2, 17, 18, \end{aligned} \tag{6.213}$$

$$\begin{aligned} &[\lambda_1^3(1)]^{(bl)} = [\rho_1^3(1)]^{(bl)} \cdot \lambda_1^0(1) = 1.296 \cdot 0.015873 = 0.020571, \\ &[\lambda_1^3(2)]^{(bl)} = [\rho_1^3(2)]^{(bl)} \cdot \lambda_1^0(2) = 1.308 \cdot 0.021739 = 0.028435, \\ &b = 3, 5, \quad l = 33, 34, 65, 66, 81, 82, \end{aligned} \tag{6.214}$$

$$\begin{aligned} &[\lambda_1^3(1)]^{(bl)} = [\rho_1^3(1)]^{(bl)} \cdot \lambda_1^0(1) = 1.236 \cdot 0.015873 = 0.019619, \\ &[\lambda_1^3(2)]^{(bl)} = [\rho_1^3(2)]^{(bl)} \cdot \lambda_1^0(2) = 1.248 \cdot 0.021739 = 0.027130, \\ &b = 3, 5, \quad l = 38, 39, 42, 43, 70, 71, 74, 75, 86, 87, 90, 91, \end{aligned} \tag{6.215}$$

$$\begin{aligned} &[\lambda_1^3(1)]^{(bl)} = [\rho_1^3(1)]^{(bl)} \cdot \lambda_1^0(1) = 1.20 \cdot 0.015873 = 0.019048, \\ &[\lambda_1^3(2)]^{(bl)} = [\rho_1^3(2)]^{(bl)} \cdot \lambda_1^0(2) = 1.20 \cdot 0.021739 = 0.026087, \\ &b = 3, 5, \quad l = 6, 7, 10, 11, 22, 23, 26, 27, \end{aligned} \tag{6.216}$$

$$\begin{aligned}
&\left[\lambda_1^3(1)\right]^{(bl)} = \left[\rho_1^3(1)\right]^{(bl)} \cdot \lambda_1^0(1) = 1.365 \cdot 0.015873 = 0.021666,\\
&\left[\lambda_1^3(2)\right]^{(bl)} = \left[\rho_1^3(2)\right]^{(bl)} \cdot \lambda_1^0(2) = 1.365 \cdot 0.021739 = 0.029673,\\
&b = 4, 6, \quad l = 1, 2, 17, 18,
\end{aligned} \tag{6.217}$$

$$\begin{aligned}
&\left[\lambda_1^3(1)\right]^{(bl)} = \left[\rho_1^3(1)\right]^{(bl)} \cdot \lambda_1^0(1) = 1.404 \cdot 0.015873 = 0.022286,\\
&\left[\lambda_1^3(2)\right]^{(bl)} = \left[\rho_1^3(2)\right]^{(bl)} \cdot \lambda_1^0(2) = 1.417 \cdot 0.021739 = 0.030804,\\
&b = 4, 6, \quad l = 33, 34, 65, 66, 81, 82,
\end{aligned} \tag{6.218}$$

$$\begin{aligned}
&\left[\lambda_1^3(1)\right]^{(bl)} = \left[\rho_1^3(1)\right]^{(bl)} \cdot \lambda_1^0(1) = 1.339 \cdot 0.015873 = 0.021254,\\
&\left[\lambda_1^3(2)\right]^{(bl)} = \left[\rho_1^3(2)\right]^{(bl)} \cdot \lambda_1^0(2) = 1.352 \cdot 0.021739 = 0.029391,\\
&b = 4, 6, \quad l = 38, 39, 42, 43, 70, 71, 74, 75, 86, 87, 90, 91,
\end{aligned} \tag{6.219}$$

$$\begin{aligned}
&\left[\lambda_1^3(1)\right]^{(bl)} = \left[\rho_1^3(1)\right]^{(bl)} \cdot \lambda_1^0(1) = 1.30 \cdot 0.015873 = 0.020634,\\
&\left[\lambda_1^3(2)\right]^{(bl)} = \left[\rho_1^3(2)\right]^{(bl)} \cdot \lambda_1^0(2) = 1.30 \cdot 0.021739 = 0.028260,\\
&b = 4, 6, \quad l = 6, 7, 10, 11, 22, 23, 26, 27;
\end{aligned} \tag{6.220}$$

– for asset $A_2$

$$\begin{aligned}
&\left[\lambda_2^3(1)\right]^{(bl)} = \left[\rho_2^3(1)\right]^{(bl)} \cdot \lambda_2^0(1) = 1.10 \cdot 0.0125 = 0.01375,\\
&\left[\lambda_2^3(2)\right]^{(bl)} = \left[\rho_2^3(2)\right]^{(bl)} \cdot \lambda_2^0(2) = 1.10 \cdot 0.021739 = 0.023913, \quad b = 1, 2, 7,\\
&l = 6, 7, 10, 11, 22, 23, 26, 27, 38, 39, 42, 43, 70, 71, 74, 75, 86, 87, 90, 91,
\end{aligned} \tag{6.221}$$

$$\begin{aligned}
&\left[\lambda_2^3(1)\right]^{(bl)} = \left[\rho_2^3(1)\right]^{(bl)} \cdot \lambda_2^0(1) = 1.21 \cdot 0.0125 = 0.015125,\\
&\left[\lambda_2^3(2)\right]^{(bl)} = \left[\rho_2^3(2)\right]^{(bl)} \cdot \lambda_2^0(2) = 1.21 \cdot 0.021739 = 0.026304,\\
&b = 1, 2, 7, \quad l = 1, 2, 17, 18, 33, 34, 65, 66, 81, 82,
\end{aligned} \tag{6.222}$$

$$\begin{aligned}
&\left[\lambda_2^3(1)\right]^{(bl)} = \left[\rho_2^3(1)\right]^{(bl)} \cdot \lambda_2^0(1) = 1.20 \cdot 0.0125 = 0.015,\\
&\left[\lambda_2^3(2)\right]^{(bl)} = \left[\rho_2^3(2)\right]^{(bl)} \cdot \lambda_2^0(2) = 1.20 \cdot 0.021739 = 0.026086, \quad b = 3, 5,\\
&l = 6, 7, 10, 11, 22, 23, 26, 27, 38, 39, 42, 43, 70, 71, 74, 75, 86, 87, 90, 91,
\end{aligned} \tag{6.223}$$

$$\begin{aligned}
&\left[\lambda_2^3(1)\right]^{(bl)} = \left[\rho_2^3(1)\right]^{(bl)} \cdot \lambda_2^0(1) = 1.32 \cdot 0.0125 = 0.0165,\\
&\left[\lambda_2^3(2)\right]^{(bl)} = \left[\rho_2^3(2)\right]^{(bl)} \cdot \lambda_2^0(2) = 1.32 \cdot 0.021739 = 0.028695,\\
&b = 3, 5, \quad l = 1, 2, 17, 18, 33, 34, 65, 66, 81, 82,
\end{aligned} \tag{6.224}$$

$$\begin{aligned}
&\left[\lambda_2^3(1)\right]^{(bl)} = \left[\rho_2^3(1)\right]^{(bl)} \cdot \lambda_2^0(1) = 1.30 \cdot 0.0125 = 0.01625,\\
&\left[\lambda_2^3(2)\right]^{(bl)} = \left[\rho_2^3(2)\right]^{(bl)} \cdot \lambda_2^0(2) = 1.30 \cdot 0.021739 = 0.028261, \quad b = 4, 6,\\
&l = 6, 7, 10, 11, 22, 23, 26, 27, 38, 39, 42, 43, 70, 71, 74, 75, 86, 87, 90, 91,
\end{aligned} \tag{6.225}$$

$$\begin{aligned} &\left[\lambda_2^3(1)\right]^{(bl)} = \left[\rho_2^3(1)\right]^{(bl)} \cdot \lambda_2^0(1) = 1.43 \cdot 0.0125 = 0.017875, \\ &\left[\lambda_2^3(2)\right]^{(bl)} = \left[\rho_2^3(2)\right]^{(bl)} \cdot \lambda_2^0(2) = 1.43 \cdot 0.021739 = 0.031086, \\ &b = 4, 6, \quad l = 1, 2, 17, 18, 33, 34, 65, 66, 81, 82; \end{aligned} \tag{6.226}$$

– for asset $A_3$

$$\begin{aligned} &\left[\lambda_3^3(1)\right]^{(bl)} = \left[\rho_3^3(1)\right]^{(bl)} \cdot \lambda_3^0(1) = 1 \cdot 0.0125 = 0.0125, \\ &\left[\lambda_3^3(2)\right]^{(bl)} = \left[\rho_3^3(2)\right]^{(bl)} \cdot \lambda_3^0(2) = 1 \cdot 0.021739 = 0.021739, \quad b = 1, 2, 3, 7, \\ &l = 6, 7, 10, 11, 22, 23, 26, 27, 38, 39, 42, 43, 70, 71, 74, 75, 86, 87, 90, 91, \end{aligned} \tag{6.227}$$

$$\begin{aligned} &\left[\lambda_3^3(1)\right]^{(bl)} = \left[\rho_3^3(1)\right]^{(bl)} \cdot \lambda_3^0(1) = 1.10 \cdot 0.0125 = 0.01375, \\ &\left[\lambda_3^3(2)\right]^{(bl)} = \left[\rho_3^3(2)\right]^{(bl)} \cdot \lambda_3^0(2) = 1.10 \cdot 0.021739 = 0.023913, \\ &b = 1, 2, 3, 7, \quad l = 1, 2, 17, 18, 33, 34, 65, 66, 81, 82, \end{aligned} \tag{6.228}$$

$$\begin{aligned} &\left[\lambda_3^3(1)\right]^{(bl)} = \left[\rho_3^3(1)\right]^{(bl)} \cdot \lambda_3^0(1) = 1.20 \cdot 0.0125 = 0.015, \\ &\left[\lambda_3^3(2)\right]^{(bl)} = \left[\rho_3^3(2)\right]^{(bl)} \cdot \lambda_3^0(2) = 1.20 \cdot 0.021739 = 0.026086, \\ &b = 5, \quad l = 6, 7, 10, 11, 22, 23, 26, 27, 38, 39, 42, 43, 70, 71, 74, 75, 86, 87, 90, 91, \end{aligned} \tag{6.229}$$

$$\begin{aligned} &\left[\lambda_3^3(1)\right]^{(bl)} = \left[\rho_3^3(1)\right]^{(bl)} \cdot \lambda_3^0(1) = 1.32 \cdot 0.0125 = 0.0165, \\ &\left[\lambda_3^3(2)\right]^{(bl)} = \left[\rho_3^3(2)\right]^{(bl)} \cdot \lambda_3^0(2) = 1.32 \cdot 0.021739 = 0.028695, \\ &b = 5, \quad l = 1, 2, 17, 18, 33, 34, 65, 66, 81, 82, \end{aligned} \tag{6.230}$$

$$\begin{aligned} &\left[\lambda_3^3(1)\right]^{(bl)} = \left[\rho_3^3(1)\right]^{(bl)} \cdot \lambda_3^0(1) = 1.30 \cdot 0.0125 = 0.01625, \\ &\left[\lambda_3^3(2)\right]^{(bl)} = \left[\rho_3^3(2)\right]^{(bl)} \cdot \lambda_3^0(2) = 1.30 \cdot 0.021739 = 0.028261, \quad b = 4, 6, \\ &l = 6, 7, 10, 11, 22, 23, 26, 27, 38, 39, 42, 43, 70, 71, 74, 75, 86, 87, 90, 91, \end{aligned} \tag{6.231}$$

$$\begin{aligned} &\left[\lambda_3^3(1)\right]^{(bl)} = \left[\rho_3^3(1)\right]^{(bl)} \cdot \lambda_3^0(1) = 1.43 \cdot 0.0125 = 0.017875, \\ &\left[\lambda_3^3(2)\right]^{(bl)} = \left[\rho_3^3(2)\right]^{(bl)} \cdot \lambda_3^0(2) = 1.43 \cdot 0.021739 = 0.031086, \\ &b = 4, 6, \quad l = 1, 2, 17, 18, 33, 34, 65, 66, 81, 82; \end{aligned} \tag{6.232}$$

– for asset $A_4$

$$\begin{aligned} &\left[\lambda_4^3(1)\right]^{(bl)} = \left[\rho_4^3(1)\right]^{(bl)} \cdot \lambda_4^0(1) = 1.10 \cdot 0.0125 = 0.01375, \\ &\left[\lambda_4^3(2)\right]^{(bl)} = \left[\rho_4^3(2)\right]^{(bl)} \cdot \lambda_4^0(2) = 1.10 \cdot 0.021739 = 0.023913, \quad b = 1, 2, 7, \\ &l = 6, 7, 10, 11, 22, 23, 26, 27, 38, 39, 42, 43, 70, 71, 74, 75, 86, 87, 90, 91, \end{aligned} \tag{6.233}$$

$$\begin{aligned}&\left[\lambda_4^3(1)\right]^{(bl)} = \left[\rho_4^3(1)\right]^{(bl)} \cdot \lambda_4^0(1) = 1.21 \cdot 0.0125 = 0.015125,\\&\left[\lambda_4^3(2)\right]^{(bl)} = \left[\rho_4^3(2)\right]^{(bl)} \cdot \lambda_4^0(2) = 1.21 \cdot 0.021739 = 0.026304,\\&b = 1, 2, 7, \quad l = 1, 2, 17, 18, 33, 34, 65, 66, 81, 82,\end{aligned} \tag{6.234}$$

$$\begin{aligned}&\left[\lambda_4^3(1)\right]^{(bl)} = \left[\rho_4^3(1)\right]^{(bl)} \cdot \lambda_4^0(1) = 1.20 \cdot 0.0125 = 0.015,\\&\left[\lambda_4^3(2)\right]^{(bl)} = \left[\rho_4^3(2)\right]^{(bl)} \cdot \lambda_4^0(2) = 1.20 \cdot 0.021739 = 0.026086, \quad b = 3, 5,\\&l = 6, 7, 10, 11, 22, 23, 26, 27, 38, 39, 42, 43, 70, 71, 74, 75, 86, 87, 90, 91,\end{aligned} \tag{6.235}$$

$$\begin{aligned}&\left[\lambda_4^3(1)\right]^{(bl)} = \left[\rho_4^3(1)\right]^{(bl)} \cdot \lambda_4^0(1) = 1.32 \cdot 0.0125 = 0.0165,\\&\left[\lambda_4^3(2)\right]^{(bl)} = \left[\rho_4^3(2)\right]^{(bl)} \cdot \lambda_4^0(2) = 1.32 \cdot 0.021739 = 0.028695,\\&b = 3, 5, \quad l = 1, 2, 17, 18, 33, 34, 65, 66, 81, 82,\end{aligned} \tag{6.236}$$

$$\begin{aligned}&\left[\lambda_4^3(1)\right]^{(bl)} = \left[\rho_4^3(1)\right]^{(bl)} \cdot \lambda_4^0(1) = 1.30 \cdot 0.0125 = 0.01625,\\&\left[\lambda_4^3(2)\right]^{(bl)} = \left[\rho_4^3(2)\right]^{(bl)} \cdot \lambda_4^0(2) = 1.30 \cdot 0.021739 = 0.028261, \quad b = 4, 6,\\&l = 6, 7, 10, 11, 22, 23, 26, 27, 38, 39, 42, 43, 70, 71, 74, 75, 86, 87, 90, 91,\end{aligned} \tag{6.237}$$

$$\begin{aligned}&\left[\lambda_4^3(1)\right]^{(bl)} = \left[\rho_4^3(1)\right]^{(bl)} \cdot \lambda_4^0(1) = 1.43 \cdot 0.0125 = 0.017875,\\&\left[\lambda_4^3(2)\right]^{(bl)} = \left[\rho_4^3(2)\right]^{(bl)} \cdot \lambda_4^0(2) = 1.43 \cdot 0.021739 = 0.031086,\\&b = 4, 6, \quad l = 1, 2, 17, 18, 33, 34, 65, 66, 81, 82;\end{aligned} \tag{6.238}$$

– for asset $A_5$

$$\begin{aligned}&\left[\lambda_5^3(1)\right]^{(bl)} = \left[\rho_5^3(1)\right]^{(bl)} \cdot \lambda_5^0(1) = 1 \cdot 0.0125 = 0.0125,\\&\left[\lambda_5^3(2)\right]^{(bl)} = \left[\rho_5^3(2)\right]^{(bl)} \cdot \lambda_5^0(2) = 1 \cdot 0.021739 = 0.021739, \quad b = 1, 2, 7,\\&l = 6, 7, 10, 11, 22, 23, 26, 27, 38, 39, 42, 43, 70, 71, 74, 75, 86, 87, 90, 91,\end{aligned} \tag{6.239}$$

$$\begin{aligned}&\left[\lambda_5^3(1)\right]^{(bl)} = \left[\rho_5^3(1)\right]^{(bl)} \cdot \lambda_5^0(1) = 1.10 \cdot 0.0125 = 0.01375,\\&\left[\lambda_5^3(2)\right]^{(bl)} = \left[\rho_5^3(2)\right]^{(bl)} \cdot \lambda_5^0(2) = 1.10 \cdot 0.021739 = 0.023913,\\&b = 1, 2, 7, \quad l = 1, 2, 17, 18, 33, 34, 65, 66, 81, 82,\end{aligned} \tag{6.240}$$

$$\begin{aligned}&\left[\lambda_5^3(1)\right]^{(bl)} = \left[\rho_5^3(1)\right]^{(bl)} \cdot \lambda_5^0(1) = 1.20 \cdot 0.0125 = 0.015,\\&\left[\lambda_5^3(2)\right]^{(bl)} = \left[\rho_5^3(2)\right]^{(bl)} \cdot \lambda_5^0(2) = 1.20 \cdot 0.021739 = 0.026087, \quad b = 3, 5,\\&l = 6, 7, 10, 11, 22, 23, 26, 27, 38, 39, 42, 43, 70, 71, 74, 75, 86, 87, 90, 91,\end{aligned} \tag{6.241}$$

$$\begin{aligned}&\left[\lambda_5^3(1)\right]^{(bl)} = \left[\rho_5^3(1)\right]^{(bl)} \cdot \lambda_5^0(1) = 1.32 \cdot 0.0125 = 0.0165,\\&\left[\lambda_5^3(2)\right]^{(bl)} = \left[\rho_5^3(2)\right]^{(bl)} \cdot \lambda_5^0(2) = 1.32 \cdot 0.021739 = 0.028695,\\&b = 3, 5, \quad l = 1, 2, 17, 18, 33, 34, 65, 66, 81, 82,\end{aligned} \tag{6.242}$$

$$\begin{aligned}&\left[\lambda_5^3(1)\right]^{(bl)} = \left[\rho_5^3(1)\right]^{(bl)} \cdot \lambda_5^0(1) = 1.30 \cdot 0.0125 = 0.01625,\\ &\left[\lambda_5^3(2)\right]^{(bl)} = \left[\rho_5^3(2)\right]^{(bl)} \cdot \lambda_5^0(2) = 1.30 \cdot 0.021739 = 0.028261, \quad b = 4, 6,\\ &l = 6, 7, 10, 11, 22, 23, 26, 27, 38, 39, 42, 43, 70, 71, 74, 75, 86, 87, 90, 91,\end{aligned} \tag{6.243}$$

$$\begin{aligned}&\left[\lambda_5^3(1)\right]^{(bl)} = \left[\rho_5^3(1)\right]^{(bl)} \cdot \lambda_5^0(1) = 1.43 \cdot 0.0125 = 0.017875,\\ &\left[\lambda_5^3(2)\right]^{(bl)} = \left[\rho_5^3(2)\right]^{(bl)} \cdot \lambda_5^0(2) = 1.43 \cdot 0.021739 = 0.031087,\\ &b = 4, 6, \quad l = 1, 2, 17, 18, 33, 34, 65, 66, 81, 82;\end{aligned} \tag{6.244}$$

– for asset $A_6$

$$\begin{aligned}&\left[\lambda_6^3(1)\right]^{(bl)} = \left[\rho_6^3(1)\right]^{(bl)} \cdot \lambda_6^0(1) = 1 \cdot 0.0125 = 0.0125,\\ &\left[\lambda_5^3(2)\right]^{(bl)} = \left[\rho_6^3(2)\right]^{(bl)} \cdot \lambda_6^0(2) = 1 \cdot 0.021739 = 0.021739, \quad b = 2, 5,\\ &l = 6, 7, 10, 11, 22, 23, 26, 27, 38, 39, 42, 43, 70, 71, 74, 75, 86, 87, 90, 91,\end{aligned} \tag{6.245}$$

$$\begin{aligned}&\left[\lambda_6^3(1)\right]^{(bl)} = \left[\rho_6^3(1)\right]^{(bl)} \cdot \lambda_6^0(1) = 1.10 \cdot 0.0125 = 0.01375,\\ &\left[\lambda_6^3(2)\right]^{(bl)} = \left[\rho_6^3(2)\right]^{(bl)} \cdot \lambda_6^0(2) = 1.10 \cdot 0.021739 = 0.023913,\\ &b = 2, 5, \quad l = 1, 2, 17, 18, 33, 34, 65, 66, 81, 82,\end{aligned} \tag{6.246}$$

$$\begin{aligned}&\left[\lambda_6^3(1)\right]^{(bl)} = \left[\rho_6^3(1)\right]^{(bl)} \cdot \lambda_6^0(1) = 1.10 \cdot 0.0125 = 0.01375,\\ &\left[\lambda_6^3(2)\right]^{(bl)} = \left[\rho_6^3(2)\right]^{(bl)} \cdot \lambda_6^0(2) = 1.10 \cdot 0.021739 = 0.023913, \quad b = 1, 7,\\ &l = 6, 7, 10, 11, 22, 23, 26, 27, 38, 39, 42, 43, 70, 71, 74, 75, 86, 87, 90, 91,\end{aligned} \tag{6.247}$$

$$\begin{aligned}&\left[\lambda_6^3(1)\right]^{(bl)} = \left[\rho_6^3(1)\right]^{(bl)} \cdot \lambda_6^0(1) = 1.21 \cdot 0.0125 = 0.015125,\\ &\left[\lambda_6^3(2)\right]^{(bl)} = \left[\rho_6^3(2)\right]^{(bl)} \cdot \lambda_6^0(2) = 1.21 \cdot 0.021739 = 0.026304,\\ &b = 1, 7, \quad l = 1, 2, 17, 18, 33, 34, 65, 66, 81, 82,\end{aligned} \tag{6.248}$$

$$\begin{aligned}&\left[\lambda_6^3(1)\right]^{(bl)} = \left[\rho_6^3(1)\right]^{(bl)} \cdot \lambda_6^0(1) = 1.20 \cdot 0.0125 = 0.015,\\ &\left[\lambda_6^3(2)\right]^{(bl)} = \left[\rho_6^3(2)\right]^{(bl)} \cdot \lambda_6^0(2) = 1.20 \cdot 0.021739 = 0.026087,\\ &b = 3, \quad l = 6, 7, 10, 11, 22, 23, 26, 27, 38, 39, 42, 43, 70, 71, 74, 75, 86, 87, 90, 91,\end{aligned} \tag{6.249}$$

$$\begin{aligned}&\left[\lambda_6^3(1)\right]^{(bl)} = \left[\rho_6^3(1)\right]^{(bl)} \cdot \lambda_6^0(1) = 1.32 \cdot 0.0125 = 0.0165,\\ &\left[\lambda_6^3(2)\right]^{(bl)} = \left[\rho_6^3(2)\right]^{(bl)} \cdot \lambda_6^0(2) = 1.32 \cdot 0.021739 = 0.028695,\\ &b = 3, \quad l = 1, 2, 17, 18, 33, 34, 65, 66, 81, 82,\end{aligned} \tag{6.250}$$

$$\left[\lambda_6^3(1)\right]^{(bl)} = \left[\rho_6^3(1)\right]^{(bl)} \cdot \lambda_6^0(1) = 1.30 \cdot 0.0125 = 0.01625,$$
$$\left[\lambda_6^3(2)\right]^{(bl)} = \left[\rho_6^3(2)\right]^{(bl)} \cdot \lambda_6^0(2) = 1.30 \cdot 0.021739 = 0.028261, \quad b = 4, 6,$$
$$l = 6, 7, 10, 11, 22, 23, 26, 27, 38, 39, 42, 43, 70, 71, 74, 75, 86, 87, 90, 91, \tag{6.251}$$

$$\left[\lambda_6^3(1)\right]^{(bl)} = \left[\rho_6^3(1)\right]^{(bl)} \cdot \lambda_6^0(1) = 1.43 \cdot 0.0125 = 0.017875,$$
$$\left[\lambda_6^3(2)\right]^{(bl)} = \left[\rho_6^3(2)\right]^{(bl)} \cdot \lambda_6^0(2) = 1.43 \cdot 0.021739 = 0.031087, \tag{6.252}$$
$$b = 4, 6, \quad l = 1, 2, 17, 18, 33, 34, 65, 66, 81, 82;$$

– for asset $A_7$

$$\left[\lambda_7^3(1)\right]^{(bl)} = \left[\rho_7^3(1)\right]^{(bl)} \cdot \lambda_7^0(1) = 1 \cdot 0.0125 = 0.0125,$$
$$\left[\lambda_7^3(2)\right]^{(bl)} = \left[\rho_7^3(2)\right]^{(bl)} \cdot \lambda_7^0(2) = 1 \cdot 0.021739 = 0.021739, \quad b = 2, 3, 5,$$
$$l = 6, 7, 10, 11, 22, 23, 26, 27, 38, 39, 42, 43, 70, 71, 74, 75, 86, 87, 90, 91, \tag{6.253}$$

$$\left[\lambda_7^3(1)\right]^{(bl)} = \left[\rho_7^3(1)\right]^{(bl)} \cdot \lambda_7^0(1) = 1.10 \cdot 0.0125 = 0.01375,$$
$$\left[\lambda_7^3(2)\right]^{(bl)} = \left[\rho_7^3(2)\right]^{(bl)} \cdot \lambda_7^0(2) = 1.10 \cdot 0.021739 = 0.023913, \tag{6.254}$$
$$b = 2, 3, 5, \quad l = 1, 2, 17, 18, 33, 34, 65, 66, 81, 82,$$

$$\left[\lambda_7^3(1)\right]^{(bl)} = \left[\rho_7^3(1)\right]^{(bl)} \cdot \lambda_7^0(1) = 1.10 \cdot 0.0125 = 0.01375,$$
$$\left[\lambda_7^3(2)\right]^{(bl)} = \left[\rho_7^3(2)\right]^{(bl)} \cdot \lambda_7^0(2) = 1.10 \cdot 0.021739 = 0.023913, \quad b = 1, 7,$$
$$l = 6, 7, 10, 11, 22, 23, 26, 27, 38, 39, 42, 43, 70, 71, 74, 75, 86, 87, 90, 91, \tag{6.255}$$

$$\left[\lambda_7^3(1)\right]^{(bl)} = \left[\rho_7^3(1)\right]^{(bl)} \cdot \lambda_7^0(1) = 1.21 \cdot 0.0125 = 0.015125,$$
$$\left[\lambda_7^3(2)\right]^{(bl)} = \left[\rho_7^3(2)\right]^{(bl)} \cdot \lambda_7^0(2) = 1.21 \cdot 0.021739 = 0.026304, \tag{6.256}$$
$$b = 1, 7, \quad l = 1, 2, 17, 18, 33, 34, 65, 66, 81, 82,$$

$$\left[\lambda_7^3(1)\right]^{(bl)} = \left[\rho_7^3(1)\right]^{(bl)} \cdot \lambda_7^0(1) = 1.30 \cdot 0.0125 = 0.01625,$$
$$\left[\lambda_7^3(2)\right]^{(bl)} = \left[\rho_7^3(2)\right]^{(bl)} \cdot \lambda_7^0(2) = 1.30 \cdot 0.021739 = 0.028261,$$
$$b = 4, 6, \quad l = 6, 7, 10, 11, 22, 23, 26, 27, 38, 39, 42, 43, 70, 71, 74, 75, 86, 87, 90, 91, \tag{6.257}$$

$$\begin{aligned}&\left[\lambda_7^3(1)\right]^{(bl)} = \left[\rho_7^3(1)\right]^{(bl)} \cdot \lambda_7^0(1) = 1.43 \cdot 0.0125 = 0.017875,\\&\left[\lambda_7^3(2)\right]^{(bl)} = \left[\rho_7^3(2)\right]^{(bl)} \cdot \lambda_7^0(2) = 1.43 \cdot 0.021739 = 0.031087,\\&b = 4, 6, \quad l = 1, 2, 17, 18, 33, 34, 65, 66, 81, 82;\end{aligned} \tag{6.258}$$

- for asset $A_8$

$$\begin{aligned}&\left[\lambda_8^3(1)\right]^{(bl)} = \left[\rho_8^3(1)\right]^{(bl)} \cdot \lambda_8^0(1) = 1 \cdot 0.0125 = 0.0125,\\&\left[\lambda_8^3(2)\right]^{(bl)} = \left[\rho_8^3(2)\right]^{(bl)} \cdot \lambda_8^0(2) = 1 \cdot 0.021739 = 0.021739, \quad b = 1, 3, 4, 5, 6,\\&l = 6, 7, 10, 11, 22, 23, 26, 27, 38, 39, 42, 43, 70, 71, 74, 75, 86, 87, 90, 91,\end{aligned} \tag{6.259}$$

$$\begin{aligned}&\left[\lambda_8^3(1)\right]^{(bl)} = \left[\rho_8^3(1)\right]^{(bl)} \cdot \lambda_8^0(1) = 1.10 \cdot 0.0125 = 0.01375,\\&\left[\lambda_8^3(2)\right]^{(bl)} = \left[\rho_8^3(2)\right]^{(bl)} \cdot \lambda_8^0(2) = 1.10 \cdot 0.021739 = 0.023913,\\&b = 1, 3, 4, 5, 6, \quad l = 1, 2, 17, 18, 33, 34, 65, 66, 81, 82,\end{aligned} \tag{6.260}$$

$$\begin{aligned}&\left[\lambda_8^3(1)\right]^{(bl)} = \left[\rho_8^3(1)\right]^{(bl)} \cdot \lambda_8^0(1) = 1.10 \cdot 0.0125 = 0.01375,\\&\left[\lambda_8^3(2)\right]^{(bl)} = \left[\rho_8^3(2)\right]^{(bl)} \cdot \lambda_8^0(2) = 1.10 \cdot 0.021739 = 0.023913, \quad b = 2, 7,\\&l = 6, 7, 10, 11, 22, 23, 26, 27, 38, 39, 42, 43, 70, 71, 74, 75, 86, 87, 90, 91,\end{aligned} \tag{6.261}$$

$$\begin{aligned}&\left[\lambda_8^3(1)\right]^{(bl)} = \left[\rho_8^3(1)\right]^{(bl)} \cdot \lambda_8^0(1) = 1.21 \cdot 0.0125 = 0.015125,\\&\left[\lambda_8^3(2)\right]^{(bl)} = \left[\rho_8^3(2)\right]^{(bl)} \cdot \lambda_8^0(2) = 1.21 \cdot 0.021739 = 0.026304,\\&b = 2, 7, \quad l = 1, 2, 17, 18, 33, 34, 65, 66, 81, 82;\end{aligned} \tag{6.262}$$

- for asset $A_9$

$$\begin{aligned}&\left[\lambda_9^3(1)\right]^{(bl)} = \left[\rho_9^3(1)\right]^{(bl)} \cdot \lambda_9^0(1) = 1.10 \cdot 0.0125 = 0.01375,\\&\left[\lambda_9^3(2)\right]^{(bl)} = \left[\rho_9^3(2)\right]^{(bl)} \cdot \lambda_9^0(2) = 1.10 \cdot 0.021739 = 0.023913, \quad b = 1, 2, 7,\\&l = 6, 7, 10, 11, 22, 23, 26, 27, 38, 39, 42, 43, 70, 71, 74, 75, 86, 87, 90, 91,\end{aligned} \tag{6.263}$$

$$\begin{aligned}&\left[\lambda_9^3(1)\right]^{(bl)} = \left[\rho_9^3(1)\right]^{(bl)} \cdot \lambda_9^0(1) = 1.21 \cdot 0.0125 = 0.015125,\\&\left[\lambda_9^3(2)\right]^{(bl)} = \left[\rho_9^3(2)\right]^{(bl)} \cdot \lambda_9^0(2) = 1.21 \cdot 0.021739 = 0.026304,\\&b = 1, 2, 7, \quad l = 1, 2, 17, 18, 33, 34, 65, 66, 81, 82,\end{aligned} \tag{6.264}$$

$$\begin{aligned}&\left[\lambda_9^3(1)\right]^{(bl)} = \left[\rho_9^3(1)\right]^{(bl)} \cdot \lambda_9^0(1) = 1.20 \cdot 0.0125 = 0.015,\\&\left[\lambda_9^3(2)\right]^{(bl)} = \left[\rho_9^3(2)\right]^{(bl)} \cdot \lambda_9^0(2) = 1.20 \cdot 0.021739 = 0.026087, \quad b = 3, 5,\\&l = 6, 7, 10, 11, 22, 23, 26, 27, 38, 39, 42, 43, 70, 71, 74, 75, 86, 87, 90, 91,\end{aligned} \tag{6.265}$$

$$\left[\lambda_9^3(1)\right]^{(bl)} = \left[\rho_9^3(1)\right]^{(bl)} \cdot \lambda_9^0(1) = 1.32 \cdot 0.0125 = 0.0165,$$
$$\left[\lambda_9^3(2)\right]^{(bl)} = \left[\rho_9^3(2)\right]^{(bl)} \cdot \lambda_9^0(2) = 1.32 \cdot 0.021739 = 0.028695, \quad (6.266)$$
$$b = 3, 5, \quad l = 1, 2, 17, 18, 33, 34, 65, 66, 81, 82,$$

$$\left[\lambda_9^3(1)\right]^{(bl)} = \left[\rho_9^3(1)\right]^{(bl)} \cdot \lambda_9^0(1) = 1.30 \cdot 0.0125 = 0.01625,$$
$$\left[\lambda_9^3(2)\right]^{(bl)} = \left[\rho_9^3(2)\right]^{(bl)} \cdot \lambda_9^0(2) = 1.30 \cdot 0.021739 = 0.028261, \quad b = 4, 6,$$
$$l = 6, 7, 10, 11, 22, 23, 26, 27, 38, 39, 42, 43, 70, 71, 74, 75, 86, 87, 90, 91, \quad (6.267)$$

$$\left[\lambda_9^3(1)\right]^{(bl)} = \left[\rho_9^3(1)\right]^{(bl)} \cdot \lambda_9^0(1) = 1.43 \cdot 0.0125 = 0.017875,$$
$$\left[\lambda_9^3(2)\right]^{(bl)} = \left[\rho_9^3(2)\right]^{(bl)} \cdot \lambda_9^0(2) = 1.43 \cdot 0.021739 = 0.031087, \quad (6.268)$$
$$b = 4, 6, \quad l = 1, 2, 17, 18, 33, 34, 65, 66, 81, 82.$$

### 6.5.3 Safety and Resilience Indicators of Port Oil Terminal Critical Infrastructure Impacted by Operation Process Related to Climate-Weather Change

Considering that the coordinates $[S_i^3(t,u)]^{(bl)}$, $t \geq 0$, $u = 1, 2$, $i = 1, 2, \ldots 9$, $b = 1, 2, \ldots 7, l = 1, 2, 6, 7, 10, 11, 17, 18, 22, 23, 26, 27, 33, 34, 38, 39, 42, 43, 65, 66, 70, 71, 74, 75, 81, 82, 86, 87, 90, 91$, of the conditional safety functions (6.122), (6.123) for the port oil terminal critical infrastructure assets $A_i$, $i = 1, 2, \ldots 9$, are of the form (6.130), i.e. they are given by the vectors

$$\left[S_i^3(t,\cdot)\right]^{(bl)} = \left[1, \left[S_i^3(t,1)\right]^{(bl)}, \left[S_i^3(t,2)\right]^{(bl)}\right] t \geq 0, u = 1, 2, i = 1, 2, \ldots 9,$$
$$b = 1, 2, \ldots 7, \quad l = 1, 2, 6, 7, 10, 11, 17, 18, 22, 23, 26, 27, 33, 34, 38, 39, 42, 43, 65, 66, 70, 71, 74, 75, 81, 82, 86, 87, 90, 91,$$

with the coordinates

$$\left[S_i^3(t,u)\right]^{(bl)} = \exp\left[-\left[\lambda_i^3(u)\right]^{(bl)} t\right], \quad t \geq 0, \quad u = 1, 2, \quad i = 1, 2, \ldots 9,$$
$$b = 1, 2, \ldots 7, \quad l = 1, 2, 6, 7, 10, 11, 17, 18, 22, 23, 26, 27, 33, 34, 38, 39, 42, 43, 65, 66, 70, 71, 74, 75, 81, 82, 86, 87, 90, 91,$$

and the intensities of ageing $[\lambda_i^3(u)]^{(bl)}$, $t \geq 0$, $u = 1, 2$, $i = 1, 2, \ldots 9$, $b = 1, 2, \ldots 7, l = 1, 2, 6, 7, 10, 11, 17, 18, 22, 23, 26, 27, 33, 34, 38, 39, 42, 43, 65, 66, 70, 71, 74, 75, 81, 82, 86, 87, 90, 91$, at the operation and climate-weather states $zc_{bl}$, $b = 1, 2, \ldots, 7$, $l = 1, 2, 6, 7, 10, 11, 17, 18, 22, 23, 26, 27, 33, 34, 38, 39, 42, 43, 65, 66, 70, 71, 74, 75, 81, 82, 86, 87, 90, 91$, are given respectively by (6.209)–(6.268), we get their coordinates following final forms:

- for asset $A_1$

$$[S_1^3(1)]^{(bl)} = \exp[-0.018333t], \quad [S_1^3(2)]^{(bl)} = \exp[-0.025109t], \quad b = 1, 2, 7, \quad l = 1, 2, 17, 18, \tag{6.269}$$

$$[S_1^3(1)]^{(bl)} = \exp[-0.018857t], \quad [S_1^3(2)]^{(bl)} = \exp[-0.026065t], \quad b = 1, 2, 7, \quad l = 33, 34, 65, 66, 81, 82, \tag{6.270}$$

$$[S_1^3(1)]^{(bl)} = \exp[-0.017984t], \quad [S_1^3(2)]^{(bl)} = \exp[-0.024869t], \quad b = 1, 2, 7, \quad l = 38, 39, 42, 43, 70, 71, 74, 75, 86, 87, 90, 91, \tag{6.271}$$

$$[S_1^3(1)]^{(bl)} = \exp[-0.017460t], \quad [S_1^3(2)]^{(bl)} = \exp[-0.023913t], \quad b = 1, 2, 7, \quad l = 6, 7, 10, 11, 22, 23, 26, 27, \tag{6.272}$$

$$[S_1^3(1)]^{(bl)} = \exp[-0.019999t], \quad [S_1^3(2)]^{(bl)} = \exp[-0.027391t], \quad b = 3, 5, \quad l = 1, 2, 17, 18, \tag{6.273}$$

$$[S_1^3(1)]^{(bl)} = \exp[-0.020571t], \quad [S_1^3(2)]^{(bl)} = \exp[-0.028435t], \quad b = 3, 5, \quad l = 33, 34, 65, 66, 81, 82, \tag{6.274}$$

$$[S_1^3(1)]^{(bl)} = \exp[-0.019619t], \quad [S_1^3(2)]^{(bl)} = \exp[-0.027130t], \quad b = 3, 5, \quad l = 38, 39, 42, 43, 70, 71, 74, 75, 86, 87, 90, 91, \tag{6.275}$$

$$[S_1^3(1)]^{(bl)} = \exp[-0.019048t], \quad [S_1^3(2)]^{(bl)} = \exp[-0.026087t], \quad b = 3, 5, \quad l = 6, 7, 10, 11, 22, 23, 26, 27, \tag{6.276}$$

$$[S_1^3(1)]^{(bl)} = \exp[-0.021666t], \quad [S_1^3(2)]^{(bl)} = \exp[-0.029673t], \quad b = 4, 6, \quad l = 1, 2, 17, 18, \tag{6.277}$$

$$[S_1^3(1)]^{(bl)} = \exp[-0.022286t], \quad [S_1^3(2)]^{(bl)} = \exp[-0.030804t], \quad b = 4, 6, \quad l = 33, 34, 65, 66, 81, 82, \tag{6.278}$$

$$[S_1^3(1)]^{(bl)} = \exp[-0.021254t], \quad [S_1^3(2)]^{(bl)} = \exp[-0.029391t], \quad b = 4, 6, \quad l = 38, 39, 42, 43, 70, 71, 74, 75, 86, 87, 90, 91, \tag{6.279}$$

$$[S_1^3(1)]^{(bl)} = \exp[-0.020634t], \quad [S_1^3(2)]^{(bl)} = \exp[-0.028260t], \quad b = 4, 6, l = 6, 7, 10, 11, 22, 23, 26, 27; \tag{6.280}$$

- for asset $A_2$

$$[S_2^3(1)]^{(bl)} = \exp[-0.01375t], \quad [S_2^3(2)]^{(bl)} = \exp[-0.023913t], \quad b = 1, 2, 7, \quad l = 6, 7, 10, 11, 22, 23, 26, 27, 38, 39, 42, 43, 70, 71, 74, 75, 86, 87, 90, 91, \tag{6.281}$$

$$[S_2^3(1)]^{(bl)} = \exp[-0.015125t], \quad [S_2^3(2)]^{(bl)} = \exp[-0.026304t],$$
$$b = 1, 2, 7, l = 1, 2, 17, 18, 33, 34, 65, 66, 81, 82, \tag{6.282}$$

$$[S_2^3(1)]^{(bl)} = \exp[-0.015t], \quad [S_2^3(2)]^{(bl)} = \exp[-0.026086t], \quad b = 3, 5,$$
$$l = 6, 7, 10, 11, 22, 23, 26, 27, 38, 39, 42, 43, 70, 71, 74, 75, 86, 87, 90, 91, \tag{6.283}$$

$$[S_2^3(1)]^{(bl)} = \exp[-0.0165t], \quad [S_2^3(2)]^{(bl)} = \exp[-0.028695t],$$
$$b = 3, 5, l = 1, 2, 17, 18, 33, 34, 65, 66, 81, 82, \tag{6.284}$$

$$[S_2^3(1)]^{(bl)} = \exp[-0.01625t], \quad [S_2^3(2)]^{(bl)} = \exp[-0.028261t], \quad b = 4, 6,$$
$$l = 6, 7, 10, 11, 22, 23, 26, 27, 38, 39, 42, 43, 70, 71, 74, 75, 86, 87, 90, 91, \tag{6.285}$$

$$[S_2^3(1)]^{(bl)} = \exp[-0.017875t], \quad [S_2^3(2)]^{(bl)} = \exp[-0.031086t],$$
$$b = 4, 6, l = 1, 2, 17, 18, 33, 34, 65, 66, 81, 82; \tag{6.286}$$

– for asset $A_3$

$$[S_3^3(1)]^{(bl)} = \exp[-0.0125t], \quad [S_3^3(2)]^{(bl)} = \exp[-0.021739t], \quad b = 1, 2, 3, 7,$$
$$l = 6, 7, 10, 11, 22, 23, 26, 27, 38, 39, 42, 43, 70, 71, 74, 75, 86, 87, 90, 91, \tag{6.287}$$

$$[S_3^3(1)]^{(bl)} = \exp[-0.01375t], \quad [S_3^3(2)]^{(bl)} = \exp[-0.023913t],$$
$$b = 1, 2, 3, 7, l = 1, 2, 17, 18, 33, 34, 65, 66, 81, 82, \tag{6.288}$$

$$[S_3^3(1)]^{(bl)} = \exp[-0.015t], \quad [S_3^3(2)]^{(bl)} = \exp[-0.026086t],$$
$$b = 5, l = 6, 7, 10, 11, 22, 23, 26, 27, 38, 39, 42, 43, 70, 71, 74, 75, 86, 87, 90, 91, \tag{6.289}$$

$$[S_3^3(1)]^{(bl)} = \exp[-0.0165t], \quad [S_3^3(2)]^{(bl)} = \exp[-0.028695t],$$
$$b = 5, l = 1, 2, 17, 18, 33, 34, 65, 66, 81, 82, \tag{6.290}$$

$$[S_3^3(1)]^{(bl)} = \exp[-0.01625t], \quad [S_3^3(2)]^{(bl)} = \exp[-0.028261t], \quad b = 4, 6,$$
$$l = 6, 7, 10, 11, 22, 23, 26, 27, 38, 39, 42, 43, 70, 71, 74, 75, 86, 87, 90, 91, \tag{6.291}$$

$$[S_3^3(1)]^{(bl)} = \exp[-0.017875t], \quad [S_3^3(2)]^{(bl)} = \exp[-0.031086t],$$
$$b = 4, 6, l = 1, 2, 17, 18, 33, 34, 65, 66, 81, 82; \quad (6.292)$$

– for asset $A_4$

$$[S_4^3(1)]^{(bl)} = \exp[-0.01375t], \quad [S_4^3(2)]^{(bl)} = \exp[-0.023913t],$$
$$b = 1, 2, 7, \quad l = 6, 7, 10, 11, 22, 23, 26, 27, 38,$$
$$39, 42, 43, 70, 71, 74, 75, 86, 87, 90, 91, \quad (6.293)$$

$$[S_4^3(1)]^{(bl)} = \exp[-0.015125t], \quad [S_4^3(2)]^{(bl)} = \exp[-0.026304t],$$
$$b = 1, 2, 7, \quad l = 1, 2, 17, 18, 33, 34, 65, 66, 81, 82, \quad (6.294)$$

$$[S_4^3(1)]^{(bl)} = \exp[-0.015t], \quad [S_4^3(2)]^{(bl)} = \exp[-0.026086t], \quad b = 3, 5,$$
$$l = 6, 7, 10, 11, 22, 23, 26, 27, 38, 39, 42, 43, 70, 71, 74, 75, 86, 87, 90, 91, \quad (6.295)$$

$$[S_4^3(1)]^{(bl)} = \exp[-0.0165t], \quad [S_4^3(2)]^{(bl)} = \exp[-0.028695t],$$
$$b = 3, 5, \quad l = 1, 2, 17, 18, 33, 34, 65, 66, 81, 82, \quad (6.296)$$

$$[S_4^3(1)]^{(bl)} = \exp[-0.01625t], \quad [S_4^3(2)]^{(bl)} = \exp[-0.028261t], \quad b = 4, 6,$$
$$l = 6, 7, 10, 11, 22, 23, 26, 27, 38, 39, 42, 43, 70, 71, 74, 75, 86, 87, 90, 91, \quad (6.297)$$

$$[S_4^3(1)]^{(bl)} = \exp[-0.017875t], \quad [S_4^3(2)]^{(bl)} = \exp[-0.031086t],$$
$$b = 4, 6, \quad l = 1, 2, 17, 18, 33, 34, 65, 66, 81, 82; \quad (6.298)$$

– for asset $A_5$

$$[S_5^3(1)]^{(bl)} = \exp[-0.0125t], \quad [S_5^3(2)]^{(bl)} = \exp[-0.021739t], \quad b = 1, 2, 7,$$
$$l = 6, 7, 10, 11, 22, 23, 26, 27, 38, 39, 42, 43, 70, 71, 74, 75, 86, 87, 90, 91, \quad (6.299)$$

$$[S_5^3(1)]^{(bl)} = \exp[-0.01375t], \quad [S_5^3(2)]^{(bl)} = \exp[-0.023913t],$$
$$b = 1, 2, 7, \quad l = 1, 2, 17, 18, 33, 34, 65, 66, 81, 82, \quad (6.300)$$

$$[S_5^3(1)]^{(bl)} = \exp[-0.015t], \quad [S_5^3(2)]^{(bl)} = \exp[-0.026087t], \quad b = 3, 5,$$
$$l = 6, 7, 10, 11, 22, 23, 26, 27, 38, 39, 42, 43, 70, 71, 74, 75, 86, 87, 90, 91, \quad (6.301)$$

$$[S_5^3(1)]^{(bl)} = \exp[-0.0165t], \quad [S_5^3(2)]^{(bl)} = \exp[-0.028695t],$$
$$b = 3, 5, \quad l = 1, 2, 17, 18, 33, 34, 65, 66, 81, 82, \quad (6.302)$$

$$[S_5^3(1)]^{(bl)} = \exp[-0.01625t], \quad [S_5^3(2)]^{(bl)} = \exp[-0.028261t], \quad b = 4, 6,$$
$$l = 6, 7, 10, 11, 22, 23, 26, 27, 38, 39, 42, 43, 70, 71, 74, 75, 86, 87, 90, 91, \tag{6.303}$$

$$[S_5^3(1)]^{(bl)} = \exp[-0.017875t], \quad [S_5^3(2)]^{(bl)} = \exp[-0.031087t],$$
$$b = 4, 6, \quad l = 1, 2, 17, 18, 33, 34, 65, 66, 81, 82; \tag{6.304}$$

– for asset $A_6$

$$[S_6^3(1)]^{(bl)} = \exp[-0.0125t], \quad [S_6^3(2)]^{(bl)} = \exp[-0.021739t], \quad b = 2, 5,$$
$$l = 6, 7, 10, 11, 22, 23, 26, 27, 38, 39, 42, 43, 70, 71, 74, 75, 86, 87, 90, 91, \tag{6.305}$$

$$[S_6^3(1)]^{(bl)} = \exp[-0.01375t], \quad [S_6^3(2)]^{(bl)} = \exp[-0.023913t],$$
$$b = 2, 5, \quad l = 1, 2, 17, 18, 33, 34, 65, 66, 81, 82, \tag{6.306}$$

$$[S_6^3(1)]^{(bl)} = \exp[-0.01375t], \quad [S_6^3(2)]^{(bl)} = \exp[-0.023913t], b = 1, 7,$$
$$l = 6, 7, 10, 11, 22, 23, 26, 27, 38, 39, 42, 43, 70, 71, 74, 75, 86, 87, 90, 91, \tag{6.307}$$

$$[S_6^3(1)]^{(bl)} = \exp[-0.015125t], \quad [S_6^3(2)]^{(bl)} = \exp[-0.026304t],$$
$$b = 1, 7, \quad l = 1, 2, 17, 18, 33, 34, 65, 66, 81, 82, \tag{6.308}$$

$$[S_6^3(1)]^{(bl)} = \exp[-0.015t], \quad [S_6^3(2)]^{(bl)} = \exp[-0.026087t], \quad b = 3,$$
$$l = 6, 7, 10, 11, 22, 23, 26, 27, 38, 39, 42, 43, 70, 71, 74, 75, 86, 87, 90, 91, \tag{6.309}$$

$$[S_6^3(1)]^{(bl)} = \exp[-0.0165t], \quad [S_6^3(2)]^{(bl)} = \exp[-0.028695t],$$
$$b = 3, \quad l = 1, 2, 17, 18, 33, 34, 65, 66, 81, 82, \tag{6.310}$$

$$[S_6^3(1)]^{(bl)} = \exp[-0.01625t], \quad [S_6^3(2)]^{(bl)} = \exp[-0.028261t], \quad b = 4, 6,$$
$$l = 6, 7, 10, 11, 22, 23, 26, 27, 38, 39, 42, 43, 70, 71, 74, 75, 86, 87, 90, 91, \tag{6.311}$$

$$[S_6^3(1)]^{(bl)} = \exp[-0.017875t], \quad [S_6^3(2)]^{(bl)} = \exp[-0.031087t],$$
$$b = 4, 6, \quad l = 1, 2, 17, 18, 33, 34, 65, 66, 81, 82; \tag{6.312}$$

– for asset $A_7$

$$[S_7^3(1)]^{(bl)} = \exp[-0.0125t], \quad [S_7^3(2)]^{(bl)} = \exp[-0.021739t], b = 2, 3, 5,$$
$$l = 6, 7, 10, 11, 22, 23, 26, 27, 38, 39, 42, 43, 70, 71, 74, 75, 86, 87, 90, 91, \tag{6.313}$$

$$[S_7^3(1)]^{(bl)} = \exp[-0.01375t], \quad [S_7^3(2)]^{(bl)} = \exp[-0.023913t],$$
$$b = 2, 3, 5, \quad l = 1, 2, 17, 18, 33, 34, 65, 66, 81, 82, \tag{6.314}$$

$$[S_7^3(1)]^{(bl)} = \exp[-0.01375t], \quad [S_7^3(2)]^{(bl)} = \exp[-0.023913t], \quad b = 1, 7,$$
$$l = 6, 7, 10, 11, 22, 23, 26, 27, 38, 39, 42, 43, 70, 71, 74, 75, 86, 87, 90, 91, \tag{6.315}$$

$$[S_7^3(1)]^{(bl)} = \exp[-0.015125t], \quad [S_7^3(2)]^{(bl)} = \exp[-0.026304t],$$
$$b = 1, 7, \quad l = 1, 2, 17, 18, 33, 34, 65, 66, 81, 82, \tag{6.316}$$

$$[S_7^3(1)]^{(bl)} = \exp[-0.01625t], [S_7^3(2)]^{(bl)} = \exp[-0.028261t], \quad b = 4, 6,$$
$$l = 6, 7, 10, 11, 22, 23, 26, 27, 38, 39, 42, 43, 70, 71, 74, 75, 86, 87, 90, 91, \tag{6.317}$$

$$[S_7^3(1)]^{(bl)} = \exp[-0.017875t], \quad [S_7^3(2)]^{(bl)} = \exp[-0.031087t],$$
$$b = 4, 6, \quad l = 1, 2, 17, 18, 33, 34, 65, 66, 81, 82; \tag{6.318}$$

- for asset $A_8$

$$[S_8^3(1)]^{(bl)} = \exp[-0.0125t], \quad [S_8^3(2)]^{(bl)} = \exp[-0.021739t], \quad b = 1, 3, 4, 5, 6,$$
$$l = 6, 7, 10, 11, 22, 23, 26, 27, 38, 39, 42, 43, 70, 71, 74, 75, 86, 87, 90, 91, \tag{6.319}$$

$$[S_8^3(1)]^{(bl)} = \exp[-0.01375t], \quad [S_8^3(2)]^{(bl)} = \exp[-0.023913t],$$
$$b = 1, 3, 4, 5, 6, \quad l = 1, 2, 17, 18, 33, 34, 65, 66, 81, 82, \tag{6.320}$$

$$[S_8^3(1)]^{(bl)} = \exp[-0.01375t], \quad [S_8^3(2)]^{(bl)} = \exp[-0.023913t],$$
$$b = 2, 7, \quad l = 6, 7, 10, 11, 22, 23, 26, 27, 38, 39, 42, 43, 70, 71, 74, 75, 86, 87, 90, 91, \tag{6.321}$$

$$[S_8^3(1)]^{(bl)} = \exp[-0.015125t], \quad [S_8^3(2)]^{(bl)} = \exp[-0.026304t],$$
$$b = 2, 7, \quad l = 1, 2, 17, 18, 33, 34, 65, 66, 81, 82; \tag{6.322}$$

- for asset $A_9$

$$[S_9^3(1)]^{(bl)} = \exp[-0.01375t], \quad [S_9^3(2)]^{(bl)} = \exp[-0.023913t], \quad b = 1, 2, 7,$$
$$l = 6, 7, 10, 11, 22, 23, 26, 27, 38, 39, 42, 43, 70, 71, 74, 75, 86, 87, 90, 91, \tag{6.323}$$

$$[S_9^3(1)]^{(bl)} = \exp[-0.015125t], \quad [S_9^3(2)]^{(bl)} = \exp[-0.026304t],$$
$$b = 1, 2, 7, \quad l = 1, 2, 17, 18, 33, 34, 65, 66, 81, 82, \tag{6.324}$$

$$[S_9^3(1)]^{(bl)} = \exp[-0.015t], \quad [S_9^3(2)]^{(bl)} = \exp[-0.026087t], \quad b = 3, 5,$$
$$l = 6, 7, 10, 11, 22, 23, 26, 27, 38, 39, 42, 43, 70, 71, 74, 75, 86, 87, 90, 91, \tag{6.325}$$

$$[S_1^3(1)]^{(bl)} = \exp[-0.0165t], \quad [S_9^3(2)]^{(bl)} = \exp[-0.028695t],$$
$$b = 3, 5, \quad l = 1, 2, 17, 18, 33, 34, 65, 66, 81, 82, \tag{6.326}$$

$$[S_9^3(1)]^{(bl)} = \exp[-0.01625t], \quad [S_9^3(2)]^{(bl)} = \exp[-0.028261t], b = 4, 6,$$
$$l = 6, 7, 10, 11, 22, 23, 26, 27, 38, 39, 42, 43, 70, 71, 74, 75, 86, 87, 90, 91, \tag{6.327}$$

$$[S_9^3(1)]^{(bl)} = \exp[-0.017875t], \quad [S_9^3(2)]^{(bl)} = \exp[-0.031087t],$$
$$b = 4, 6, \quad l = 1, 2, 17, 18, 33, 34, 65, 66, 81, 82. \tag{6.328}$$

Since the oil terminal critical infrastructure is a three-state (z = 2) series system, then considering (6.269)–(6.328) and applying (2.8), (2.9) its conditional safety functions at the operation and climate-weather states $zc_{bl}$, $b = 1, 2, \ldots, 7$, $l = 1, 2, 6, 7, 10, 11, 17, 18, 22, 23, 26, 27, 33, 34, 38, 39, 42, 43, 65, 66, 70, 71, 74, 75, 81, 82, 86, 87, 90, 91$, are given by:

$$\left[\boldsymbol{S}^3(t,\cdot)\right]^{(bl)} = \left[1, \left[\boldsymbol{S}^3(t,1)\right]^{(bl)}, \left[\boldsymbol{S}^3(t,2)\right]^{(bl)}\right] t \geq 0, \tag{6.329}$$

where

$$\begin{aligned}
\left[\boldsymbol{S}^3(t,1)\right]^{(bl)} &= \prod_{i=1}^{9} [\boldsymbol{S}_i^3(t,1)]^{(bl)} \\
&= [\boldsymbol{S}_1^3(t,1)]^{(bl)} \cdot [\boldsymbol{S}_2^3(t,1)]^{(bl)} \cdot [\boldsymbol{S}_3^3(t,1)]^{(bl)} \\
&\quad \cdot [\boldsymbol{S}_4^3(t,1)]^{(bl)} \cdot [\boldsymbol{S}_5^3(t,1)]^{(bl)} \cdot [\boldsymbol{S}_6^3(t,1)]^{(bl)} \\
&\quad \cdot [\boldsymbol{S}_7^3(t,1)]^{(bl)} \cdot [\boldsymbol{S}_8^3(t,1)]^{(bl)} \cdot [\boldsymbol{S}_9^3(t,1)]^{(bl)} \\
&= \exp[-0.018333t] \exp[-0.015125t] \exp[-0.01375t] \\
&\quad \exp[-0.015125t] \exp[-0.01375t] \exp[-0.015125t] \\
&\quad \exp[-0.015125t] \exp[-0.01375t] \exp[-0.015125t] \\
&= \exp[-0.135208t],
\end{aligned} \tag{6.330}$$

$$\begin{aligned}
\left[\boldsymbol{S}^3(t,2)\right]^{(bl)} &= \prod_{i=1}^{9} [\boldsymbol{S}_i^3(t,2)]^{(bl)} \\
&= [\boldsymbol{S}_1^3(t,2)]^{(bl)} \cdot [\boldsymbol{S}_2^3(t,2)]^{(bl)} \cdot [\boldsymbol{S}_3^3(t,2)]^{(bl)} \\
&\quad \cdot [\boldsymbol{S}_4^3(t,2)]^{(bl)} \cdot [\boldsymbol{S}_5^3(t,2)]^{(bl)} \cdot [\boldsymbol{S}_6^3(t,2)]^{(bl)} \\
&\quad \cdot [\boldsymbol{S}_7^3(t,2)]^{(bl)} \cdot [\boldsymbol{S}_8^3(t,2)]^{(bl)} \cdot [\boldsymbol{S}_9^3(t,2)]^{(bl)} \\
&= \exp[-0.025105t] \exp[-0.026304t] \exp[-0.023913t]
\end{aligned}$$

$$
\begin{aligned}
&\exp[-0.026304t]\exp[-0.023913t]\exp[-0.026304t]\\
&\exp[-0.026304t]\exp[-0.023913t]\exp[-0.026304t]\\
&= \exp[-0.228364t],
\end{aligned}
\tag{6.331}
$$

for operation and climate-weather states $zc_{bl}$, $b = 1$, $l = 1, 2, 17, 18$;

$$
\left[\boldsymbol{S}^3(t,\cdot)\right]^{(bl)} = \left[1, \left[\boldsymbol{S}^3(t,1)\right]^{(bl)}, \left[\boldsymbol{S}^3(t,2)\right]^{(bl)}\right] t \geq 0, \tag{6.332}
$$

where

$$
\begin{aligned}
\left[\boldsymbol{S}^3(t,1)\right]^{(bl)} &= \prod_{i=1}^{9}\left[\boldsymbol{S}_i^3(t,1)\right]^{(bl)}\\
&= \left[S_1^3(t,1)\right]^{(bl)} \cdot \left[S_2^3(t,1)\right]^{(bl)} \cdot S_3^0(t,1) \cdot \left[S_4^3(t,1)\right]^{(bl)} \cdot S_5^0(t,1)\\
&\quad \cdot \left[S_6^3(t,1)\right]^{(bl)} \cdot \left[S_7^3(t,1)\right]^{(bl)} \cdot S_8^0(t,1) \cdot \left[S_9^3(t,1)\right]^{(bl)}\\
&= \exp[-0.017460t]\exp[-0.01375t]\exp[-0.0125t]\\
&\quad \exp[-0.01375t]\exp[-0.0125t]\exp[-0.01375t]\\
&\quad \exp[-0.013755t]\exp[-0.0125t]\exp[-0.01375t]\\
&= \exp[-0.12371t],
\end{aligned}
\tag{6.333}
$$

$$
\begin{aligned}
\left[\boldsymbol{S}^3(t,2)\right]^{(bl)} &= \prod_{i=1}^{9}\left[\boldsymbol{S}_i^3(t,2)\right]^{(bl)}\\
&= \left[S_1^3(t,2)\right]^{(bl)} \cdot \left[S_2^3(t,2)\right]^{(bl)} \cdot S_3^0(t,2) \cdot \left[S_4^3(t,2)\right]^{(bl)} \cdot S_5^0(t,2)\\
&\quad \cdot \left[S_6^3(t,2)\right]^{(bl)} \cdot \left[S_7^3(t,2)\right]^{(bl)} \cdot S_8^0(t,2) \cdot \left[S_9^3(t,2)\right]^{(bl)}\\
&= \exp[-0.023913t]\exp[-0.023913t]\exp[-0.021739t]\\
&\quad \exp[-0.023913t]\exp[-0.021739t]\exp[-0.023913t]\\
&\quad \exp[-0.023913t]\exp[-0.021739t]\exp[-0.023913t]\\
&= \exp[-0.208695t],
\end{aligned}
\tag{6.334}
$$

for operation and climate-weather states $zc_{bl}$, $b = 1$, $l = 6, 7, 10, 11, 22, 23, 26, 27$;

$$
\left[\boldsymbol{S}^3(t,\cdot)\right]^{(bl)} = \left[1, \left[\boldsymbol{S}^3(t,1)\right]^{(bl)}, \left[\boldsymbol{S}^3(t,2)\right]^{(bl)}\right] t \geq 0, \tag{6.335}
$$

where

$$
\begin{aligned}
\left[\boldsymbol{S}^3(t,1)\right]^{(bl)} &= \prod_{i=1}^{9}\left[\boldsymbol{S}_i^3(t,1)\right]^{(bl)}\\
&= \left[S_1^3(t,1)\right]^{(bl)} \cdot \left[S_2^3(t,1)\right]^{(bl)} \cdot \left[S_3^3(t,1)\right]^{(bl)}
\end{aligned}
$$

$$\cdot \left[S_4^3(t,1)\right]^{(bl)} \cdot \left[S_5^3(t,1)\right]^{(bl)} \cdot \left[S_6^3(t,1)\right]^{(bl)}$$
$$\cdot \left[S_7^3(t,1)\right]^{(bl)} \cdot \left[S_8^3(t,1)\right]^{(bl)} \cdot \left[S_9^3(t,1)\right]^{(bl)}$$
$$= \exp[-0.018857t]\exp[-0.015125t]\exp[-0.01375t]$$
$$\exp[-0.015125t]\exp[-0.01375t]\exp[-0.015125t]$$
$$\exp[-0.015125t]\exp[-0.01375t]\exp[-0.015125t]$$
$$= \exp[-0.135732t], \tag{6.336}$$

$$\left[\boldsymbol{S}^3(t,2)\right]^{(bl)} = \prod_{i=1}^{9}\left[S_i^3(t,2)\right]^{(bl)}$$
$$= \left[S_1^3(t,2)\right]^{(bl)} \cdot \left[S_2^3(t,2)\right]^{(bl)} \cdot \left[S_3^3(t,2)\right]^{(bl)}$$
$$\cdot \left[S_4^3(t,2)\right]^{(bl)} \cdot \left[S_5^3(t,2)\right]^{(bl)} \cdot \left[S_6^3(t,2)\right]^{(bl)}$$
$$\cdot \left[S_7^3(t,2)\right]^{(bl)} \cdot \left[S_8^3(t,2)\right]^{(bl)} \cdot \left[S_9^3(t,2)\right]^{(bl)}$$
$$= \exp[-0.026065t]\exp[-0.026304t]\exp[-0.023913t]$$
$$\exp[-0.026304t]\exp[-0.023913t]\exp[-0.026304t]$$
$$\exp[-0.026304t]\exp[-0.023913t]\exp[-0.026304t]$$
$$= \exp[-0.229324t], \tag{6.337}$$

for operation and climate-weather states $zc_{bl}$, $b = 1$, $l = 33, 34, 65, 66, 81, 82$;

$$\left[\boldsymbol{S}^3(t,\cdot)\right]^{(bl)} = \left[1, \left[\boldsymbol{S}^3(t,1)\right]^{(bl)}, \left[\boldsymbol{S}^3(t,2)\right]^{(bl)}\right] t \geq 0, \tag{6.338}$$

where

$$\left[\boldsymbol{S}^3(t,1)\right]^{(bl)} = \prod_{i=1}^{9}\left[S_i^3(t,1)\right]^{(bl)}$$
$$= \left[S_1^3(t,1)\right]^{(bl)} \cdot \left[S_2^3(t,1)\right]^{(bl)} \cdot S_3^0(t,1) \cdot \left[S_4^3(t,1)\right]^{(bl)} \cdot S_5^0(t,1)$$
$$\cdot \left[S_6^3(t,1)\right]^{(bl)} \cdot \left[S_7^3(t,1)\right]^{(bl)} \cdot S_8^0(t,1) \cdot \left[S_9^3(t,1)\right]^{(bl)}$$
$$= \exp[-0.017984t]\exp[-0.01375t]\exp[-0.0125t]$$
$$\exp[-0.01375t]\exp[-0.0125t]\exp[-0.01375t]$$
$$\exp[-0.01375t]\exp[-0.0125t]\exp[-0.01375t]$$
$$= \exp[-0.124234t], \tag{6.339}$$

$$\begin{aligned}
\left[\boldsymbol{S}^{3}(t,2)\right]^{(bl)} &= \prod_{i=1}^{9}\left[\boldsymbol{S}_{i}^{3}(t,2)\right]^{(bl)} \\
&= \left[S_{\mathbf{1}}^{\mathbf{3}}(t,2)\right]^{(bl)}\cdot\left[S_{\mathbf{2}}^{\mathbf{3}}(t,2)\right]^{(bl)}\cdot S_{\mathbf{3}}^{\mathbf{0}}(t,2)\cdot\left[S_{\mathbf{4}}^{\mathbf{3}}(t,2)\right]^{(bl)}\cdot S_{\mathbf{5}}^{\mathbf{0}}(t,2) \\
&\quad\cdot\left[S_{\mathbf{6}}^{\mathbf{3}}(t,2)\right]^{(bl)}\cdot\left[S_{\mathbf{7}}^{\mathbf{3}}(t,2)\right]^{(bl)}\cdot S_{\mathbf{8}}^{\mathbf{0}}(t,2)\cdot\left[S_{\mathbf{9}}^{\mathbf{3}}(t,2)\right]^{(bl)} \\
&= \exp[-0.024869t]\exp[-0.023913t]\exp[-0.021739t] \\
&\quad\exp[-0.023913t]\exp[-0.021739t]\exp[-0.023913t] \\
&\quad\exp[-0.023913t]\exp[-0.021739t]\exp[-0.023913t] \\
&= \exp[-0.209651t],
\end{aligned} \tag{6.340}$$

for operation and climate-weather states $zc_{bl}$, $b = 1$, $l = 38, 39, 42, 43, 70, 71, 74, 75, 86, 87, 90, 91$;

$$\left[\boldsymbol{S}^{3}(t,\cdot)\right]^{(bl)} = \left[1,\left[\boldsymbol{S}^{3}(t,1)\right]^{(bl)},\left[\boldsymbol{S}^{3}(t,2)\right]^{(bl)}\right]t\geq 0, \tag{6.341}$$

where

$$\begin{aligned}
\left[\boldsymbol{S}^{3}(t,1)\right]^{(bl)} &= \prod_{i=1}^{9}\left[\boldsymbol{S}_{i}^{3}(t,1)\right]^{(bl)} \\
&= \left[S_{\mathbf{1}}^{\mathbf{3}}(t,1)\right]^{(bl)}\cdot\left[S_{\mathbf{2}}^{\mathbf{3}}(t,1)\right]^{(bl)}\cdot\left[S_{\mathbf{3}}^{\mathbf{3}}(t,1)\right]^{(bl)} \\
&\quad\cdot\left[S_{\mathbf{4}}^{\mathbf{3}}(t,1)\right]^{(bl)}\cdot\left[S_{\mathbf{5}}^{\mathbf{3}}(t,1)\right]^{(bl)}\cdot\left[S_{\mathbf{6}}^{\mathbf{3}}(t,1)\right]^{(bl)} \\
&\quad\cdot\left[S_{\mathbf{7}}^{\mathbf{3}}(t,1)\right]^{(bl)}\cdot\left[S_{\mathbf{8}}^{\mathbf{3}}(t,1)\right]^{(bl)}\cdot\left[S_{\mathbf{9}}^{\mathbf{3}}(t,1)\right]^{(bl)} \\
&= \exp[-0.018333t]\exp[-0.015125t]\exp[-0.01375t] \\
&\quad\exp[-0.015125t]\exp[-0.01375t]\exp[-0.01375t] \\
&\quad\exp[-0.01375t]\exp[-0.015125t]\exp[-0.015125t] \\
&= \exp[-0.133833t],
\end{aligned} \tag{6.342}$$

$$\begin{aligned}
\left[\boldsymbol{S}^{3}(t,2)\right]^{(bl)} &= \prod_{i=1}^{9}\left[\boldsymbol{S}_{i}^{3}(t,2)\right]^{(bl)} \\
&= \left[S_{\mathbf{1}}^{\mathbf{3}}(t,2)\right]^{(bl)}\cdot\left[S_{\mathbf{2}}^{\mathbf{3}}(t,2)\right]^{(bl)}\cdot\left[S_{\mathbf{3}}^{\mathbf{3}}(t,2)\right]^{(bl)} \\
&\quad\cdot\left[S_{\mathbf{4}}^{\mathbf{3}}(t,2)\right]^{(bl)}\cdot\left[S_{\mathbf{5}}^{\mathbf{3}}(t,2)\right]^{(bl)}\cdot\left[S_{\mathbf{6}}^{\mathbf{3}}(t,2)\right]^{(bl)} \\
&\quad\cdot\left[S_{\mathbf{7}}^{\mathbf{3}}(t,2)\right]^{(bl)}\cdot\left[S_{\mathbf{8}}^{\mathbf{3}}(t,2)\right]^{(bl)}\cdot\left[S_{\mathbf{9}}^{\mathbf{3}}(t,2)\right]^{(bl)} \\
&= \exp[-0.025105t]\exp[-0.026304t]\exp[-0.023913t] \\
&\quad\exp[-0.026304t]\exp[-0.023913t]\exp[-0.023913t] \\
&\quad\exp[-0.023913t]\exp[-0.026304t]\exp[-0.026304t] \\
&= \exp[-0.225977t],
\end{aligned} \tag{6.343}$$

for operation and climate-weather states $zc_{bl}$, $b = 2$, $l = 1, 2, 17, 18$;

$$\left[\boldsymbol{S}^3(t,\cdot)\right]^{(bl)} = \left[1, \left[\boldsymbol{S}^3(t,1)\right]^{(bl)}, \left[\boldsymbol{S}^3(t,2)\right]^{(bl)}\right] t \geq 0, \tag{6.344}$$

where

$$\begin{aligned}\left[\boldsymbol{S}^3(t,1)\right]^{(bl)} &= \prod_{i=1}^{9}\left[S_i^3(t,1)\right]^{(bl)} \\ &= \left[S_{\mathbf{1}}^{\mathbf{3}}(t,1)\right]^{(bl)} \cdot \left[S_{\mathbf{2}}^{\mathbf{3}}(t,1)\right]^{(bl)} \cdot S_{\mathbf{3}}^{\mathbf{0}}(t,1) \cdot \left[S_{\mathbf{4}}^{\mathbf{3}}(t,1)\right]^{(bl)} \cdot S_{\mathbf{5}}^{\mathbf{0}}(t,1) \\ &\quad \cdot S_{\mathbf{6}}^{\mathbf{0}}(t,1) \cdot S_{\mathbf{7}}^{\mathbf{0}}(t,1) \cdot \left[S_{\mathbf{8}}^{\mathbf{3}}(t,1)\right]^{(bl)} \cdot \left[S_{\mathbf{9}}^{\mathbf{3}}(t,1)\right]^{(bl)} \\ &= \exp[-0.017460t]\exp[-0.01375t]\exp[-0.0125t] \\ &\quad \exp[-0.01375t]\exp[-0.0125t]\exp[-0.0125t] \\ &\quad \exp[-0.0125t]\exp[-0.01375t]\exp[-0.01375t] \\ &= \exp[-0.12246t],\end{aligned} \tag{6.345}$$

$$\begin{aligned}\left[\boldsymbol{S}^3(t,2)\right]^{(bl)} &= \prod_{i=1}^{9}\left[S_i^3(t,2)\right]^{(bl)} \\ &= \left[S_{\mathbf{1}}^{\mathbf{3}}(t,2)\right]^{(bl)} \cdot \left[S_{\mathbf{2}}^{\mathbf{3}}(t,2)\right]^{(bl)} \cdot S_{\mathbf{3}}^{\mathbf{0}}(t,2) \cdot \left[S_{\mathbf{4}}^{\mathbf{3}}(t,2)\right]^{(bl)} \cdot S_{\mathbf{5}}^{\mathbf{0}}(t,2) \\ &\quad \cdot S_{\mathbf{6}}^{\mathbf{0}}(t,2) \cdot S_{\mathbf{7}}^{\mathbf{0}}(t,2) \cdot \left[S_{\mathbf{8}}^{\mathbf{3}}(t,2)\right]^{(bl)} \cdot \left[S_{\mathbf{9}}^{\mathbf{3}}(t,2)\right]^{(bl)} \\ &= \exp[-0.023913t]\exp[-0.023913t]\exp[-0.021739t] \\ &\quad \exp[-0.023913t]\exp[-0.021739t]\exp[-0.021739t] \\ &\quad \exp[-0.021739t]\exp[-0.023913t]\exp[-0.023913t] \\ &= \exp[-0.206521t];\end{aligned} \tag{6.346}$$

for operation and climate-weather states $zc_{bl}$, $b = 2$, $l = 6, 7, 10, 11, 22, 23, 26, 27$;

$$\left[\boldsymbol{S}^3(t,\cdot)\right]^{(bl)} = \left[1, \left[\boldsymbol{S}^3(t,1)\right]^{(bl)}, \left[\boldsymbol{S}^3(t,2)\right]^{(bl)}\right] t \geq 0, \tag{6.347}$$

where

$$\begin{aligned}\left[\boldsymbol{S}^3(t,1)\right]^{(bl)} &= \prod_{i=1}^{9}\left[S_i^3(t,1)\right]^{(bl)} \\ &= \left[S_{\mathbf{1}}^{\mathbf{3}}(t,1)\right]^{(bl)} \cdot \left[S_{\mathbf{2}}^{\mathbf{3}}(t,1)\right]^{(bl)} \cdot \left[S_{\mathbf{3}}^{\mathbf{3}}(t,1)\right]^{(bl)} \\ &\quad \cdot \left[S_{\mathbf{4}}^{\mathbf{3}}(t,1)\right]^{(bl)} \cdot \left[S_{\mathbf{5}}^{\mathbf{3}}(t,1)\right]^{(bl)} \cdot \left[S_{\mathbf{6}}^{\mathbf{3}}(t,1)\right]^{(bl)} \\ &\quad \cdot \left[S_{\mathbf{7}}^{\mathbf{3}}(t,1)\right]^{(bl)} \cdot \left[S_{\mathbf{8}}^{\mathbf{3}}(t,1)\right]^{(bl)} \cdot \left[S_{\mathbf{9}}^{\mathbf{3}}(t,1)\right]^{(bl)} \\ &= \exp[-0.018857t]\exp[-0.015125t]\exp[-0.01375t]\end{aligned}$$

$$\exp[-0.015125t]\exp[-0.01375t]\exp[-0.01375t]$$
$$\exp[-0.01375t]\exp[-0.015125t]\exp[-0.015125t]$$
$$= \exp[-0.134357t], \quad (6.348)$$

$$\begin{aligned}[\boldsymbol{S}^3(t,2)]^{(bl)} &= \prod_{i=1}^{9}[\boldsymbol{S}_i^3(t,2)]^{(bl)} \\ &= [S_1^3(t,2)]^{(bl)} \cdot [S_2^3(t,2)]^{(bl)} \cdot [S_3^3(t,2)]^{(bl)} \\ &\cdot [S_4^3(t,2)]^{(bl)} \cdot [S_5^3(t,2)]^{(bl)} \cdot [S_6^3(t,2)]^{(bl)} \\ &\cdot [S_7^3(t,2)]^{(bl)} \cdot [S_8^3(t,2)]^{(bl)} \cdot [S_9^3(t,2)]^{(bl)} \\ &= \exp[-0.026065t]\exp[-0.026304t]\exp[-0.023913t] \\ &\quad \exp[-0.026304t]\exp[-0.023913t]\exp[-0.023913t] \\ &\quad \exp[-0.023913t]\exp[-0.026304t]\exp[-0.026304t] \\ &= \exp[-0.226933t], \end{aligned} \quad (6.349)$$

for operation and climate-weather states $zc_{bl}$, $b = 2$, $l = 33, 34, 65, 66, 81, 82$;

$$[\boldsymbol{S}^3(t,\cdot)]^{(bl)} = \left[1, [\boldsymbol{S}^3(t,1)]^{(bl)}, [\boldsymbol{S}^3(t,2)]^{(bl)}\right] t \geq 0, \quad (6.350)$$

where

$$\begin{aligned}[\boldsymbol{S}^3(t,1)]^{(bl)} &= \prod_{i=1}^{9}[\boldsymbol{S}_i^3(t,1)]^{(bl)} \\ &= [S_1^3(t,1)]^{(bl)} \cdot [S_2^3(t,1)]^{(bl)} \cdot S_3^0(t,1) \cdot [S_4^3(t,1)]^{(bl)} \\ &\cdot S_5^0(t,1) \cdot S_6^0(t,1) \cdot S_7^0(t,1) \cdot [S_8^3(t,1)]^{(bl)} \cdot [S_9^3(t,1)]^{(bl)} \\ &= \exp[-0.017984t]\exp[-0.01375t]\exp[-0.0125t] \\ &\quad \exp[-0.01375t]\exp[-0.0125t]\exp[-0.0125t] \\ &\quad \exp[-0.0125t]\exp[-0.01375t]\exp[-0.01375t] \\ &= \exp[-0.122984t], \end{aligned} \quad (6.351)$$

$$\begin{aligned}[\boldsymbol{S}^3(t,2)]^{(bl)} &= \prod_{i=1}^{9}[\boldsymbol{S}_i^3(t,2)]^{(bl)} \\ &= [S_1^3(t,2)]^{(bl)} \cdot [S_2^3(t,2)]^{(bl)} \cdot S_3^0(t,2) \cdot [S_4^3(t,2)]^{(bl)} \cdot S_5^0(t,2) \\ &\cdot S_6^0(t,2) \cdot S_7^0(t,2) \cdot [S_8^3(t,3)]^{(bl)} \cdot [S_9^3(t,2)]^{(bl)} \\ &= \exp[-0.024869t]\exp[-0.023913t]\exp[-0.021739t] \\ &\quad \exp[-0.023913t]\exp[-0.021739t]\exp[-0.021739t] \end{aligned}$$

$$\exp[-0.021739t]\exp[-0.023913t]\exp[-0.023913t]$$
$$= \exp[-0.207477t]; \tag{6.352}$$

for operation and climate-weather states $zc_{bl}$, $b = 2$, $l = 38, 39, 42, 43, 70, 71, 74, 75, 86, 87, 90, 91$;

$$[\boldsymbol{S}^3(t,\cdot)]^{(bl)} = \left[1, [\boldsymbol{S}^3(t,1)]^{(bl)}, [\boldsymbol{S}^3(t,2)]^{(bl)}\right] t \geq 0, \tag{6.353}$$

where

$$\begin{aligned}
[\boldsymbol{S}^3(t,1)]^{(bl)} &= \prod_{i=1}^{9}[S_i^3(t,1)]^{(bl)} \\
&= [S_1^3(t,1)]^{(bl)} \cdot [S_2^3(t,1)]^{(bl)} \cdot [S_3^3(t,1)]^{(bl)} \\
&\quad \cdot [S_4^3(t,1)]^{(bl)} \cdot [S_5^3(t,1)]^{(bl)} \cdot [S_6^3(t,1)]^{(bl)} \\
&\quad \cdot [S_7^3(t,1)]^{(bl)} \cdot [S_8^3(t,1)]^{(bl)} \cdot [S_9^3(t,1)]^{(bl)} \\
&= \exp[-0.019999t]\exp[-0.0165t]\exp[-0.01375t] \\
&\quad \exp[-0.0165t]\exp[-0.0165t]\exp[-0.0165t] \\
&\quad \exp[-0.01375t]\exp[-0.01375t]\exp[-0.0165t] \\
&= \exp[-0.143749t],
\end{aligned} \tag{6.354}$$

$$\begin{aligned}
[\boldsymbol{S}^3(t,2)]^{(bl)} &= \prod_{i=1}^{9}[S_i^3(t,2)]^{(bl)} \\
&= [S_1^3(t,2)]^{(bl)} \cdot [S_2^3(t,2)]^{(bl)} \cdot [S_3^3(t,2)]^{(bl)} \\
&\quad \cdot [S_4^3(t,2)]^{(bl)} \cdot [S_5^3(t,2)]^{(bl)} \cdot [S_6^3(t,2)]^{(bl)} \\
&\quad \cdot [S_7^3(t,2)]^{(bl)} \cdot [S_8^3(t,2)]^{(bl)} \cdot [S_9^3(t,2)]^{(bl)} \\
&= \exp[-0.027391t]\exp[-0.028695t]\exp[-0.023913t] \\
&\quad \exp[-0.028695t]\exp[-0.028695t]\exp[-0.028695t] \\
&\quad \exp[-0.023913t]\exp[-0.023913t]\exp[-0.028695t] \\
&= \exp[-0.242605t];
\end{aligned} \tag{6.355}$$

for operation and climate-weather states $zc_{bl}$, $b = 3$, $l = 1, 2, 17, 18$;

$$[\boldsymbol{S}^3(t,\cdot)]^{(bl)} = \left[1, [\boldsymbol{S}^3(t,1)]^{(bl)}, [\boldsymbol{S}^3(t,2)]^{(bl)}\right] t \geq 0, \tag{6.356}$$

where

$$\begin{aligned}\left[\boldsymbol{S}^3(t,1)\right]^{(bl)} &= \prod_{i=1}^{9}\left[S_i^3(t,1)\right]^{(bl)}\\ &= \left[S_1^3(t,1)\right]^{(bl)}\cdot\left[S_2^3(t,1)\right]^{(bl)}\cdot S_3^0(t,1)\cdot\left[S_4^3(t,1)\right]^{(bl)}\cdot\left[S_5^3(t,1)\right]^{(bl)}\\ &\quad\cdot\left[S_6^3(t,1)\right]^{(bl)}\cdot S_7^0(t,1)\cdot S_8^0(t,1)\cdot\left[S_9^3(t,1)\right]^{(bl)}\\ &= \exp[-0.019048t]\exp[-0.015t]\exp[-0.0125t]\\ &\quad\exp[-0.015t]\exp[-0.015t]\exp[-0.015t]\\ &\quad\exp[-0.0125t]\exp[-0.0125t]\exp[-0.015t]\\ &= \exp[-0.131548t],\end{aligned} \tag{6.357}$$

$$\begin{aligned}\left[\boldsymbol{S}^3(t,2)\right]^{(bl)} &= \prod_{i=1}^{9}\left[S_i^3(t,2)\right]^{(bl)}\\ &= \left[S_1^3(t,2)\right]^{(bl)}\cdot\left[S_2^3(t,2)\right]^{(bl)}\cdot S_3^0(t,2)\\ &\quad\cdot\left[S_4^3(t,2)\right]^{(bl)}\cdot\left[S_5^3(t,2)\right]^{(bl)}\cdot\left[S_6^3(t,2)\right]^{(bl)}\\ &\quad\cdot S_7^0(t,2)\cdot S_8^0(t,2)\cdot\left[S_9^3(t,2)\right]^{(bl)}\\ &= \exp[-0.026087t]\exp[-0.026086t]\exp[-0.021739t]\\ &\quad\exp[-0.026086t]\exp[-0.026087t]\exp[-0.026087t]\\ &\quad\exp[-0.021739t]\exp[-0.021739t]\exp[-0.026087t]\\ &= \exp[-0.221737t];\end{aligned} \tag{6.358}$$

for operation and climate-weather states $zc_{bl}$, $b = 3$, $l = 6, 7, 10, 11, 22, 23, 26, 27$;

$$\left[\boldsymbol{S}^3(t,\cdot)\right]^{(bl)} = \left[1, \left[\boldsymbol{S}^3(t,1)\right]^{(bl)}, \left[\boldsymbol{S}^3(t,2)\right]^{(bl)}\right] t \geq 0, \tag{6.359}$$

where

$$\begin{aligned}\left[\boldsymbol{S}^3(t,1)\right]^{(bl)} &= \prod_{i=1}^{9}\left[S_i^3(t,1)\right]^{(bl)}\\ &= \left[S_1^3(t,1)\right]^{(bl)}\cdot\left[S_2^3(t,1)\right]^{(bl)}\cdot\left[S_3^3(t,1)\right]^{(bl)}\\ &\quad\cdot\left[S_4^3(t,1)\right]^{(bl)}\cdot\left[S_5^3(t,1)\right]^{(bl)}\cdot\left[S_6^3(t,1)\right]^{(bl)}\\ &\quad\cdot\left[S_7^3(t,1)\right]^{(bl)}\cdot\left[S_8^3(t,1)\right]^{(bl)}\cdot\left[S_9^3(t,1)\right]^{(bl)}\\ &= \exp[-0.020571t]\exp[-0.0165t]\exp[-0.01375t]\\ &\quad\exp[-0.0165t]\exp[-0.0165t]\exp[-0.0165t]\\ &\quad\exp[-0.01375t]\exp[-0.01375t]\exp[-0.0165t]\\ &= \exp[-0.144321t],\end{aligned} \tag{6.360}$$

$$\begin{aligned}
[S^3(t,2)]^{(bl)} &= \prod_{i=1}^{9}[S_i^3(t,2)]^{(bl)} \\
&= [S_1^3(t,2)]^{(bl)} \cdot [S_2^3(t,2)]^{(bl)} \cdot [S_3^3(t,2)]^{(bl)} \\
&\quad \cdot [S_4^3(t,2)]^{(bl)} \cdot [S_5^3(t,2)]^{(bl)} \cdot [S_6^3(t,2)]^{(bl)} \\
&\quad \cdot [S_7^3(t,2)]^{(bl)} \cdot [S_8^3(t,2)]^{(bl)} \cdot [S_9^3(t,2)]^{(bl)} \\
&= \exp[-0.028435t]\exp[-0.028695t]\exp[-0.023913t] \\
&\quad \exp[-0.028695t]\exp[-0.028695t]\exp[-0.028695t] \\
&\quad \exp[-0.023913t]\exp[-0.023913t]\exp[-0.028695t] \\
&= \exp[-0.243649t]; \qquad (6.361)
\end{aligned}$$

for operation and climate-weather states $zc_{bl}$, $b = 3$, $l = 33, 34, 65, 66, 81, 82$;

$$[S^3(t,\cdot)]^{(bl)} = \left[1, [S^3(t,1)]^{(bl)}, [S^3(t,2)]^{(bl)}\right] t \geq 0, \qquad (6.362)$$

where

$$\begin{aligned}
[S^3(t,1)]^{(bl)} &= \prod_{i=1}^{0}[S_i^3(t,1)]^{(bl)} \\
&= [S_1^3(t,1)]^{(bl)} \cdot [S_2^3(t,1)]^{(bl)} \cdot S_3^0(t,1) \\
&\quad \cdot [S_4^3(t,1)]^{(bl)} \cdot [S_5^3(t,1)]^{(bl)} \cdot [S_6^3(t,1)]^{(bl)} \\
&\quad \cdot S_7^0(t,1) \cdot S_8^0(t,1) \cdot [S_9^3(t,1)]^{(bl)} \\
&= \exp[-0.019619t]\exp[-0.015t]\exp[-0.0125t] \\
&\quad \exp[-0.015t]\exp[-0.015t]\exp[-0.015t] \\
&\quad \exp[-0.0125t]\exp[-0.0125t]\exp[-0.015t] \\
&= \exp[-0.124234t], \qquad (6.363)
\end{aligned}$$

$$\begin{aligned}
[S^3(t,2)]^{(bl)} &= \prod_{i=1}^{9}[S_i^3(t,2)]^{(bl)} \\
&= [S_1^3(t,2)]^{(bl)} \cdot [S_2^3(t,2)]^{(bl)} \cdot S_3^0(t,2) \\
&\quad \cdot [S_4^3(t,2)]^{(bl)} \cdot [S_5^3(t,2)]^{(bl)} \cdot [S_6^3(t,2)]^{(bl)} \\
&\quad \cdot S_7^0(t,2) \cdot S_8^0(t,2) \cdot [S_9^3(t,2)]^{(bl)} \\
&= \exp[-0.027130t]\exp[-0.026086t]\exp[-0.021739t] \\
&\quad \exp[-0.026086t]\exp[-0.026087t]\exp[-0.026087t] \\
&\quad \exp[-0.021739t]\exp[-0.021739t]\exp[-0.026087t]
\end{aligned}$$

$$= \exp[-0.22278t]; \tag{6.364}$$

for operation and climate-weather states $zc_{bl}$, $b = 3$, $l = 38, 39, 42, 43, 70, 71, 74, 75, 86, 87, 90, 91$;

$$[\boldsymbol{S}^3(t,\cdot)]^{(bl)} = \left[1, [\boldsymbol{S}^3(t,1)]^{(bl)}, [\boldsymbol{S}^3(t,2)]^{(bl)}\right] t \geq 0, \tag{6.365}$$

where

$$\begin{aligned}
[\boldsymbol{S}^3(t,1)]^{(bl)} &= \prod_{i=1}^{9} [S_i^3(t,1)]^{(bl)} \\
&= [S_1^3(t,1)]^{(bl)} \cdot [S_2^3(t,1)]^{(bl)} \cdot [S_3^3(t,1)]^{(b)} \\
&\quad \cdot [S_4^3(t,1)]^{(bl)} \cdot [S_5^3(t,1)]^{(bl)} \cdot [S_6^3(t,1)]^{(bl)} \\
&\quad \cdot [S_7^3(t,1)]^{(bl)} \cdot [S_8^3(t,1)]^{(bl)} \cdot [S_9^3(t,1)]^{(bl)} \\
&= \exp[-0.021666t] \exp[-0.017875t] \exp[-0.017875t] \\
&\quad \exp[-0.017875t] \exp[-0.017875t] \exp[-0.017875t] \\
&\quad \exp[-0.017875t] \exp[-0.01375t] \exp[-0.017875t] \\
&= \exp[-0.160541t],
\end{aligned} \tag{6.366}$$

$$\begin{aligned}
[\boldsymbol{S}^3(t,2)]^{(bl)} &= \prod_{i=1}^{9} [S_i^3(t,2)]^{(bl)} \\
&= [S_1^3(t,2)]^{(bl)} \cdot [S_2^3(t,2)]^{(bl)} \cdot [S_3^3(t,2)]^{(bl)} \\
&\quad \cdot [S_4^3(t,2)]^{(bl)} \cdot [S_5^3(t,2)]^{(b))} \cdot [S_6^3(t,2)]^{(bl)} \\
&\quad \cdot [S_7^3(t,2)]^{(bl)} \cdot [S_8^3(t,2)]^{(bl)} \cdot [S_9^3(t,2)]^{(bl)} \\
&= \exp[-0.029673t] \exp[-0.031086t] \exp[-0.031086t] \\
&\quad \exp[-0.031086t] \exp[-0.031086t] \exp[-0.031086t] \\
&\quad \exp[-0.031086t] \exp[-0.023913t] \exp[-0.031087t] \\
&= \exp[-0.271189t],
\end{aligned} \tag{6.367}$$

for operation and climate-weather states $zc_{bl}$, $b = 4$, $l = 1, 2, 17, 18$;

$$[\boldsymbol{S}^3(t,\cdot)]^{(bl)} = \left[1, [\boldsymbol{S}^3(t,1)]^{(bl)}, [\boldsymbol{S}^3(t,2)]^{(bl)}\right] t \geq 0, \tag{6.368}$$

where

$$\begin{aligned}\left[\boldsymbol{S}^3(t,1)\right]^{(bl)} &= \prod_{i=1}^{9}\left[\boldsymbol{S}_i^3(t,1)\right]^{(bl)}\\ &= \left[S_{\mathbf{1}}^{\mathbf{3}}(t,1)\right]^{(bl)}\cdot\left[S_{\mathbf{2}}^{\mathbf{3}}(t,1)\right]^{(bl)}\cdot\left[S_{\mathbf{3}}^{\mathbf{3}}(t,1)\right]^{(bl)}\\ &\quad\cdot\left[S_{\mathbf{4}}^{\mathbf{3}}(t,1)\right]^{(bl)}\cdot\left[S_{\mathbf{5}}^{\mathbf{3}}(t,1)\right]^{(bl)}\cdot\left[S_{\mathbf{6}}^{\mathbf{3}}(t,1)\right]^{(bl)}\\ &\quad\cdot\left[S_{\mathbf{7}}^{\mathbf{3}}(t,1)\right]^{(bl)}\cdot S_{\mathbf{8}}^{\mathbf{0}}(t,1)\cdot\left[S_{\mathbf{9}}^{\mathbf{3}}(t,1)\right]^{(bl)}\\ &= \exp[-0.020634t]\exp[-0.01625t]\exp[-0.01625t]\\ &\quad\exp[-0.01625t]\exp[-0.01625t]\exp[-0.01625t]\\ &\quad\exp[-0.01625t]\exp[-0.0125t]\exp[-0.01625t]\\ &= \exp[-0.146884t],\end{aligned} \tag{6.369}$$

$$\begin{aligned}\left[\boldsymbol{S}^3(t,2)\right]^{(bl)} &= \prod_{i=1}^{9}\left[\boldsymbol{S}_i^3(t,2)\right]^{(bl)}\\ &= \left[S_{\mathbf{1}}^{\mathbf{3}}(t,2)\right]^{(bl)}\cdot\left[S_{\mathbf{2}}^{\mathbf{3}}(t,2)\right]^{(bl)}\cdot\left[S_{\mathbf{3}}^{\mathbf{3}}(t,2)\right]^{(bl)}\\ &\quad\cdot\left[S_{\mathbf{4}}^{\mathbf{3}}(t,2)\right]^{(bl)}\cdot\left[S_{\mathbf{5}}^{\mathbf{3}}(t,2)\right]^{(bl)}\cdot\left[S_{\mathbf{6}}^{\mathbf{3}}(t,2)\right]^{(bl)}\\ &\quad\cdot\left[S_{\mathbf{7}}^{\mathbf{3}}(t,2)\right]^{(bl)}\cdot S_{\mathbf{8}}^{\mathbf{0}}(t,2)\cdot\left[S_{\mathbf{9}}^{\mathbf{3}}(t,2)\right]^{(bl)}\\ &= \exp[-0.028260t]\exp[-0.028261t]\exp[-0.028261t]\\ &\quad\exp[-0.028261t]\exp[-0.028261t]\exp[-0.028261t]\\ &\quad\exp[-0.028261t]\exp[-0.021739t]\exp[-0.028261t]\\ &= \exp[-0.247826t],\end{aligned} \tag{6.370}$$

for operation and climate-weather states $zc_{bl}$, $b = 4$, $l = 6, 7, 10, 11, 22, 23, 26, 27$;

$$\left[\boldsymbol{S}^3(t,\cdot)\right]^{(bl)} = \left[1, \left[\boldsymbol{S}^3(t,1)\right]^{(bl)}, \left[\boldsymbol{S}^3(t,2)\right]^{(bl)}\right] t \geq 0, \tag{6.371}$$

where

$$\begin{aligned}\left[\boldsymbol{S}^3(t,1)\right]^{(bl)} &= \prod_{i=1}^{9}\left[\boldsymbol{S}_i^3(t,1)\right]^{(bl)}\\ &= \left[S_{\mathbf{1}}^{\mathbf{3}}(t,1)\right]^{(bl)}\cdot\left[S_{\mathbf{2}}^{\mathbf{3}}(t,1)\right]^{(bl)}\cdot\left[S_{\mathbf{3}}^{\mathbf{3}}(t,1)\right]^{(bl)}\\ &\quad\cdot\left[S_{\mathbf{4}}^{\mathbf{3}}(t,1)\right]^{(bl)}\cdot\left[S_{\mathbf{5}}^{\mathbf{3}}(t,1)\right]^{(bl)}\cdot\left[S_{\mathbf{6}}^{\mathbf{3}}(t,1)\right]^{(bl)}\\ &\quad\cdot\left[S_{\mathbf{7}}^{\mathbf{3}}(t,1)\right]^{(bl)}\cdot\left[S_{\mathbf{8}}^{\mathbf{3}}(t,1)\right]^{(bl)}\cdot\left[S_{\mathbf{9}}^{\mathbf{3}}(t,1)\right]^{(bl)}\\ &= \exp[-0.022286t]\exp[-0.017875t]\exp[-0.017875t]\\ &\quad\exp[-0.017875t]\exp[-0.017875t]\exp[-0.017875t]\end{aligned}$$

$$\exp[-0.017875t]\exp[-0.01375t]\exp[-0.017875t]$$
$$= \exp[-0.161161t], \tag{6.372}$$

$$\begin{aligned}
[\boldsymbol{S}^3(t,2)]^{(bl)} &= \prod_{i=1}^{9}[\boldsymbol{S}_i^3(t,2)]^{(bl)} \\
&= [S_1^3(t,2)]^{(bl)} \cdot [S_2^3(t,2)]^{(bl)} \cdot [S_3^3(t,2)]^{(bl)} \\
&\cdot [S_4^3(t,2)]^{(bl)} \cdot [S_5^3(t,2)]^{(bl)} \cdot [S_6^3(t,2)]^{(bl)} \\
&\cdot [S_7^3(t,2)]^{(bl)} \cdot [S_8^3(t,2)]^{(bl)} \cdot [S_9^3(t,2)]^{(bl)} \\
&= \exp[-0.030804t]\exp[-0.031086t]\exp[-0.031086t] \\
&\quad \exp[-0.031086t]\exp[-0.031087t]\exp[-0.031087t] \\
&\quad \exp[-0.031087t]\exp[-0.023913t]\exp[-0.031087t] \\
&= \exp[-0.272323t];
\end{aligned} \tag{6.373}$$

for operation and climate-weather states $zc_{bl}$, $b = 4$, $l = 33, 34, 65, 66, 81, 82$;

$$[\boldsymbol{S}^3(t,\cdot)]^{(bl)} = \left[1, [\boldsymbol{S}^3(t,1)]^{(bl)}, [\boldsymbol{S}^3(t,2)]^{(bl)}\right] t \geq 0, \tag{6.374}$$

where

$$\begin{aligned}
[\boldsymbol{S}^3(t,1)]^{(bl)} &= \prod_{i=1}^{9}[\boldsymbol{S}_i^3(t,1)]^{(bl)} \\
&= [S_1^3(t,1)]^{(bl)} \cdot [S_2^3(t,1)]^{(bl)} \cdot [S_3^3(t,1)]^{(bl)} \\
&\cdot [S_4^3(t,1)]^{(bl)} \cdot [S_5^3(t,1)]^{(bl)} \cdot [S_6^3(t,1)]^{(bl)} \\
&\cdot [S_7^3(t,1)]^{(bl)} \cdot S_8^0(t,1) \cdot [S_9^3(t,1)]^{(bl)} \\
&= \exp[-0.021254t]\exp[-0.01625t]\exp[-0.01625t] \\
&\quad \exp[-0.01625t]\exp[-0.01625t]\exp[-0.01625t] \\
&\quad \exp[-0.01625t]\exp[-0.0125t]\exp[-0.01625t] \\
&= \exp[-0.147504t],
\end{aligned} \tag{6.375}$$

$$\begin{aligned}
\left[\boldsymbol{S}^3(t,2)\right]^{(bl)} &= \prod_{i=1}^{9}\left[\boldsymbol{S}_i^3(t,2)\right]^{(bl)} \\
&= \left[S_1^3(t,2)\right]^{(bl)} \cdot \left[S_2^3(t,2)\right]^{(bl)} \cdot \left[S_3^3(t,2)\right]^{(bl)} \cdot \left[S_4^3(t,2)\right]^{(bl)} \cdot \left[S_5^3(t,2)\right]^{(bl)} \\
&\cdot \left[S_6^3(t,2)\right]^{(bl)} \cdot \left[S_7^3(t,2)\right]^{(bl)} \cdot S_8^0(t,2) \cdot \left[S_9^3(t,2)\right]^{(bl)} \\
&= \exp[-0.029391t]\exp[-0.028261t]\exp[-0.028261t] \\
&\quad \exp[-0.028261t]\exp[-0.028261t]\exp[-0.028261t]
\end{aligned}$$

$$\exp[-0.028261t]\exp[-0.021739t]\exp[-0.028261t]$$
$$= \exp[-0.248957t], \tag{6.376}$$

for operation and climate-weather states $zc_{bl}$, $b = 4$, $l = 38, 39, 42, 43, 70, 71, 74, 75, 86, 87, 90, 91$;

$$\left[\boldsymbol{S}^3(t,\cdot)\right]^{(bl)} = \left[1, \left[\boldsymbol{S}^3(t,1)\right]^{(bl)}, \left[\boldsymbol{S}^3(t,2)\right]^{(bl)}\right] t \geq 0, \tag{6.377}$$

where

$$\begin{aligned}
\left[\boldsymbol{S}^3(t,1)\right]^{(bl)} &= \prod_{i=1}^{9}\left[S_i^3(t,1)\right]^{(bl)} \\
&= \left[S_1^3(t,1)\right]^{(bl)}\cdot\left[S_2^3(t,1)\right]^{(bl)}\cdot\left[S_3^3(t,1)\right]^{(bl)} \\
&\quad\cdot\left[S_4^3(t,1)\right]^{(bl)}\cdot\left[S_5^3(t,1)\right]^{(bl)}\cdot\left[S_6^3(t,1)\right]^{(bl)} \\
&\quad\cdot\left[S_7^3(t,1)\right]^{(bl)}\cdot\left[S_8^3(t,1)\right]^{(bl)}\cdot\left[S_9^3(t,1)\right]^{(bl)} \\
&= \exp[-0.019999t]\exp[-0.0165t]\exp[-0.0165t] \\
&\quad \exp[-0.0165t]\exp[-0.0165t]\exp[-0.01375t] \\
&\quad \exp[-0.01375t]\exp[-0.01375t]\exp[-0.0165t] \\
&= \exp[-0.143749t],
\end{aligned} \tag{6.378}$$

$$\begin{aligned}
\left[\boldsymbol{S}^3(t,2)\right]^{(bl)} &= \prod_{i=1}^{0}\left[S_i^3(t,2)\right]^{(bl)} \\
&= \left[S_1^3(t,2)\right]^{(bl)}\cdot\left[S_2^3(t,2)\right]^{(bl)}\cdot\left[S_3^3(t,2)\right]^{(bl)} \\
&\quad\cdot\left[S_4^3(t,2)\right]^{(bl)}\cdot\left[S_5^3(t,2)\right]^{(bl)}\cdot\left[S_6^3(t,2)\right]^{(bl)} \\
&\quad\cdot\left[S_7^3(t,2)\right]^{(bl)}\cdot\left[S_8^3(t,2)\right]^{(bl)}\cdot\left[S_9^3(t,2)\right]^{(bl)} \\
&= \exp[-0.027391t]\exp[-0.028695t]\exp[-0.028695t] \\
&\quad \exp[-0.028695t]\exp[-0.028695t]\exp[-0.023913t] \\
&\quad \exp[-0.023913t]\exp[-0.023913t]\exp[-0.028695t] \\
&= \exp[-0.242605t],
\end{aligned} \tag{6.379}$$

for operation and climate-weather states $zc_{bl}$, $b = 5$, $l = 1, 2, 17, 18$;

$$\left[\boldsymbol{S}^3(t,\cdot)\right]^{(bl)} = \left[1, \left[\boldsymbol{S}^3(t,1)\right]^{(bl)}, \left[\boldsymbol{S}^3(t,2)\right]^{(bl)}\right] t \geq 0, \tag{6.380}$$

where

$$\begin{aligned}
\left[\boldsymbol{S}^3(t,1)\right]^{(bl)} &= \prod_{i=1}^{9}\left[\boldsymbol{S}_i^3(t,1)\right]^{(bl)} \\
&= \left[S_1^3(t,1)\right]^{(bl)}\cdot\left[S_2^3(t,1)\right]^{(bl)}\cdot\left[S_3^3(t,1)\right]^{(bl)} \\
&\quad\cdot\left[S_4^3(t,1)\right]^{(bl)}\cdot\left[S_5^3(t,1)\right]^{(bl)}\cdot\left[S_6^3(t,1)\right]^{(bl)} \\
&\quad\cdot\left[S_7^3(t,1)\right]^{(bl)}\cdot S_8^0(t,1)\cdot\left[S_9^3(t,1)\right]^{(bl)} \\
&= \exp[-0.019048t]\exp[-0.015t]\exp[-0.015t] \\
&\quad \exp[-0.015t]\exp[-0.015t]\exp[-0.01375t] \\
&\quad \exp[-0.01375t]\exp[-0.0125t]\exp[-0.015t] \\
&= \exp[-0.134048t],
\end{aligned} \tag{6.381}$$

$$\begin{aligned}
\left[\boldsymbol{S}^3(t,2)\right]^{(bl)} &= \prod_{i=1}^{9}\left[\boldsymbol{S}_i^3(t,2)\right]^{(bl)} \\
&= \left[S_1^3(t,2)\right]^{(bl)}\cdot\left[S_2^3(t,2)\right]^{(bl)}\cdot\left[S_3^3(t,2)\right]^{(bl)} \\
&\quad\cdot\left[S_4^3(t,2)\right]^{(bl)}\cdot\left[S_5^3(t,2)\right]^{(bl)}\cdot\left[S_6^3(t,2)\right]^{(bl)} \\
&\quad\cdot\left[S_7^3(t,2)\right]^{(bl)}\cdot S_8^0(t,2)\cdot\left[S_9^3(t,2)\right]^{(bl)} \\
&= \exp[-0.026087t]\exp[-0.026086t]\exp[-0.026086t] \\
&\quad \exp[-0.026086t]\exp[-0.026086t]\exp[-0.023913t] \\
&\quad \exp[-0.023913t]\exp[-0.021739t]\exp[-0.026087t] \\
&= \exp[-0.226083t],
\end{aligned} \tag{6.382}$$

for operation and climate-weather states $zc_{bl}$, $b = 5$, $l = 6, 7, 10, 11, 22, 23, 26, 27$;

$$\left[\boldsymbol{S}^3(t,\cdot)\right]^{(bl)} = \left[1, \left[\boldsymbol{S}^3(t,1)\right]^{(bl)}, \left[\boldsymbol{S}^3(t,2)\right]^{(bl)}\right]t \geq 0, \tag{6.383}$$

where

$$\begin{aligned}
\left[\boldsymbol{S}^3(t,1)\right]^{(bl)} &= \prod_{i=1}^{0}\left[\boldsymbol{S}_i^3(t,1)\right]^{(bl)} \\
&= \left[S_1^3(t,1)\right]^{(bl)}\cdot\left[S_2^3(t,1)\right]^{(bl)}\cdot\left[S_3^3(t,1)\right]^{(bl)} \\
&\quad\cdot\left[S_4^3(t,1)\right]^{(bl)}\cdot\left[S_5^3(t,1)\right]^{(bl)}\cdot\left[S_6^3(t,1)\right]^{(bl)} \\
&\quad\cdot\left[S_7^3(t,1)\right]^{(bl)}\cdot\left[S_8^3(t,1)\right]^{(bl)}\cdot\left[S_9^3(t,1)\right]^{(bl)} \\
&= \exp[-0.020571t]\exp[-0.0165t]\exp[-0.0165t] \\
&\quad \exp[-0.0165t]\exp[-0.0165t]\exp[-0.01375t] \\
&\quad \exp[-0.01375t]\exp[-0.01375t]\exp[-0.0165t]
\end{aligned}$$

$$= \exp[-0.144321t], \tag{6.384}$$

$$\begin{aligned}
[S^3(t,2)]^{(bl)} &= \prod_{i=1}^{9}[S_i^3(t,2)]^{(bl)} \\
&= [S_1^3(t,2)]^{(bl)} \cdot [S_2^3(t,2)]^{(bl)} \cdot [S_3^3(t,2)]^{(bl)} \\
&\cdot [S_4^3(t,2)]^{(bl)} \cdot [S_5^3(t,2)]^{(bl)} \cdot [S_6^3(t,2)]^{(bl)} \\
&\cdot [S_7^3(t,2)]^{(bl)} \cdot [S_8^3(t,2)]^{(bl)} \cdot [S_9^3(t,2)]^{(bl)} \\
&= \exp[-0.028435t]\exp[-0.028695t]\exp[-0.028695t] \\
&\quad \exp[-0.028695t]\exp[-0.028695t]\exp[-0.023913t] \\
&\quad \exp[-0.023913t]\exp[-0.023913t]\exp[-0.028695t] \\
&= \exp[-0.243649t],
\end{aligned} \tag{6.385}$$

for operation and climate-weather states $zc_{bl}$, $b = 5$, $l = 33, 34, 65, 66, 81, 82$;

$$[S^3(t,\cdot)]^{(bl)} = \left[1, [S^3(t,1)]^{(bl)}, [S^3(t,2)]^{(bl)}\right] t \geq 0, \tag{6.386}$$

where

$$\begin{aligned}
[S^3(t,1)]^{(bl)} &= \prod_{i=1}^{0}[S_i^3(t,1)]^{(bl)} \\
&= [S_1^3(t,1)]^{(bl)} \cdot [S_2^3(t,1)]^{(bl)} \cdot [S_3^3(t,1)]^{(bl)} \\
&\cdot [S_4^3(t,1)]^{(bl)} \cdot [S_5^3(t,1)]^{(bl)} \cdot S_6^0(t,1) \\
&\cdot S_7^0(t,1) \cdot S_8^0(t,1) \cdot [S_9^3(t,1)]^{(bl)} \\
&= \exp[-0.019619t]\exp[-0.015t]\exp[-0.015t] \\
&\quad \exp[-0.015t]\exp[-0.015t]\exp[-0.0125t] \\
&\quad \exp[-0.0125t]\exp[-0.0125t]\exp[-0.015t] \\
&= \exp[-0.132119t],
\end{aligned} \tag{6.387}$$

$$\begin{aligned}
[S^3(t,2)]^{(bl)} &= \prod_{i=1}^{9}[S_i^3(t,2)]^{(bl)} \\
&= [S_1^3(t,2)]^{(bl)} \cdot [S_2^3(t,2)]^{(bl)} \cdot [S_3^3(t,2)]^{(bl)} \\
&\cdot [S_4^3(t,2)]^{(bl)} \cdot [S_5^3(t,2)]^{(bl)} \cdot S_6^0(t,2) \\
&\cdot S_7^0(t,2) \cdot S_8^0(t,2) \cdot [S_9^3(t,2)]^{(bl)} \\
&= \exp[-0.027130t]\exp[-0.026086t]\exp[-0.026086t]
\end{aligned}$$

$$\exp[-0.0260861t]\exp[-0.026087t]\exp[-0.021739t]$$
$$\exp[-0.021739t]\exp[-0.021739t]\exp[-0.026087t]$$
$$= \exp[-0.222779t], \quad (6.388)$$

for operation and climate-weather states $zc_{bl}$, $b = 5$, $l = 38, 39, 42, 43, 70, 71, 74, 75, 86, 87, 90, 91$;

$$\left[\boldsymbol{S}^3(t, \cdot)\right]^{(bl)} = \left[1, \left[\boldsymbol{S}^3(t, 1)\right]^{(bl)}, \left[\boldsymbol{S}^3(t, 2)\right]^{(bl)}\right] t \geq 0, \quad (6.389)$$

where

$$\begin{aligned}
\left[\boldsymbol{S}^3(t, 1)\right]^{(bl)} &= \prod_{i=1}^{0}\left[\boldsymbol{S}_i^3(t, 1)\right]^{(bl)} \\
&= \left[S_1^3(t, 1)\right]^{(bl)} \cdot \left[S_2^3(t, 1)\right]^{(bl)} \cdot \left[S_3^3(t, 1)\right]^{(bl)} \\
&\quad \cdot \left[S_4^3(t, 1)\right]^{(bl)} \cdot \left[S_5^3(t, 1)\right]^{(bl)} \cdot \left[S_6^3(t, 1)\right]^{(bl)} \\
&\quad \cdot \left[S_7^3(t, 1)\right]^{(bl)} \cdot \left[S_8^3(t, 1)\right]^{(bl)} \cdot \left[S_9^3(t, 1)\right]^{(bl)} \\
&= \exp[-0.021666t]\exp[-0.017875t]\exp[-0.017875t] \\
&\quad \exp[-0.017875t]\exp[-0.017875t]\exp[-0.017875t] \\
&\quad \exp[-0.017875t]\exp[-0.01375t]\exp[-0.017875t] \\
&= \exp[-0.160541t], \quad (6.390)
\end{aligned}$$

$$\begin{aligned}
\left[\boldsymbol{S}^3(t, 2)\right]^{(bl)} &= \prod_{i=1}^{0}\left[\boldsymbol{S}_i^3(t, 2)\right]^{(bl)} \\
&= \left[S_1^3(t, 2)\right]^{(bl)} \cdot \left[S_2^3(t, 2)\right]^{(bl)} \\
&\quad \cdot \left[S_3^3(t, 2)\right]^{(bl)} \cdot \left[S_4^3(t, 2)\right]^{(bl)} \cdot \left[S_5^3(t, 2)\right]^{(bl)} \cdot \left[S_6^3(t, 2)\right]^{(bl)} \\
&\quad \cdot \left[S_7^3(t, 2)\right]^{(bl)} \cdot \left[S_8^3(t, 2)\right]^{(bl)} \cdot \left[S_9^3(t, 2)\right]^{(bl)} \\
&= \exp[-0.029673t]\exp[-0.031086t]\exp[-0.031086t] \\
&\quad \exp[-0.031086t]\exp[-0.031086t]\exp[-0.031087t] \\
&\quad \exp[-0.031087t]\exp[-0.023913t]\exp[-0.031087t] \\
&= \exp[-0.271192t], \quad (6.391)
\end{aligned}$$

for operation and climate-weather states $zc_{bl}$, $b = 6$, $l = 1, 2, 17, 18$;

$$\left[\boldsymbol{S}^3(t, \cdot)\right]^{(bl)} = \left[1, \left[\boldsymbol{S}^3(t, 1)\right]^{(bl)}, \left[\boldsymbol{S}^3(t, 2)\right]^{(bl)}\right] t \geq 0, \quad (6.392)$$

where

$$
\begin{aligned}
\left[\boldsymbol{S}^3(t,1)\right]^{(bl)} &= \prod_{i=1}^{9}\left[\boldsymbol{S}_i^3(t,1)\right]^{(bl)} \\
&= \left[S_1^3(t,1)\right]^{(bl)} \cdot \left[S_2^3(t,1)\right]^{(bl)} \\
&\cdot \left[S_3^3(t,1)\right]^{(bl)} \cdot \left[S_4^3(t,1)\right]^{(bl)} \cdot \left[S_5^3(t,1)\right]^{(bl)} \cdot \left[S_6^3(t,1)\right]^{(bl)} \\
&\cdot \left[S_7^3(t,1)\right]^{(bl)} \cdot S_8^0(t,1) \cdot \left[S_9^3(t,1)\right]^{(bl)} \\
&= \exp[-0.020634t]\exp[-0.016 \downarrow 25t]\exp[-0.01625t] \\
&\quad \exp[-0.01625t]\exp[-0.01625t]\exp[-0.01625t] \\
&\quad \exp[-0.01625t]\exp[-0.0125t]\exp[-0.01625t] \\
&= \exp[-0.146884t],
\end{aligned} \tag{6.393}
$$

$$
\begin{aligned}
\left[\boldsymbol{S}^3(t,2)\right]^{(bl)} &= \prod_{i=1}^{9}\left[\boldsymbol{S}_i^3(t,2)\right]^{(bl)} \\
&= \left[S_1^3(t,2)\right]^{(bl)} \cdot \left[S_2^3(t,2)\right]^{(bl)} \cdot \left[S_3^3(t,2)\right]^{(bl)} \\
&\cdot \left[S_4^3(t,2)\right]^{(bl)} \cdot \left[S_5^3(t,2)\right]^{(bl)} \cdot \left[S_6^3(t,2)\right]^{(bl)} \\
&= \left[S_7^3(t,2)\right]^{(bl)} \cdot S_8^0(t,2) \cdot \left[S_9^3(t,2)\right]^{(bl)} \\
&= \exp[-0.028260t]\exp[-0.028261t]\exp[-0.028261t] \\
&\quad \exp[-0.028261t]\exp[-0.028261t]\exp[-0.028261t] \\
&\quad \exp[-0.028261t]\exp[-0.021739t]\exp[-0.028261t] \\
&= \exp[-0.247826t],
\end{aligned} \tag{6.394}
$$

for operation and climate-weather states $zc_{bl}$, $b = 6$, $l = 6, 7, 10, 11, 22, 23, 26, 27$;

$$
\left[\boldsymbol{S}^3(t,\cdot)\right]^{(bl)} = \left[1, \left[\boldsymbol{S}^3(t,1)\right]^{(bl)}, \left[\boldsymbol{S}^3(t,2)\right]^{(bl)}\right] t \geq 0, \tag{6.395}
$$

where

$$
\begin{aligned}
\left[\boldsymbol{S}^3(t,1)\right]^{(bl)} &= \prod_{i=1}^{9}\left[\boldsymbol{S}_i^3(t,1)\right]^{(bl)} \\
&= \left[S_1^3(t,1)\right]^{(bl)} \cdot \left[S_2^3(t,1)\right]^{(bl)} \cdot \left[S_3^3(t,1)\right]^{(bl)} \\
&\cdot \left[S_4^3(t,1)\right]^{(bl)} \cdot \left[S_5^3(t,1)\right]^{(bl)} \cdot \left[S_6^3(t,1)\right]^{(bl)} \\
&\cdot \left[S_7^3(t,1)\right]^{(bl)} \cdot \left[S_8^3(t,1)\right]^{(bl)} \cdot \left[S_9^3(t,1)\right]^{(bl)} \\
&= \exp[-0.022286t]\exp[-0.017875t]\exp[-0.017875t] \\
&\quad \exp[-0.017875t]\exp[-0.01375t]\exp[-0.017875t] \\
&\quad \exp[-0.017875t]\exp[-0.01375t]\exp[-0.017875t]
\end{aligned}
$$

$$= \exp[-0.157036t], \tag{6.396}$$

$$\begin{aligned}
[S^3(t,2)]^{(bl)} &= \prod_{i=1}^{9} [S_i^3(t,2)]^{(bl)} \\
&= [S_1^3(t,2)]^{(bl)} \cdot [S_2^3(t,2)]^{(bl)} \cdot [S_3^3(t,2)]^{(bl)} \\
&\quad \cdot [S_4^3(t,2)]^{(bl)} \cdot [S_5^3(t,2)]^{(bl)} \cdot [S_6^3(t,2)]^{(bl)} \\
&\quad \cdot [S_7^3(t,2)]^{(bl)} \cdot [S_8^3(t,2)]^{(bl)} \cdot [S_9^3(t,2)]^{(bl)} \\
&= \exp[-0.030804t]\exp[-0.031086t]\exp[-0.031086t] \\
&\quad \exp[-0.031086t]\exp[-0.023913t]\exp[-0.031087t] \\
&\quad \exp[-0.031087t]\exp[-0.023913t]\exp[-0.031087t] \\
&= \exp[-0.265149t],
\end{aligned} \tag{6.397}$$

for operation and climate-weather states $zc_{bl}$, $b = 6$, $l = 33, 34, 65, 66, 81, 82$;

$$[S^3(t,\cdot)]^{(bl)} = \left[1, [S^3(t,1)]^{(bl)}, [S^3(t,2)]^{(bl)}\right] t \geq 0, \tag{6.398}$$

where

$$\begin{aligned}
[S^3(t,1)]^{(bl)} &= \prod_{i=1}^{9} [S_i^3(t,1)]^{(bl)} \\
&= [S_1^3(t,1)]^{(bl)} \cdot [S_2^3(t,1)]^{(bl)} \cdot [S_3^3(t,1)]^{(bl)} \\
&\quad \cdot [S_4^3(t,1)]^{(bl)} \cdot [S_5^3(t,1)]^{(bl)} \cdot [S_6^3(t,1)]^{(bl)} \\
&\quad \cdot [S_7^3(t,1)]^{(bl)} \cdot S_8^0(t,1) \cdot [S_9^3(t,1)]^{(bl)} \\
&= \exp[-0.021254t]\exp[-0.01625t]\exp[-0.01625t] \\
&\quad \exp[-0.01625t]\exp[-0.01625t]\exp[-0.01625t] \\
&\quad \exp[-0.01625t]\exp[-0.0125t]\exp[-0.01625t] \\
&= \exp[-0.147504t],
\end{aligned} \tag{6.399}$$

$$\begin{aligned}
[S^3(t,2)]^{(bl)} &= \prod_{i=1}^{9} [S_i^3(t,2)]^{(bl)} \\
&= [S_1^3(t,2)]^{(bl)} \cdot [S_2^3(t,2)]^{(bl)} \cdot [S_3^3(t,2)]^{(bl)} \\
&\quad \cdot [S_4^3(t,2)]^{(bl)} \cdot [S_5^3(t,2)]^{(bl)} \cdot [S_6^3(t,2)]^{(bl)} \\
&\quad \cdot [S_7^3(t,2)]^{(bl)} \cdot S_8^0(t,2) \cdot [S_9^3(t,2)]^{(bl)} \\
&= \exp[-0.029391t]\exp[-0.028261t]\exp[-0.028261t]
\end{aligned}$$

$$\begin{aligned}
&\exp[-0.028261t]\exp[-0.028261t]\exp[-0.028261t]\\
&\exp[-0.028261t]\exp[-0.021739t]\exp[-0.028261t]\\
&= \exp[-0.248957t],
\end{aligned} \tag{6.400}$$

for operation and climate-weather states $zc_{bl}$, $b = 6$, $l = 38, 39, 42, 43, 70, 71, 74, 75, 86, 87, 90, 91$;

$$\left[\boldsymbol{S}^3(t,\cdot)\right]^{(bl)} = \left[1, \left[\boldsymbol{S}^3(t,1)\right]^{(bl)}, \left[\boldsymbol{S}^3(t,2)\right]^{(bl)}\right] t \geq 0, \tag{6.401}$$

where

$$\begin{aligned}
\left[\boldsymbol{S}^3(t,1)\right]^{(bl)} &= \prod_{i=1}^{9}\left[S_i^3(t,1)\right]^{(bl)}\\
&= \left[S_1^3(t,1)\right]^{(bl)} \cdot \left[S_2^3(t,1)\right]^{(bl)} \cdot \left[S_3^3(t,1)\right]^{(bl)}\\
&\quad \cdot \left[S_4^3(t,1)\right]^{(bl)} \cdot \left[S_5^3(t,1)\right]^{(bl)} \cdot \left[S_6^3(t,1)\right]^{(bl)}\\
&\quad \cdot \left[S_7^3(t,1)\right]^{(bl)} \cdot \left[S_8^3(t,1)\right]^{(bl)} \cdot \left[S_9^3(t,1)\right]^{(bl)}\\
&= \exp[-0.018333t]\exp[-0.015125t]\exp[-0.01375t]\\
&\quad \exp[-0.015125t]\exp[-0.01375t]\exp[-0.015125t]\\
&\quad \exp[-0.015125t]\exp[-0.015125t]\exp[-0.015125t]\\
&= \exp[-0.136583t],
\end{aligned} \tag{6.402}$$

$$\begin{aligned}
\left[\boldsymbol{S}^3(t,2)\right]^{(bl)} &= \prod_{i=1}^{9}\left[S_i^3(t,2)\right]^{(bl)}\\
&= \left[S_1^3(t,2)\right]^{(bl)} \cdot \left[S_2^3(t,2)\right]^{(bl)} \cdot \left[S_3^3(t,2)\right]^{(bl)}\\
&\quad \cdot \left[S_4^3(t,2)\right]^{(bl)} \cdot \left[S_5^3(t,2)\right]^{(bl)} \cdot \left[S_6^3(t,2)\right]^{(bl)}\\
&\quad \cdot \left[S_7^3(t,2)\right]^{(bl)} \cdot \left[S_8^3(t,2)\right]^{(bl)} \cdot \left[S_9^3(t,2)\right]^{(bl)}\\
&= \exp[-0.025109t]\exp[-0.026304t]\exp[-0.023913t]\\
&\quad \exp[-0.026304t]\exp[-0.023913t]\exp[-0.026304t]\\
&\quad \exp[-0.026304t]\exp[-0.026304t]\exp[-0.026304t]\\
&= \exp[-0.230759t],
\end{aligned} \tag{6.403}$$

for operation and climate-weather states $zc_{bl}$, $b = 7$, $l = 1, 2, 17, 18$;

$$\left[\boldsymbol{S}^3(t,\cdot)\right]^{(bl)} = \left[1, \left[\boldsymbol{S}^3(t,1)\right]^{(bl)}, \left[\boldsymbol{S}^3(t,2)\right]^{(bl)}\right] t \geq 0, \tag{6.404}$$

where

$$
\begin{aligned}
\left[\boldsymbol{S}^3(t,1)\right]^{(bl)} &= \prod_{i=1}^{9}\left[\boldsymbol{S}_i^3(t,1)\right]^{(bl)} \\
&= \left[S_1^3(t,1)\right]^{(bl)}\cdot\left[S_2^3(t,1)\right]^{(bl)}\cdot S_3^0(t,1) \\
&\quad \cdot\left[S_4^3(t,1)\right]^{(bl)}\cdot S_5^0(t,1)\cdot\left[S_6^3(t,1)\right]^{(bl)} \\
&\quad \cdot\left[S_7^3(t,1)\right]^{(bl)}\cdot\left[S_8^3(t,1)\right]^{(bl)}\cdot\left[S_9^3(t,1)\right]^{(bl)} \\
&= \exp[-0.017460t]\exp[-0.01375t]\exp[-0.0125t] \\
&\quad \exp[-0.01375t]\exp[-0.0125t]\exp[-0.01375t] \\
&\quad \exp[-0.01375t]\exp[-0.01375t]\exp[-0.01375t] \\
&= \exp[-0.124960t],
\end{aligned}
\tag{6.405}
$$

$$
\begin{aligned}
\left[\boldsymbol{S}^3(t,2)\right]^{(bl)} &= \prod_{i=1}^{9}\left[\boldsymbol{S}_i^3(t,2)\right]^{(bl)} \\
&= \left[S_1^3(t,2)\right]^{(bl)}\cdot\left[S_2^3(t,2)\right]^{(bl)}\cdot S_3^0(t,2) \\
&\quad \cdot\left[S_4^3(t,2)\right]^{(bl)}\cdot S_5^0(t,2)\cdot\left[S_6^3(t,2)\right]^{(bl)} \\
&\quad \cdot\left[S_7^3(t,2)\right]^{(bl)}\cdot\left[S_8^3(t,2)\right]^{(bl)}\cdot\left[S_9^3(t,2)\right]^{(bl)} \\
&= \exp[-0.023913t]\exp[-0.023913t]\exp[-0.021739t] \\
&\quad \exp[-0.023913t]\exp[-0.021739t]\exp[-0.023913t] \\
&\quad \exp[-0.023913t]\exp[-0.023913t]\exp[-0.023913t] \\
&= \exp[-0.210869t],
\end{aligned}
\tag{6.406}
$$

for operation and climate-weather states $zc_{bl}$, $b = 7$, $l = 6, 7, 10, 11, 22, 23, 26, 27$;

$$
\left[\boldsymbol{S}^3(t,\cdot)\right]^{(bl)} = \left[1, \left[\boldsymbol{S}^3(t,1)\right]^{(bl)}, \left[\boldsymbol{S}^3(t,2)\right]^{(bl)}\right] t \ge 0, \tag{6.407}
$$

where

$$
\begin{aligned}
\left[\boldsymbol{S}^3(t,1)\right]^{(bl)} &= \prod_{i=1}^{9}\left[\boldsymbol{S}_i^3(t,1)\right]^{(bl)} \\
&= \left[S_1^3(t,1)\right]^{(bl)}\cdot\left[S_2^3(t,1)\right]^{(bl)}\cdot\left[S_3^3(t,1)\right]^{(bl)} \\
&\quad \cdot\left[S_4^3(t,1)\right]^{(bl)}\cdot\left[S_5^3(t,1)\right]^{(bl)}\cdot\left[S_6^3(t,1)\right]^{(bl)} \\
&\quad \cdot\left[S_7^3(t,1)\right]^{(bl)}\cdot\left[S_8^3(t,1)\right]^{(bl)}\cdot\left[S_9^3(t,1)\right]^{(bl)} \\
&= \exp[-0.018857t]\exp[-0.015125t]\exp[-0.01375t] \\
&\quad \exp[-0.015125t]\exp[-0.01375t]\exp[-0.015125t] \\
&\quad \exp[-0.015125t]\exp[-0.015125t]\exp[-0.015125t]
\end{aligned}
$$

$$= \exp[-0.137107t], \tag{6.408}$$

$$\begin{aligned}[S^3(t,2)]^{(bl)} &= \prod_{i=1}^{9}[S_i^3(t,2)]^{(bl)} \\ &= [S_1^3(t,2)]^{(bl)} \cdot [S_2^3(t,2)]^{(bl)} \cdot [S_3^3(t,2)]^{(bl)} \\ &\quad \cdot [S_4^3(t,2)]^{(bl)} \cdot [S_5^3(t,2)]^{(bl)} \cdot [S_6^3(t,2)]^{(bl)} \\ &\quad \cdot [S_7^3(t,2)]^{(bl)} \cdot [S_8^3(t,2)]^{(bl)} \cdot [S_9^3(t,2)]^{(bl)} \\ &= \exp[-0.026065t]\exp[-0.026304t]\exp[-0.023913t] \\ &\quad \exp[-0.026304t]\exp[-0.023913t]\exp[-0.026304t] \\ &\quad \exp[-0.026304t]\exp[-0.026304t]\exp[-0.026304t] \\ &= \exp[-0.231715t], \end{aligned} \tag{6.409}$$

for operation and climate-weather states $zc_{bl}$, $b = 7$, $l = 33, 34, 65, 66, 81, 82$;

$$[S^3(t,\cdot)]^{(bl)} = \left[1, [S^3(t,1)]^{(bl)}, [S^3(t,2)]^{(bl)}\right] t \geq 0, \tag{6.410}$$

where

$$\begin{aligned}[S^3(t,1)]^{(bl)} &= \prod_{i=1}^{9}[S_i^3(t,1)]^{(bl)} \\ &= [S_1^3(t,1)]^{(bl)} \cdot [S_2^3(t,1)]^{(bl)} \cdot S_3^0(t,1) \\ &\quad \cdot [S_4^4(t,1)]^{(bl)} \cdot S_5^0(t,1) \cdot [S_6^3(t,1)]^{(bl)} \\ &\quad \cdot [S_7^3(t,1)]^{(bl)} \cdot [S_8^3(t,1)]^{(bl)} \cdot [S_9^3(t,1)]^{(bl)} \\ &= \exp[-0.017984t]\exp[-0.01375t]\exp[-0.0125t] \\ &\quad \exp[-0.01375t]\exp[-0.0125t]\exp[-0.01375t] \\ &\quad \exp[-0.01625t]\exp[-0.01375t]\exp[-0.015t] \\ &= \exp[-0.129234t], \end{aligned} \tag{6.411}$$

$$\begin{aligned}[S^3(t,2)]^{(bl)} &= \prod_{i=1}^{0}[S_i^3(t,2)]^{(bl)} \\ &= [S_1^3(t,2)]^{(bl)} \cdot [S_2^3(t,2)]^{(bl)} \cdot S_3^0(t,2) \\ &\quad \cdot [S_4^3(t,2)]^{(bl)} \cdot S_5^0(t,2) \cdot [S_6^3(t,2)]^{(bl)} \\ &\quad \cdot [S_7^3(t,2)]^{(bl)} \cdot [S_8^3(t,2)]^{(bl)} \cdot [S_9^3(t,2)]^{(bl)} \\ &= \exp[-0.024869t]\exp[-0.023913t]\exp[-0.021739t] \end{aligned}$$

$$\exp[-0.023913t]\exp[-0.021739t]\exp[-0.023913t]$$
$$\exp[-0.028261t]\exp[-0.023913t]\exp[-0.026087t]$$
$$= \exp[-0.218347t], \quad (6.412)$$

for operation and climate-weather states $zc_{bl}$, $b = 7$, $l = 38, 39, 42, 43, 70, 71, 74, 75, 86, 87, 90, 91$.

Hence, applying (6.95) or (6.100), to (6.329)–(6.412) the expected values of the port oil terminal critical infrastructure lifetimes in the safety state subsets {1, 2}, at the operation and climate-weather states $zc_{bl}$, $b = 1, 2, \ldots 7, l = 1, 2, 6, 7, 10, 11, 17, 18, 22, 23, 26, 27, 33, 34, 38, 39, 42, 43, 65, 66, 70, 71, 74, 75, 81, 82, 86, 87, 90, 91$, respectively are:

$$[\mu^3(1)]^{(bl)} = \int_0^\infty [S^3(t,1)]^{(bl)}dt = \int_0^\infty \exp[-0.135208t]dt \cong 7.40 \text{ years},$$
$$[\mu^3(2)]^{(bl)} = \int_0^\infty [S^3(t,2)]^{(b)}dt = \int_0^\infty \exp[-0.228364t]dt \cong 4.39 \text{ years}, \quad (6.413)$$

at the operation and climate-weather states $zc_{bl}$, $b = 1$, $l = 1, 2, 17, 18$;

$$[\mu^3(1)]^{(bl)} = \int_0^\infty [S^3(t,1)]^{(bl)}dt = \int_0^\infty \exp[-0.12371t]dt \cong 8.08 \text{ years},$$
$$[\mu^3(2)]^{(bl)} = \int_0^\infty [S^3(t,2)]^{(b)}dt = \int_0^\infty \exp[-0.208695t]dt \cong 4.79 \text{ years}, \quad (6.414)$$

at the operation and climate-weather states $zc_{bl}$, $b = 1$, $l = 6, 7, 10, 11, 22, 23, 26, 27$;

$$[\mu^3(1)]^{(bl)} = \int_0^\infty [S^3(t,1)]^{(bl)}dt = \int_0^\infty \exp[-0.135732t]dt \cong 7.38 \text{ years},$$
$$[\mu^3(2)]^{(bl)} = \int_0^\infty [S^3(t,2)]^{(b)}dt = \int_0^\infty \exp[-0.229324t]dt \cong 4.36 \text{ years}, \quad (6.415)$$

at the operation and climate-weather states $zc_{bl}$, $b = 1$, $l = 33, 34, 65, 66, 81, 82$;

$$[\mu^3(1)]^{(bl)} = \int_0^\infty [S^3(t,1)]^{(bl)}dt = \int_0^\infty \exp[-0.124234t]dt \cong 8.05 \text{ years},$$
$$[\mu^3(2)]^{(bl)} = \int_0^\infty [S^3(t,2)]^{(b)}dt = \int_0^\infty \exp[-0.209651t]dt \cong 4.77 \text{ years}, \quad (6.416)$$

at the operation and climate-weather states $zc_{bl}$, $b = 1$, $l = 38, 39, 42, 43, 70, 71, 74, 75, 86, 87, 90, 91$;

$$[\mu^3(1)]^{(bl)} = \int_0^\infty [S^3(t,1)]^{(bl)} dt = \int_0^\infty \exp[-0.133833t]dt \cong 7.47 \text{ years},$$
$$[\mu^3(2)]^{(bl)} = \int_0^\infty [S^3(t,2)]^{(b)} dt = \int_0^\infty \exp[-0.225977t]dt \cong 4.42 \text{ years}, \tag{6.417}$$

at the operation and climate-weather states $zc_{bl}$, $b = 2$, $l = 1, 2, 17, 18$;

$$[\mu^3(1)]^{(bl)} = \int_0^\infty [S^3(t,1)]^{(bl)} dt = \int_0^\infty \exp[-0.12246t]dt \cong 8.16 \text{ years},$$
$$[\mu^3(2)]^{(bl)} = \int_0^\infty [S^3(t,2)]^{(b)} dt = \int_0^\infty \exp[-0.206521t]dt \cong 4.84 \text{ years}, \tag{6.418}$$

at the operation and climate-weather states $zc_{bl}$, $b = 2$, $l = 6, 7, 10, 11, 22, 23, 26, 27$;

$$[\mu^3(1)]^{(bl)} = \int_0^\infty [S^3(t,1)]^{(bl)} dt = \int_0^\infty \exp[-0.134357t]dt \cong 7.44 \text{ years},$$

$$[\mu^3(2)]^{(bl)} = \int_0^\infty [S^3(t,2)]^{(b)} dt = \int_0^\infty \exp[-0.226933t]dt \cong 4.41 \text{ years}, \tag{6.419}$$

at the operation and climate-weather states $zc_{bl}$, $b = 2$, $l = 33, 34, 65, 66, 81, 82$;

$$[\mu^3(1)]^{(bl)} = \int_0^\infty [S^3(t,1)]^{(b)} dt = \int_0^\infty \exp[-0.122984t]dt \cong 8.13 \text{ years},$$
$$[\mu^3(2)]^{(bl)} = \int_0^\infty [S^3(t,2)]^{(b)} dt = \int_0^\infty \exp[-0.207477t]dt \cong 4.82 \text{ years}, \tag{6.420}$$

at the operation and climate-weather states $zc_{bl}$, $b = 2$, $l = 38, 39, 42, 43, 70, 71, 74, 75, 86, 87, 90, 91$;

$$[\mu^3(1)]^{(bl)} = \int_0^\infty [S^3(t,1)]^{(bl)} dt = \int_0^\infty \exp[-0.14374t]dt \cong 6.96 \text{ years},$$

$$[\mu^3(2)]^{(bl)} = \int_0^\infty [S^3(t,2)]^{(b)} dt = \int_0^\infty \exp[-0.242605t]dt \cong 4.12 \text{ years}, \tag{6.421}$$

at the operation and climate-weather states $zc_{bl}$, $b = 3$, $l = 1, 2, 17, 18$;

$$\begin{aligned} [\mu^3(1)]^{(bl)} &= \int_0^\infty [S^3(t,1)]^{(bl)}dt = \int_0^\infty \exp[-0.131548t]dt \cong 7.60 \text{ years}, \\ [\mu^3(2)]^{(bl)} &= \int_0^\infty [S^2(t,2)]^{(b)}dt = \int_0^\infty \exp[-0.221737t]dt \cong 4.51 \text{ years}, \end{aligned} \tag{6.422}$$

at the operation and climate-weather states $zc_{bl}$, $b = 3$, $l = 6, 7, 10, 11, 22, 23, 26, 27$;

$$\begin{aligned} [\mu^3(1)]^{(bl)} &= \int_0^\infty [S^3(t,1)]^{(bl)}dt = \int_0^\infty \exp[-0.144321t]dt \cong 6.93 \text{ years}, \\ [\mu^3(2)]^{(bl)} &= \int_0^\infty [S^3(t,2)]^{(b)}dt = \int_0^\infty \exp[-0.243649t]dt \cong 4.10 \text{ years}, \end{aligned} \tag{6.423}$$

at the operation and climate-weather states $zc_{bl}$, $b = 3$, $l = 33, 34, 65, 66, 81, 82$;

$$\begin{aligned} [\mu^3(1)]^{(bl)} &= \int_0^\infty [S^3(t,1)]^{(bl)}dt = \int_0^\infty \exp[-0.132119t]dt \cong 7.57 \text{ years}, \\ [\mu^3(2)]^{(bl)} &= \int_0^\infty [S^3(t,2)]^{(b)}dt = \int_0^\infty \exp[-0.22278t]dt \cong 4.49 \text{ years}, \end{aligned} \tag{6.424}$$

at the operation and climate-weather states $zc_{bl}$, $b = 3$, $l = 38, 39, 42, 43, 70, 71, 74, 75, 86, 87, 90, 91$;

$$\begin{aligned} [\mu^3(1)]^{(bl)} &= \int_0^\infty [S^3(t,1)]^{(bl)}dt = \int_0^\infty \exp[-0.160541t]dt \cong 6.23 \text{ years}, \\ [\mu^3(2)]^{(bl)} &= \int_0^\infty [S^3(t,2)]^{(b)}dt = \int_0^\infty \exp[-0.271189t]dt \cong 3.69 \text{ years}, \end{aligned} \tag{6.425}$$

at the operation and climate-weather states $zc_{bl}$, $b = 4$, $l = 1, 2, 17, 18$;

$$\begin{aligned} [\mu^3(1)]^{(bl)} &= \int_0^\infty [S^3(t,1)]^{(bl)}dt = \int_0^\infty \exp[-0.146884t]dt \cong 6.81 \text{ years}, \\ [\mu^3(2)]^{(bl)} &= \int_0^\infty [S^3(t,2)]^{(b)}dt = \int_0^\infty \exp[-0.247826t]dt \cong 4.03 \text{ years}, \end{aligned} \tag{6.426}$$

at the operation and climate-weather states $zc_{bl}$, $b = 4$, $l = 6, 7, 10, 11, 22, 23, 26, 27$;

$$\begin{aligned} [\mu^3(1)]^{(bl)} &= \int_0^\infty [S^3(t,1)]^{(bl)}dt = \int_0^\infty \exp[-0.161161t]dt \cong 6.20 \text{ years}, \\ [\mu^3(2)]^{(bl)} &= \int_0^\infty [S^3(t,2)]^{(b)}dt = \int_0^\infty \exp[-0.272323t]dt \cong 3.67 \text{ years}, \end{aligned} \tag{6.427}$$

for $b = 4$, $l = 33, 34, 65, 66, 81, 82$;

$$\left[\mu^3(1)\right]^{(bl)} = \int_0^\infty \left[S^3(t,1)\right]^{(bl)} dt = \int_0^\infty \exp[-0.147504t]dt \cong 6.78 \text{ years},$$
$$\left[\mu^3(2)\right]^{(b)} = \int_0^\infty \left[S^3(t,2)\right]^{(b)} dt = \int_0^\infty \exp[-0.248957t]dt \cong 4.02 \text{ years}, \quad (6.428)$$

at the operation and climate-weather states $zc_{bl}$, $b = 4$, $l = 38, 39, 42, 43, 70, 71, 74, 75, 86, 87, 90, 91$;

$$\left[\mu^3(1)\right]^{(bl)} = \int_0^\infty \left[S^3(t,1)\right]^{(bl)} dt = \int_0^\infty \exp[-0.143749t]dt \cong 6.96 \text{ years},$$
$$\left[\mu^3(2)\right]^{(bl)} = \int_0^\infty \left[S^3(t,2)\right]^{(b)} dt = \int_0^\infty \exp[-0.242605t]dt \cong 4.12 \text{ years}, \quad (6.429)$$

at the operation and climate-weather states $zc_{bl}$, $b = 5$, $l = 1, 2, 17, 18$;

$$\left[\mu^3(1)\right]^{(bl)} = \int_0^\infty \left[S^3(t,1)\right]^{(bl)} dt = \int_0^\infty \exp[-0.134048t]dt \cong 7.46 \text{ years},$$
$$\left[\mu^3(2)\right]^{(bl)} = \int_0^\infty \left[S^3(t,2)\right]^{(b)} dt = \int_0^\infty \exp[-0.226083t]dt \cong 4.42 \text{ years}, \quad (6.430)$$

at the operation and climate-weather states $zc_{bl}$, $b = 5$, $l = 6, 7, 10, 11, 22, 23, 26, 27$;

$$\left[\mu^3(1)\right]^{(bl)} = \int_0^\infty \left[S^3(t,1)\right]^{(bl)} dt = \int_0^\infty \exp[-0.144321t]dt \cong 6.93 \text{ years},$$
$$\left[\mu^3(2)\right]^{(bl)} = \int_0^\infty \left[S^3(t,2)\right]^{(b)} dt = \int_0^\infty \exp[-0.243649t]dt \cong 4.10 \text{ years}, \quad (6.431)$$

at the operation and climate-weather states $zc_{bl}$, $b = 5$, $l = 33, 34, 65, 66, 81, 82$;

$$\left[\mu^3(1)\right]^{(bl)} = \int_0^\infty \left[S^3(t,1)\right]^{(bl)} dt = \int_0^\infty \exp[-0.132119t]dt \cong 7.57 \text{ years},$$
$$\left[\mu^3(2)\right]^{(bl)} = \int_0^\infty \left[S^3(t,2)\right]^{(b)} dt = \int_0^\infty \exp[-0.222779t]dt \cong 4.49 \text{ years}, \quad (6.432)$$

at the operation and climate-weather states $zc_{bl}$, $b = 5$, $l = 38, 39, 42, 43, 70, 71, 74, 75, 86, 87, 90, 91$;

$$\left[\mu^3(1)\right]^{(bl)} = \int_0^\infty \left[S^3(t,1)\right]^{(bl)} dt = \int_0^\infty \exp[-0.160541t]dt \cong 6.23 \text{ years},$$
$$\left[\mu^3(2)\right]^{(bl)} = \int_0^\infty \left[S^3(t,2)\right]^{(b)} dt = \int_0^\infty \exp[-0.271192t]dt \cong 3.69 \text{ years}, \quad (6.433)$$

at the operation and climate-weather states $zc_{bl}$, $b = 6$, $l = 1, 2, 17, 18$;

$$[\mu^3(1)]^{(bl)} = \int_0^\infty [S^3(t,1)]^{(bl)} dt = \int_0^\infty \exp[-0.146884t]dt \cong 6.81 \text{ years},$$
$$[\mu^3(2)]^{(b)} = \int_0^\infty [S^3(t,2)]^{(b)} dt = \int_0^\infty \exp[-0.247826t]dt \cong 4.03 \text{ years}, \quad (6.434)$$

at the operation and climate-weather states $zc_{bl}$, $b = 6$, $l = 6, 7, 10, 11, 22, 23, 26, 27$;

$$[\mu^3(1)]^{(bl)} = \int_0^\infty [S^3(t,1)]^{(bl)} dt = \int_0^\infty \exp[-0.157036t]dt \cong 6.37 \text{ years},$$
$$[\mu^3(2)]^{(bl)} = \int_0^\infty [S^3(t,2)]^{(b)} dt = \int_0^\infty \exp[-0.265149t]dt \cong 3.77 \text{ years}, \quad (6.435)$$

at the operation and climate-weather states $zc_{bl}$, $b = 6$, $l = 33, 34, 65, 66, 81, 82$;

$$[\mu^3(1)]^{(bl)} = \int_0^\infty [S^3(t,1)]^{(bl)} dt = \int_0^\infty \exp[-0.147504t]dt \cong 6.78 \text{ years},$$
$$[\mu^3(2)]^{(bl)} = \int_0^\infty [S^3(t,2)]^{(b)} dt = \int_0^\infty \exp[-0.248957t]dt \cong 4.02 \text{ years}, \quad (6.436)$$

at the operation and climate-weather states $zc_{bl}$, $b = 6$, $l = 38, 39, 42, 43, 70, 71, 74, 75, 86, 87, 90, 91$;

$$[\mu^3(1)]^{(bl)} = \int_0^\infty [S^3(t,1)]^{(bl)} dt = \int_0^\infty \exp[-0.136583t]dt \cong 7.32 \text{ years},$$
$$[\mu^3(2)]^{(bl)} = \int_0^\infty [S^3(t,2)]^{(b)} dt = \int_0^\infty \exp[-0.230759t]dt \cong 4.33 \text{ years}, \quad (6.437)$$

at the operation and climate-weather states $zc_{bl}$, $b = 7$, $l = 1, 2, 17, 18$;

$$[\mu^3(1)]^{(bl)} = \int_0^\infty [S^3(t,1)]^{(bl)} dt = \int_0^\infty \exp[-0.12496t]dt \cong 8.00 \text{ years},$$
$$[\mu^3(2)]^{(bl)} = \int_0^\infty [S^3(t,2)]^{(b)} dt = \int_0^\infty \exp[-0.210869t]dt \cong 4.74 \text{ years}, \quad (6.438)$$

at the operation and climate-weather states $zc_{bl}$, $b = 7$, $l = 6, 7, 10, 11, 22, 23, 26, 27$;

$$[\mu^3(1)]^{(bl)} = \int_0^\infty [S^3(t,1)]^{(bl)} dt = \int_0^\infty \exp[-0.137107t]dt \cong 7.29 \text{ years},$$
$$[\mu^3(2)]^{(bl)} = \int_0^\infty [S^3(t,2)]^{(b)} dt = \int_0^\infty \exp[-0.231715t]dt \cong 4.31 \text{ years}, \quad (6.439)$$

at the operation and climate-weather states $zc_{bl}$, $b = 7$, $l = 33, 34, 65, 66, 81, 82$;

$$\left[\mu^3(1)\right]^{(bl)} = \int_0^\infty \left[S^3(t,1)\right]^{(bl)}dt = \int_0^\infty \exp[-0.147504t]dt \cong 6.78 \text{ years},$$
$$\left[\mu^3(2)\right]^{(bl)} = \int_0^\infty \left[S^3(t,2)\right]^{(b)}dt = \int_0^\infty \exp[-0.248957t]dt \cong 4.02 \text{ years}, \quad (6.440)$$

at the operation and climate-weather states $zc_{bl}$, $b = 7$, $l = 38, 39, 42, 43, 70, 71, 74, 75, 86, 87, 90, 91$.

From the results (6.141)–(6.147) and (6.329)–(6.412), applying (6.92), the port oil terminal critical infrastructure unconditional safety function (SafI1) is given by

$$\boldsymbol{S}^3(t,\cdot) = \left[1, \boldsymbol{S}^3(t,1)\boldsymbol{S}^3(t,2)\right], t \geq 0, \quad (6.441)$$

where

$$\begin{aligned}
\boldsymbol{S}^3(t,1) &\cong \sum_{b=1}^{7}\sum_{l=1}^{96} pq_{bl}[\boldsymbol{S}^3(t,1)]^{(bl)} \\
&= 0.00036577\exp[-0.135208t] + 0.04242932\exp[-0.135208t] \\
&\quad + 0.28383752\exp[-0.12371t] + 0.00768117\exp[-0.12371t] \\
&\quad + 0.00512078\exp[-0.12371t] + 0.02633544\exp[-0.12371t] \\
&\quad + 0.00002528\exp[-0.135208t] + 0.00293248\exp[-0.135208t] \\
&\quad + 0.01961728\exp[-0.12371t] + 0.00053088\exp[-0.12371t] \\
&\quad + 0.00035392\exp[-0.12371t] + 0.00182016\exp[-0.12371t] \\
&\quad + 0.000000395\exp[-0.135732t] + 0.00004582\exp[-0.135732t] \\
&\quad + 0.00030652\exp[-0.124234t] + 0.000008295\exp[-0.124234t] \\
&\quad + 0.00000553\exp[-0.124234t] + 0.00002844\exp[-0.124234t] \\
&\quad + 0.00000237\exp[-0.135732t] + 0.00027492\exp[-0.135732t] \\
&\quad + 0.00183912\exp[-0.124234t] + 0.00004977\exp[-0.124234t] \\
&\quad + 0.00003318\exp[-0.124234t] + 0.00017064\exp[-0.124234t] \\
&\quad + 0.000001185\exp[-0.135732t] + 0.00013746\exp[-0.135732t] \\
&\quad + 0.00091956\exp[-0.124234t] + 0.000024885\exp[-0.124234t] \\
&\quad + 0.00001659\exp[-0.124234t] + 0.00008532\exp[-0.124234t] \\
&\quad + 0.00005556\exp[-0.133833t] + 0.00644496\exp[-0.133833t] \\
&\quad + 0.04311456\exp[-0.12246t] + 0.00116676\exp[-0.12246t] \\
&\quad + 0.00077784\exp[-0.12246t] + 0.00400032\exp[-0.12246t] \\
&\quad + 0.00000384\exp[-0.133833t] + 0.00044544\exp[-0.133833t] \\
&\quad + 0.00297984\exp[-0.12246t] + 0.00008064\exp[-0.12246t] \\
&\quad + 0.00005376\exp[-0.12246t] + 0.00027648\exp[-0.12246t] \\
&\quad + 0.00000006\exp[-0.134357t] + 0.00000696\exp[-0.134357t]
\end{aligned}$$

$$
\begin{aligned}
&+ 0.00004656\exp[-0.122984t] + 0.00000126\exp[-0.122984t]\\
&+ 0.00000084\exp[-0.122984t] + 0.00000432\exp[-0.122984t]\\
&+ 0.00000036\exp[-0.134357t] + 0.00004176\exp[-0.134357t]\\
&+ 0.00027936\exp[-0.122984t] + 0.00000756\exp[-0.122984t]\\
&+ 0.00000504\exp[-0.122984t] + 0.00002592\exp[-0.122984t]\\
&+ 0.00000018\exp[-0.134357t] + 0.00002088\exp[-0.134357t]\\
&+ 0.00013968\exp[-0.122984t] + 0.00000378\exp[-0.122984t]\\
&+ 0.00000252\exp[-0.122984t] + 0.00001296\exp[-0.122984t]\\
&+ 0.000002778\exp[-0.143749t] + 0.000322248\exp[-0.143749t]\\
&+ 0.002155728\exp[-0.131548t] + 0.000058338\exp[-0.131548t]\\
&+ 0.000038892\exp[-0.131548t] + 0.000200016\exp[-0.131548t]\\
&+ 0.000000192\exp[-0.143749t] + 0.0000022272\exp[-0.143749t]\\
&+ 0.000148992\exp[-0.131548t] + 0.000004032\exp[-0.131548t]\\
&+ 0.000002688\exp[-0.131548t] + 0.000013824\exp[-0.131548t]\\
&+ 0.000000003\exp[-0.144321t] + 0.000000348\exp[-0.144321t]\\
&+ 0.000002328\exp[-0.132119t] + 0.000000063\exp[-0.132119t]\\
&+ 0.000000042\exp[-0.132119t] + 0.000000216\exp[-0.132119t]\\
&+ 0.000000018\exp[-0.144321t] + 0.000002088\exp[-0.144321t]\\
&+ 0.000013968\exp[-0.132119t] + 0.000000378\exp[-0.132119t]\\
&+ 0.000000252\exp[-0.132119t] + 0.000001296\exp[-0.132119t]\\
&+ 0.000000009\exp[-0.144321t] + 0.000001044\exp[-0.144321t]\\
&+ 0.000006984\exp[-0.132119t] + 0.000000189\exp[-0.132119t]\\
&+ 0.000000126\exp[-0.132119t] + 0.000000648\exp[-0.132119t]\\
&+ 0.000001852\exp[-0.160541t] + 0.000214832\exp[-0.160541t]\\
&+ 0.001437152\exp[-0.146884t] + 0.000038892\exp[-0.146884t]\\
&+ 0.000025928\exp[-0.146884t] + 0.000133344\exp[-0.146884t]\\
&+ 0.000000128\exp[-0.160541t] + 0.000014848\exp[-0.160541t]\\
&+ 0.000099328\exp[-0.146884t] + 0.000002688\exp[-0.146884t]\\
&+ 0.000001792\exp[-0.146884t] + 0.000009216\exp[-0.146884t]\\
&+ 0.000000002\exp[-0.161161t] + 0.000000232\exp[-0.161161t]\\
&+ 0.000001552\exp[-0.147504t] + 0.000000042\exp[-0.147504t]\\
&+ 0.000000028\exp[-0.147504t] + 0.000000144\exp[-0.147504t]\\
&+ 0.000000012\exp[-0.161161t] + 0.000001392\exp[-0.161161t]\\
&+ 0.000009312\exp[-0.147504t] + 0.000000252\exp[-0.147504t]\\
&+ 0.000000168\exp[-0.147504t] + 0.000000864\exp[-0.147504t]\\
&+ 0.000000006\exp[-0.161161t] + 0.000000696\exp[-0.161161t]
\end{aligned}
$$

$$+0.000004656\exp[-0.147504t]+0.000000126\exp[-0.147504t]$$
$$+0.000000084\exp[-0.147504t]+0.000000432\exp[-0.147504t]$$
$$+0.0001852\exp[-0.143749t]+0.0214832\exp[-0.143749t]$$
$$+0.1437152\exp[-0.134048t]+0.0038892\exp[-0.134048t]$$
$$+0.0025928\exp[-0.134048t]+0.0133344\exp[-0.134048t]$$
$$+0.0000128\exp[-0.143749t]+0.0014848\exp[-0.143749t]$$
$$+0.0099328\exp[-0.134048t]+0.0002688\exp[-0.134048t]$$
$$+0.0001792\exp[-0.134048t]+0.0009216\exp[-0.134048t]$$
$$+0.0000002\exp[-0.144321t]+0.0000232\exp[-0.144321t]$$
$$+0.0001552\exp[-0.132119t]+0.0000042\exp[-0.132119t]$$
$$+0.0000028\exp[-0.132119t]+0.0000144\exp[-0.132119t]$$
$$+0.0000012\exp[-0.144321t]+0.0001392\exp[-0.144321t]$$
$$+0.0009312\exp[-0.132119t]+0.0000252\exp[-0.132119t]$$
$$+0.0000168\exp[-0.132119t]+0.0000864\exp[-0.132119t]$$
$$+0.0000006\exp[-0.144321t]+0.0000696\exp[-0.144321t]$$
$$+0.0004656\exp[-0.132119t]+0.0000126\exp[-0.132119t]$$
$$+0.0000084\exp[-0.132119t]+0.0000432\exp[-0.132119t]$$
$$+0.000053708\exp[-0.160541t]+0.006230128\exp[-0.160541t]$$
$$+0.041677408\exp[-0.146884t]+0.001127868\exp[-0.146884t]$$
$$+0.000751912\exp[-0.146884t]+0.003866976\exp[-0.146884t]$$
$$+0.000003712\exp[-0.160541t]+0.000430592\exp[-0.160541t]$$
$$+0.002880512\exp[-0.146884t]+0.000077952\exp[-0.146884t]$$
$$+0.0000051968\exp[-0.146884t]+0.000267264\exp[-0.146884t]$$
$$+0.000000058\exp[-0.157036t]+0.000006728\exp[-0.157036t]$$
$$+0.000045008\exp[-0.147504t]+0.000001218\exp[-0.147504t]$$
$$+0.000000812\exp[-0.147504t]+0.000004176\exp[-0.147504t]$$
$$+0.000000348\exp[-0.157036t]+0.000040368\exp[-0.157036t]$$
$$+0.000270048\exp[-0.147504t]+0.000007308\exp[-0.147504t]$$
$$+0.000004872\exp[-0.147504t]+0.000025056\exp[-0.147504t]$$
$$+0.000000174\exp[-0.157036t]+0.000020184\exp[-0.157036t]$$
$$+0.000135024\exp[-0.147504t]+0.000003654\exp[-0.147504t]$$
$$+0.000002436\exp[-0.147504t]+0.000012528\exp[-0.147504t]$$
$$+0.000261132\exp[-0.136583t]+0.030291312\exp[-0.136583t]$$
$$+0.202638432\exp[-0.12496t]+0.005483772\exp[-0.12496t]$$
$$+0.003655848\exp[-0.12496t]+0.018801504\exp[-0.12496t]$$
$$+0.000018048\exp[-0.136583t]+0.002093568\exp[-0.136583t]$$

$$
\begin{aligned}
&+ 0.014005248 \exp[-0.12496t] + 0.000379008 \exp[-0.12496t] \\
&+ 0.000252672 \exp[-0.12496t] + 0.001299456 \exp[-0.12496t] \\
&+ 0.000000282 \exp[-0.137107t] + 0.000032712 \exp[-0.137107t] \\
&+ 0.000218832 \exp[-0.147504t\ t] + 0.000005922 \exp[-0.147504t] \\
&+ 0.000003948 \exp[-0.147504t] + 0.000020304 \exp[-0.147504t] \\
&+ 0.000001692 \exp[-0.137107t] + 0.000196272 \exp[-0.137107t] \\
&+ 0.001312992 \exp[-0.147504t\ t] + 0.000035532 \exp[-0.147504t] \\
&+ 0.000023688 \exp[-0.147504t] + 0.000121824 \exp[-0.147504t] \\
&+ 0.000000846 \exp[-0.137107t] + 0.000098136 \exp[-0.137107t] \\
&+ 0.000656496 \exp[-0.147504t] + 0.000017766 \exp[-0.147504t\ t] \\
&+ 0.000011844 \exp[-0.147504t] + 0.000060912 \exp[-0.147504t] \\
&= 0.0524502 \exp[-0.12246t] + 0.0005298 \exp[-0.122984t] \\
&+ 0.34529715 \exp[-0.12371t] + 0.003487825 \exp[-0.124234t] \\
&+ 0.24651594 \exp[-0.12496t] + 0.00262251 \exp[-0.131548t] \\
&+ 0.00179249 \exp[-0.132119t] + 0.0069498 \exp[-0.133833t] \\
&+ 0.1748334 \exp[-0.134048t] + 0.0000702 \exp[-0.134357t] \\
&+ 0.04575277 \exp[-0.135208t] + 0.00046215 \exp[-0.135732t] \\
&+ 0.03266406 \exp[-0.136583t] + 0.000330786 \exp[-0.137107t] \\
&+ 0.023493535 \exp[-0.143749t] + 0.00023751 \exp[-0.144321t] \\
&+ 0.052403429 \exp[-0.146884t] + 0.00301986 \exp[-0.147504t] \\
&+ 0.00006786 \exp[-0.157036t] + 0.0069498 \exp[-0.160541t] \\
&+ 0.00000234 \exp[-0.161161t],
\end{aligned}
\tag{6.442}
$$

$$
\begin{aligned}
\boldsymbol{S}^3(t,2) &\cong \sum_{b=1}^{7} \sum_{l=1}^{96} pq_{bl} [\boldsymbol{S}^3(t,2)]^{(bl)} \\
&= 0.00036577 \exp[-0.228364t] + 0.04242932 \exp[-0.228364t] \\
&\quad + 0.28383752 \exp[-0.208695t] + 0.00768117 \exp[-0.208695t] \\
&\quad + 0.00512078 \exp[-0.208695t] + 0.02633544 \exp[-0.208695t] \\
&\quad + 0.00002528 \exp[-0.228364t] + 0.00293248 \exp[-0.228364t] \\
&\quad + 0.01961728 \exp[-0.208695t] + 0.00053088 \exp[-0.208695t] \\
&\quad + 0.00035392 \exp[-0.208695t] + 0.0018201 \exp[-0.208695t] \\
&\quad + 0.000000395 \exp[-0.229324t] + 0.00004582 \exp[-0.229324t] \\
&\quad + 0.00030652 \exp[-0.209651t] + 0.000008295 \exp[-0.209651t] \\
&\quad + 0.00000553 \exp[-0.209651t] + 0.00002844 \exp[-0.209651t] \\
&\quad + 0.00000237 \exp[-0.229324t] + 0.00027492 \exp[-0.229324t] \\
&\quad + 0.00183912 \exp[-0.209651t] + 0.00004977 \exp[-0.209651t]
\end{aligned}
$$

$$
\begin{aligned}
&+0.00003318\exp[-0.209651t]+0.00017064\exp[-0.209651t]\\
&+0.000001185\exp[-0.229324t]+0.00013746\exp[-0.229324t]\\
&+0.00091956\exp[-0.209651t]+0.000024885\exp[-0.209651t]\\
&+0.00001659\exp[-0.209651t]+0.00008532\exp[-0.209651t]\\
&+0.00005556\exp[-0.225977t]+0.00644496\exp[-0.225977t]\\
&+0.04311456\exp[-0.206521t]+0.00116676\exp[-0.206521t]\\
&+0.00077784\exp[-0.206521t]+0.00400032\exp[-0.206521t]\\
&+0.00000384\exp[-0.225977t]+0.00044544\exp[-0.225977t]\\
&+0.00297984\exp[-0.206521t]+0.00008064\exp[-0.206521t]\\
&+0.00005376\exp[-0.206521t]+0.00027648\exp[-0.206521t]\\
&+0.00000006\exp[-0.226933t]+0.00000696\exp[-0.226933t]\\
&+0.00004656\exp[-0.207477t]+0.00000126\exp[-0.207477t]\\
&+0.00000084\exp[-0.207477t]+0.00000432\exp[-0.207477t]\\
&+0.00000036\exp[-0.226933t]+0.00004176\exp[-0.226933t]\\
&+0.00027936\exp[-0.207477t]+0.00000756\exp[-0.207477t]\\
&+0.00000504\exp[-0.207477t]+0.00002592\exp[-0.207477t]\\
&+0.00000018\exp[-0.226933t]+0.00002088\exp[-0.226933t]\\
&+0.00013968\exp[-0.207477t]+0.00000378\exp[-0.207477t]\\
&+0.00000252\exp[-0.207477t]+0.00001296\exp[-0.207477t]\\
&+0.000002778\exp[-0.242605t]+0.000322248\exp[-0.242605t]\\
&+0.002155728\exp[-0.221737t]+0.000058338\exp[-0.221737t]\\
&+0.000038892\exp[-0.221737t]+0.000200016\exp[-0.221737t]\\
&+0.000000192\exp[-0.242605t]+0.0000022272\exp[-0.242605t]\\
&+0.000148992\exp[-0.221737t]+0.000004032\exp[-0.221737t]\\
&+0.000002688\exp[-0.221737t]+0.000013824\exp[-0.221737t]\\
&+0.000000003\exp[-0.243649t]+0.000000348\exp[-0.243649t]\\
&+0.000002328\exp[-0.22278t]+0.000000063\exp[-0.22278t]\\
&+0.000000042\exp[-0.22278t]+0.000000216\exp[-0.22278t]\\
&+0.000000018\exp[-0.243649t]+0.000002088\exp[-0.243649t]\\
&+0.000013968\exp[-0.22278t]+0.000000378\exp[-0.22278t]\\
&+0.000000252\exp[-0.22278t]+0.000001296\exp[-0.22278t]\\
&+0.000000009\exp[-0.243649t]+0.000001044\exp[-0.243649t]\\
&+0.000006984\exp[-0.22278t]+0.000000189\exp[-0.22278t]\\
&+0.000000126\exp[-0.22278t]+0.000000648\exp[-0.22278t]\\
&+0.000001852\exp[-0.271189t]+0.000214832\exp[-0.271189t]\\
&+0.001437152\exp[-0.247826t]+0.000038892\exp[-0.247826t]
\end{aligned}
$$

$$
\begin{aligned}
&+ 0.000025928 \exp[-0.247826t] + 0.000133344 \exp[-0.247826t] \\
&+ 0.000000128 \exp[-0.271189t] + 0.000014848 \exp[-0.271189t] \\
&+ 0.000099328 \exp[-0.247826t] + 0.000002688 \exp[-0.247826t] \\
&+ 0.000001792 \exp[-0.247826t] + 0.000009216 \exp[-0.247826t] \\
&+ 0.000000002 \exp[-0.272323t] + 0.000000232 \exp[-0.272323t] \\
&+ 0.000001552 \exp[-0.248957t] + 0.000000042 \exp[-0.248957t] \\
&+ 0.000000028 \exp[-0.248957t] + 0.000000144 \exp[-0.248957t] \\
&+ 0.000000012 \exp[-0.272323t] + 0.000001392 \exp[-0.272323t] \\
&+ 0.000009312 \exp[-0.248957t] + 0.000000252 \exp[-0.248957t] \\
&+ 0.000000168 \exp[-0.248957t] + 0.000000864 \exp[-0.248957t] \\
&+ 0.000000006 \exp[-0.272323t] + 0.000000696 \exp[-0.272323t] \\
&+ 0.000004656 \exp[-0.248957t] + 0.000000126 \exp[-0.248957t] \\
&+ 0.000000084 \exp[-0.248957t] + 0.000000432 \exp[-0.248957t] \\
&+ 0.0001852 \exp[-0.242605t] + 0.0214832 \exp[-0.242605t] \\
&+ 0.1437152 \exp[-0.226083t] + 0.0038892 \exp[-0.226083t] \\
&+ 0.0025928 \exp[-0.226083t] + 0.0133344 \exp[-0.226083t] \\
&+ 0.0000128 \exp[-0.242605t] + 0.0014848 \exp[-0.242605t] \\
&+ 0.0099328 \exp[-0.226083t] + 0.0002688 \exp[-0.226083t] \\
&+ 0.0001792 \exp[-0.226083t] + 0.0009216 \exp[-0.226083t] \\
&+ 0.0000002 \exp[-0.243649t] + 0.0000232 \exp[-0.243649t] \\
&+ 0.0001552 \exp[-0.222779t] + 0.0000042 \exp[-0.222779t] \\
&+ 0.0000028 \exp[-0.222779t] + 0.0000144 \exp[-0.222779t] \\
&+ 0.0000012 \exp[-0.243649t] + 0.0001392 \exp[-0.243649t] \\
&+ 0.0009312 \exp[-0.222779t] + 0.0000252 \exp[-0.222779t] \\
&+ 0.0000168 \exp[-0.222779t] + 0.0000864 \exp[-0.222779t] \\
&+ 0.0000006 \exp[-0.243649t] + 0.0000696 \exp[-0.243649t] \\
&+ 0.0004656 \exp[-0.222779t] + 0.0000126 \exp[-0.222779t] \\
&+ 0.0000084 \exp[-0.222779t] + 0.0000432 \exp[-0.222779t] \\
&+ 0.000053708 \exp[-0.271192t] + 0.006230128 \exp[-0.271192t] \\
&+ 0.041677408 \exp[-0.247826t] + 0.001127868 \exp[-0.247826t] \\
&+ 0.000751912 \exp[-0.247826t] + 0.003866976 \exp[-0.247826t] \\
&+ 0.000003712 \exp[-0.271192t] + 0.000430592 \exp[-0.271192t] \\
&+ 0.002880512 \exp[-0.247826t] + 0.000077952 \exp[-0.247826t] \\
&+ 0.0000051968 \exp[-0.247826t] + 0.000267264 \exp[-0.247826t] \\
&+ 0.000000058 \exp[-0.265149t] + 0.000006728 \exp[-0.265149t] \\
&+ 0.000045008 \exp[-0.248957t] + 0.000001218 \exp[-0.248957t]
\end{aligned}
$$

$$
\begin{aligned}
&+ 0.000000812 \exp[-0.248957t] + 0.000004176 \exp[-0.248957t] \\
&+ 0.000000348 \exp[-0.265149t] + 0.000040368 \exp[-0.265149t] \\
&+ 0.000270048 \exp[-0.248957t] + 0.000007308 \exp[-0.248957t] \\
&+ 0.000004872 \exp[-0.248957t] + 0.000025056 \exp[-0.248957t] \\
&+ 0.000000174 \exp[-0.265149t] + 0.000020184 \exp[-0.265149t] \\
&+ 0.000135024 \exp[-0.248957t] + 0.000003654 \exp[-0.248957t] \\
&+ 0.000002436 \exp[-0.248957t] + 0.000012528 \exp[-0.248957t] \\
&+ 0.000261132 \exp[-0.230759t] + 0.030291312 \exp[-0.230759t] \\
&+ 0.202638432 \exp[-0.210869t] + 0.005483772 \exp[-0.210869t] \\
&+ 0.003655848 \exp[-0.210869t] + 0.018801504 \exp[-0.210869t] \\
&+ 0.000018048 \exp[-0.230759t] + 0.002093568 \exp[-0.230759t] \\
&+ 0.014005248 \exp[-0.210869t] + 0.000379008 \exp[-0.210869t] \\
&+ 0.000252672 \exp[-0.210869t] + 0.001299456 \exp[-0.210869t] \\
&+ 0.000000282 \exp[-0.231715t] + 0.000032712 \exp[-0.231715t] \\
&+ 0.000218832 \exp[-0.248957t] + 0.000005922 \exp[-0.248957t] \\
&+ 0.000003948 \exp[-0.248957t] + 0.000020304 \exp[-0.248957t] \\
&+ 0.000001692 \exp[-0.231715t] + 0.000196272 \exp[-0.231715t] \\
&+ 0.001312992 \exp[-0.248957t] + 0.000035532 \exp[-0.248957t] \\
&+ 0.000023688 \exp[-0.248957t] + 0.000121824 \exp[-0.248957t] \\
&+ 0.000000846 \exp[-0.231715t] + 0.000098136 \exp[-0.231715t] \\
&+ 0.000656496 \exp[-0.248957t] + 0.000017766 \exp[-0.248957t] \\
&+ 0.000011844 \exp[-0.248957t] + 0.000060912 \exp[-0.248957t] \\
&= 0.0524502 \exp[-0.206521t] + 0.0005298 \exp[-0.207477t] \\
&+ 0.34529715 \exp[-0.208695t] + 0.003487825 \exp[-0.209651t] \\
&+ 0.24651594 \exp[-0.210869t] + 0.00262251 \exp[-0.221737t] \\
&+ 0.00179249 \exp[-0.22278t] + 0.0069498 \exp[-0.225977t] \\
&+ 0.1748334 \exp[-0.226083t] + 0.0000702 \exp[-0.226933t] \\
&+ 0.04575277 \exp[-0.228364t] + 0.00046215 \exp[-0.229324t] \\
&+ 0.03266406 \exp[-0.230759t] + 0.000330786 \exp[-0.231715t] \\
&+ 0.023493535 \exp[-0.242605t] + 0.00023751 \exp[-0.243649t] \\
&+ 0.052403429 \exp[-0.247826t] + 0.00301986 \exp[-0.248957t] \\
&+ 0.00006786 \exp[-0.265149t] + 0.0069498 \exp[-0.271189t] \\
&+ 0.00000234 \exp[-0.272323t]. \qquad (6.443)
\end{aligned}
$$

The graph of the three-state port oil terminal critical infrastructure safety function is shown in Fig. 6.1.

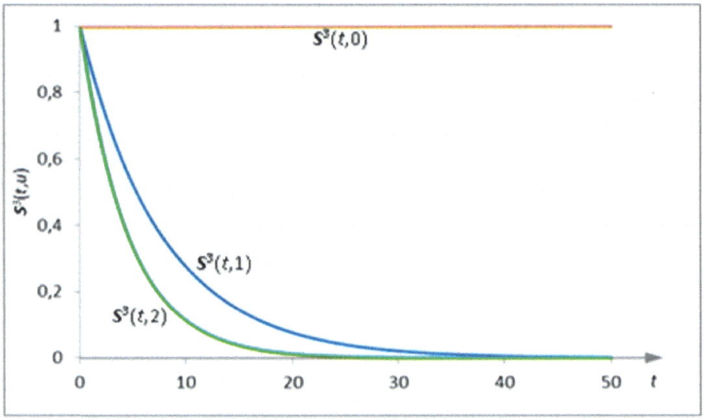


**Fig. 6.1** The graph of the port oil terminal critical infrastructure safety function $\boldsymbol{S}^3(t,\cdot)$ coordinate

As the critical safety state is $r = 1$, then by (6.93) and (6.442), the port oil terminal critical infrastructure risk function (SafI2), is given by

$$\begin{aligned}
\boldsymbol{r}^2(t) &= 1 - \boldsymbol{S}^3(t, 1) \\
&= 1 - [0.0524502 \exp[-0.12246t] + 0.0005298 \exp[-0.122984t] \\
&\quad + 0.34529715 \exp[-0.12371t] + 0.003487825 \exp[-0.124234t] \\
&\quad + 0.24651594 \exp[-0.12496t] + 0.00262251 \exp[-0.131548t] \\
&\quad + 0.00179249 \exp[-0.132119t] + 0.0069498 \exp[-0.133833t] \\
&\quad + 0.1748334 \exp[-0.134048t] + 0.0000702 \exp[-0.134357t] \\
&\quad + 0.04575277 \exp[-0.135208t] + 0,00046215 \exp[-0.135732t] \\
&\quad + 0.03266406 \exp[-0.136583t] + 0.000330786 \exp[-0.137107t] \\
&\quad + 0,023493535 \exp[-0.143749t] + 0.00023751 \exp[-0.144321t] \\
&\quad + 0.052403429 \exp[-0.146884t] + 0.00301986 \exp[-0.147504t] \\
&\quad + 0.00006786 \exp[-0.157036t] + 0.0069498 \exp[-0.160541t] \\
&\quad + 0.00000234 \exp[-0,161161t]] \\
&\text{for } t \geq 0.
\end{aligned} \tag{6.444}$$

The graph of the risk function $\boldsymbol{r}^3(t)$ of the port oil terminal critical infrastructure, the fragility curve (SafI3), is shown in Fig. 6.2.

The values of the above risk function are given in Table 6.1.

From (6.99) and (6.444), and the Table 6.1, the moment when the port oil terminal critical infrastructure risk function exceeds a permitted level $\delta = 0.05$ (SafI6), is

$$\tau^3 = \boldsymbol{r}^{3^{-1}}(\delta) \cong 0.398 \text{ years.} \tag{6.445}$$

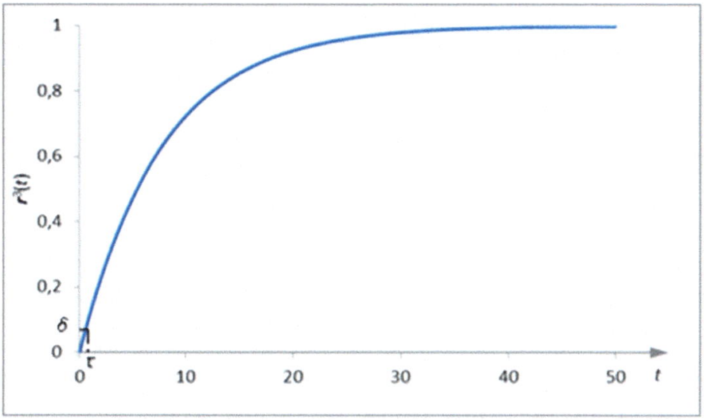


**Fig. 6.2** The graph of the risk function $r^3(t)$ (the fragility curve) of the port oil terminal critical infrastructure

**Table 6.1** The values of the port oil terminal critical infrastructure risk function

| $t$ | $r^3(t) = 1 -- S^3(t,1)$ |
|---|---|
| 0.390 | 0.0491 |
| 0.391 | 0.0492 |
| 0.392 | 0.0493 |
| 0.393 | 0.0494 |
| 0.394 | 0.0496 |
| 0.395 | 0.0497 |
| 0.396 | 0.0498 |
| 0.397 | 0.0499 |
| **0.398** | **0.0500** |
| 0.399 | 0.0502 |
| 0.400 | 0.0503 |
| 0.401 | 0.0504 |
| 0.402 | 0.0505 |
| 0.403 | 0.0507 |
| 0.404 | 0.0508 |
| 0.405 | 0.0509 |
| 0.406 | 0.0510 |
| 0.407 | 0.0511 |
| 0.408 | 0.0513 |
| 0.409 | 0.0514 |
| 0.410 | 0.0515 |

Considering (6.141)–(6.147) and (6.413)–(6.440) and applying (6.100)–(6.103), the expected values (SafI4), (SafI7) and standard deviations (SafI5), (SafI8) of the port oil terminal critical infrastructure lifetimes in the safety state subsets {1,2}, {2}, respectively are:

$$\begin{aligned}\boldsymbol{\mu}^3(1) &\cong \int_0^\infty [\boldsymbol{S}^3(t,1)]dt \cong \sum_{b=1}^{7}\sum_{l=1}^{96} pq_{bl}[\boldsymbol{\mu}^3(1)]^{(bl)}\\ &= 0.0524502 \cdot 8.16 + 0.0005298 \cdot 8.13 + 0.34529715 \cdot 8.08\\ &\quad + 0.003487825 \cdot 8.05 + 0.24651594 \cdot 8.00\\ &\quad + 0.00262251 \cdot 7.60 + 0.00179249 \cdot 7.57 + 0.0069498 \cdot 7.47\\ &\quad + 0.1748334 \cdot 7.46 + 0.0000702 \cdot 7.44 + 0.04575277 \cdot 7.40\\ &\quad + 0.00046215 \cdot 7.29 + 0.03266406 \cdot 7.32\\ &\quad + 0.000330786 \cdot 7.29 + 0.023493535 \cdot 6.96 + 0.00023751 \cdot 6.93\\ &\quad + 0.052403429 \cdot 6.81 + 0.00301986 \cdot 6.78 + 0.00006786 \cdot 6.37\\ &\quad + 0.0069498 \cdot 6.23 + 0.00000234 \cdot 6.20\\ &\cong 7.35 \text{ years,}\end{aligned} \tag{6.446}$$

$$\begin{aligned}\boldsymbol{\mu}^2(2) &\cong \int_0^\infty [\boldsymbol{S}^3(t,2)]dt \cong \sum_{b=1}^{7}\sum_{l=1}^{96} pq_{bl}[\boldsymbol{\mu}^3(2)]^{(bl)}\\ &= 0.0524502 \cdot 4.84 + 0.0005298 \cdot 4.82 + 0.34529715 \cdot 4.79\\ &\quad + 0.003487825 \cdot 4.77 + 0.24651594 \cdot 4.74\\ &\quad + 0.00262251 \cdot 4.51 + 0.00179249 \cdot 4.49 + 0.0069498 \cdot 4.42\\ &\quad + 0.1748334 \cdot 4.42 + 0.0000702 \cdot 4.41 + 0.04575277 \cdot 4.39\\ &\quad + 0.00046215 \cdot 4.36 + 0.03266406 \cdot 4.33\\ &\quad + 0.000330786 \cdot 4.31 + 0.023493535 \cdot 4.12 + 0.00023751 \cdot 4.10\\ &\quad + 0.052403429 \cdot 4.03 + 0.00301986 \cdot 4.02 + 0.00006786 \cdot 3.77\\ &\quad + 0.0069498 \cdot 3.69 + 0.00000234 \cdot 3.67\\ &\cong 4.61 \text{ years,}\end{aligned} \tag{6.447}$$

$$\boldsymbol{\sigma}^3(1) \cong 8.21, \quad \boldsymbol{\sigma}^3(2) \cong 4.63 \text{ years.} \tag{6.448}$$

The detailed way of finding the above standard deviations (6.448) is illustrated below.

According to (6.103) using (6.442), (6.443)

$$n^3(1) = 2\int_0^\infty tS^3(t, 1)dt = 121.477, \quad (6.449)$$

and

$$n^3(2) = 2\int_0^\infty tS^3(t, 2)dt = 42.686. \quad (6.450)$$

Hence, after applying (6.102) we get

$$\sigma^3(1) = \sqrt{n^3(1) - [\mu^3(1)]^2} = \sqrt{121.477 - [7.35]^2} \cong 8.21,$$
$$\sigma^3(2) = \sqrt{n^3(2) - [\mu^3(2)]^2} = \sqrt{42.688 - [4.61]^2} \cong 4.63.$$

Further, it follows that the mean values of the port oil terminal critical infrastructure lifetimes in the particular safety states (SafI9), according (6.104) and using (6.445), (6.446) are:

$$\overline{\mu}^3(1) = \mu^3(1) - \mu^3(2) \cong 2.74, \overline{\mu}^3(2) = \mu^3(2) \cong 4.61 \text{ years.} \quad (6.451)$$

From (6.446) and (6.447), applying (6.116) the approximate values of the port oil terminal critical infrastructure intensities of ageing are:

$$\lambda^3(t, 1) \cong \frac{1}{7.35} \cong 0.13605, \quad (6.452)$$

$$\lambda^3(t, 2) \cong \frac{1}{4.61} \cong 0.21692. \quad (6.453)$$

By (6.442), (6.443), applying (6.106), the port oil terminal critical infrastructure intensities of ageing (SafI10) are:

$$\begin{aligned}
\lambda^3(t, 1) = \frac{-\frac{dS^3(t,1)}{dt}}{S^3(t, 1)} \\
&= \{0.12246 \cdot 0.0524502 \exp[-0.12246t] \\
&+ 0.122984 \cdot 0.0005298 \exp[-0.122984t] \\
&+ 0.12371 \cdot 0.34529715 \exp[-0.12371t] \\
&+ 0.124234 \cdot 0.003487825 \exp[-0.124234t] \\
&+ 0.12496 \cdot 0.24651594 \exp[-0.12496t] \\
&+ 0.131548 \cdot 0.00262251 \exp[-0.131548t] \\
&+ 0.132119 \cdot 0.00179249 \exp[-0.132119t]
\end{aligned}$$

$$
\begin{aligned}
&+ 0.133833 \cdot 0.0069498 \exp[-0.133833t] \\
&+ 0.134048 \cdot 0.1748334 \exp[-0.134048t] \\
&+ 0.134357 \cdot 0.0000702 \exp[-0.134357t] \\
&+ 0.135208 \cdot 0.04575277 \exp[-0.135208t] \\
&+ 0.135732 \cdot 0.00046215 \exp[-0.135732t] \\
&+ 0.136583 \cdot 0.03266406 \exp[-0.136583t] \\
&+ 0.137107 \cdot 0.000330786 \exp[-0.137107t] \\
&+ 0.143749 \cdot 0.023493535 \exp[-0.143749t] \\
&+ 0.144321 \cdot 0.00023751 \exp[-0.144321t] \\
&+ 0.146884 \cdot 0.052403429 \exp[-0.146884t] \\
&+ 0.147504 \cdot 0.00301986 \exp[-0.147504t] \\
&+ 0.157036 \cdot 0.00006786 \exp[-0.157036t] \\
&+ 0.160541 \cdot 0.0069498 \exp[-0.160541t] \\
&+ 0.161161 \cdot 0.00000234 \exp[-0.161161t]\}\{0.0524502 \exp[-0.12246t] \\
&+ 0.0005298 \exp[-0.122984t] + 0.34529715 \exp[-0.12371t] \\
&+ 0.003487825 \exp[-0.124234t] + 0.24651594 \exp[-0.12496t] \\
&+ 0.00262251 \exp[-0.131548t] + 0.00179249 \exp[-0.132119t] \\
&+ 0.0069498 \exp[-0.133833t] + 0.1748334 \exp[-0.134048t] \\
&+ 0.0000702 \exp[-0.134357t] + 0.04575277 \exp[-0.135208t] \\
&+ 0.00046215 \exp[-0.135732t] + 0.03266406 \exp[-0.136583t] \\
&+ 0.000330786 \exp[-0.137107t] + 0.023493535 \exp[-0.143749t] \\
&+ 0.00023751 \exp[-0.144321t] \\
&+ 0.052403429 \exp[-0.146884t] \\
&+ 0.00301986 \exp[-0.147504t] + 0.00006786 \exp[-0.157036t] \\
&+ 0.0069498 \exp[-0.160541t] + 0.00000234 \exp[-0.161161t]\}, \quad t \geq 0,
\end{aligned} \tag{6.454}
$$

$$
\begin{aligned}
\lambda^3(t,2) &= \frac{-\frac{d\boldsymbol{S}^3(t,2)}{dt}}{\boldsymbol{S}^3(t,2)} \\
&= \{0.206521 \cdot 0.0524502 \exp[-0.206521t] \\
&\quad + 0.207477 \cdot 0.0005298 \exp[-0.207477t] \\
&\quad + 0.208695 \cdot 0.34529715 \exp[-0.208695t] \\
&\quad + 0.209651 \cdot 0.003487825 \exp[-0.209651t] \\
&\quad + 0.210869 \cdot 0.24651594 \exp[-0.210869t] \\
&\quad + 0.221737 \cdot 0.00262251 \exp[-0.221737t] \\
&\quad + 0.22278 \cdot 0.00179249 \exp[-0.22278t]
\end{aligned}
$$

$$
\begin{aligned}
&+ 0.225977 \cdot 0.0069498 \exp[-0.225977t] \\
&+ 0.226083 \cdot 0.1748334 \exp[-0.226083t] \\
&+ 0.226933 \cdot 0.0000702 \exp[-0.226933t] \\
&+ 0.22836 \cdot 0.04575277 \exp[-0.228364t] \\
&+ 0.229324 \cdot 0.00046215 \exp[-0.229324t] \\
&+ 0.230759 \cdot 0.03266406 \exp[-0.230759t] \\
&+ 0.231715 \cdot 0.000330786 \exp[-0.231715t] \\
&+ 0.242605 \cdot 0.023493535 \exp[-0.242605t] \\
&+ 0.243649 \cdot 0.00023751 \exp[-0.243649t] \\
&+ 0.247826 \cdot 0.052403429 \exp[-0.247826t] \\
&+ 0.248957 \cdot 0.00301986 \exp[-0.248957t] \\
&+ 0.265149 \cdot 0.00006786 \exp[-0.265149t] \\
&+ 0.271189 \cdot 0.0069498 \exp[-0.271189t] \\
&+ 0.272323 \cdot 0.00000234 \exp[-0.272323t]\} \\
&/\{0.0524502 \exp[-0.206521t] + 0.0005298 \exp[-0.207477t] \\
&+ 0.34529715 \exp[-0.208695t] + 0.003487825 \exp[-0.209651t] \\
&+ 0.24651594 \exp[-0.210869t] \\
&+ 0.00262251 \exp[-0.221737t] + 0.00179249 \exp[-0.22278t] \\
&+ 0.0069498 \exp[-0.225977t] \\
&+ 0.1748334 \exp[-0.226083t] + 0.0000702 \exp[-0.226933t] \\
&+ 0.04575277 \exp[-0.228364t] \\
&+ 0.00046215 \exp[-0.229324t] + 0.03266406 \exp[-0.230759t] \\
&+ 0.000330786 \exp[-0.231715t] \\
&+ 0.023493535 \exp[-0.242605t] + 0.00023751 \exp[-0.243649t] \\
&+ 0.052403429 \exp[-0.247826t] \\
&+ 0.00301986 \exp[-0.248957t] + 0.00006786 \exp[-0.265149t] \\
&+ 0.0069498 \exp[-0.271189t] \\
&+ 0.00000234 \exp[-0.272323t]\}, \quad t \geq 0. \qquad (6.455)
\end{aligned}
$$

From (6.454) and (6.455), the limit values of the port oil terminal critical infrastructure intensities of ageing are:

$$
\begin{aligned}
\boldsymbol{\lambda}^3(1) &= \lim_{t\to\infty} \boldsymbol{\lambda}^3(t,1) = \lim_{t\to\infty} \frac{[0.12246 \cdot 0.0524502 \exp[-0.12246t]][1+o(1)]}{0.0524502 \exp[-0.12246t][1+o(1)]} \\
&= 0.12246, \qquad (6.456)
\end{aligned}
$$

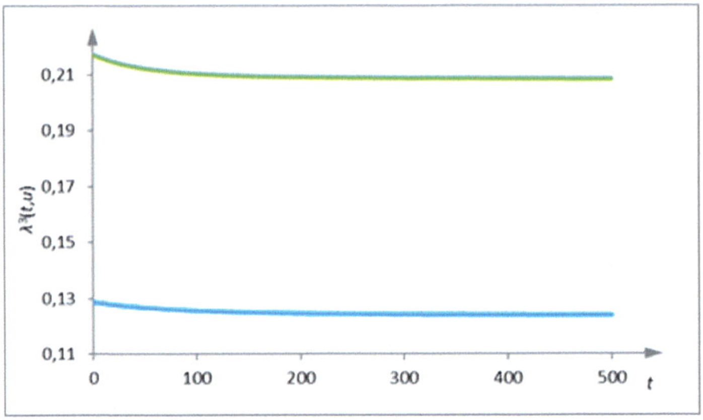


**Fig. 6.3** The graph of the intensieties of ageing of the port oil terminal critical infrastructure

$$\boldsymbol{\lambda}^3(2) = \lim_{t\to\infty} \boldsymbol{\lambda}^3(t,2) = \lim_{t\to\infty} \frac{[0.206521 \cdot 0.87417 \exp[-0.206521]][1+o(1)]}{0.87417 \exp[-0.206521][1+o(1)]}$$
$$= 0.206521. \tag{6.457}$$

The graphs of the intensities of ageing of the port oil terminal critical infrastructure are shown in Fig. 6.3.

Considering (6.452), (6.453), (6.456), (6.457) and the values of the port oil terminal critical infrastructure intensities of ageing without of operation and climate-weather impact, determined in Chap. 3,

$$\boldsymbol{\lambda}^0(t,1) = 0.115873, \quad \boldsymbol{\lambda}^0(t,2) = 0.181739, \tag{6.458}$$

and applying (6.109), the coefficients of the operation process impact on the port oil terminal critical infrastructure intensities of ageing (ResI1) are:

$$\boldsymbol{\rho}^3(t,1) = \frac{\lambda^3(t,1)}{\lambda^0(t,1)} \cong \frac{0.13605}{0.115873} \cong 1.17413, \tag{6.459}$$

$$\boldsymbol{\rho}^3(t,1) = \frac{\lambda^3(t,1)}{\lambda^0(t,1)} \cong \frac{0.12246}{0.115873} \cong 1.05685 \tag{6.460}$$

$$\boldsymbol{\rho}^3(t,2) = \frac{\lambda^3(t,2)}{\lambda^0(t,2)} \cong \frac{0.21692}{0.181739} \cong 1.19358, \tag{6.461}$$

$$\boldsymbol{\rho}^3(t,2) = \frac{\lambda^3(t,2)}{\lambda^0(t,2)} \cong \frac{0.206521}{0.181739} \cong 1.13636. \tag{6.462}$$

Finally, by (6.112) and (6.460), the port oil terminal critical infrastructure resilence indicator (ResI2), i.e. the coefficient of the port oil terminal critical infrastructure resilience to operation process impact, is

$$\boldsymbol{RI}^3(t, 1) = \frac{1}{\rho^3(t, 1)} \cong 0.946 = 94.6\%, \quad t \in < 0, +\infty. \quad (6.463)$$

The comparison of safety indicators (4.124)–(4.136), (5.109)–(5.123) and (6.441)–(6.455) proves that the influence of the operation process is significantly then the influence of the climate-weather change process and the influence of the operation process related to climate-weather change process on the port oil terminal safety what is also clearly expressed in its resilience indicators to operation process impact given by (4.137)–(4.139) and in its resilience indicators to climate-weather change process impact given by (5.125)–(5.129) and operation process related to climate-weather change process impact given by (6.459)–(6.463).

## 6.6 Conclusion

This book presents a comprehensive approach to the safety analysis, evaluation and prediction of critical infrastructures, showing the role of their operation process and the climate-weather change process at their operating area. This approach is based on two-state and multistate system reliability modelling. At first, safety of ageing multistate critical infrastructure without outside impacts is modeled. Then, the approach is modified and developed for safety analysis of critical infrastructures impacted, separately and then simultaneously, by their operation process, climate-weather change process and by their operation process influenced by the climate-weather change process at the operating area.

Starting from the critical infrastructure simplest safety model, defined as a multistate ageing system without considering outside impacts, this book shows how to combine that model with the model of critical infrastructure operation process, and to apply the integrated model to analyze safety of critical infrastructures. Furthermore, we define safety and resilience indicators, useful in the critical infrastructure examination. Another step in this work was to develop an integrated safety model of critical infrastructures considering the influence of climate-weather change on infrastructure's safety.

Thus, a comprehensive critical infrastructure safety model is proposed, which, simultaneously, considers the operation process and the influence of the climate-weather change process, on the safety of a critical infrastructure.

All proposed safety and resilience indicators are defined for any critical infrastructures with varying in time safety structures and assets safety parameters. It's been shown how changes of operation process and climate-weather process affect assets' safety parameters. The model has been validated on a port oil terminal.

The methods and procedures of identification and prediction of operation process and climate-weather change process and their influence on critical infrastructure safety presented in the book may also be extended to the estimation of safety of other complex technical systems. Moreover, they can be used in maintenance optimization and operation cost analysis of other critical infrastructures. They can be applied in the safety assessment of a wide class of real technical systems in varying operation and climate-weather conditions.

In our future research activity, we will focus on investigating and solving different problems relating to safety of critical infrastructure impacted by operation and climate-weather change, and to methods for strengthening critical infrastructure resilience upon the influence of operation and weather changes. We will also focus on methods for critical infrastructure safety optimization, as well as on studying their degradation, identifying accident consequences and mitigating them.

Finding optimal values of safety and resilience indicators, as well as analyzing resilience and strengthening the critical infrastructure against climate-weather change will be the aim of our future research.

## References

1. Ferreira F, Pacheco A (2007) Comparison of level-crossing times for Markov and semi-Markov processes. Stat Probab Lett 77(2):151–157
2. Glynn PW, Haas PJ (2006) Laws of large numbers and functional central limit theorems for generalized semi-Markov processes. Stoch Model 22(2):201–231
3. Grabski F (2002) Semi-Markov models of systems reliability and operations analysis. System Research Institute, Polish Academy of Science (in Polish)
4. Jakusik E, Kołowrocki K, Kuligowska E, Soszyńska-Budny J, Torbicki M (2016) Modelling climate-weather change process including extreme weather hazards for port oil piping transportation system. J Pol Saf Reliab Assoc, Summer Safety Reliab Semin 7(3):31–40
5. Jakusik E, Kołowrocki K, Kuligowska E, Soszyńska-Budny J, Torbicki M (2016) Identification methods and procedures of climate-weather change process including extreme weather hazards for port oil piping transportation system operating at under water Baltic Sea area. J Pol Saf Reliab Assoc, Summer Safety Reliab Semin 7(3):47–56
6. Jakusik E, Kołowrocki K, Kuligowska E, Soszyńska-Budny J, Torbicki M (2016) Identification methods and procedures of climate-weather change process including extreme weather hazards of port oil piping transportation system operating at land Baltic seaside area. J Pol Saf Reliab Assoc, Summer Safety Reliab Seminars 7(3):57–64
7. Kołowrocki K (2005) Reliability of large systems. Elsevier Amsterdam—Boston—Heidelberg—London—New York—Oxford—Paris—San Diego—San Francisco—Singapore—Sydney—Tokyo
8. Kołowrocki K (2014) Reliability of Large and Complex Systems (2nd edn), Elsevier Amsterdam—Boston—Heidelberg—London—New York—Oxford—Paris—San Diego—San Francisco—Singapore—Sydney—Tokyo
9. Kołowrocki K, Kuligowska E, Soszyńska-Budny J, Torbicki M (2018) Oil transport in port Part 1 Port oil piping transportation system safety and resilience impacted by its operation process. J Pol Saf Reliab Assoc, Summer Safety and Reliability Seminars 9(2):11–28
10. Kołowrocki K, Soszyńska-Budny J (2011) Reliability and safety of complex technical systems and processes. Modeling – identification – prediction—optimization. London, Dordrecht, Heildeberg, New York, Springer
11. Kołowrocki K, Soszyńska-Budny J (2015) Modelling Safety of Multistate Systems with Ageing Components. Proc. The International Conference of Numerical Analysis and Applied Mathematics—ICNAAM 2015 Symposium/Workshop on Safety of Critical Infrastructures, (ISBN: 978–0–7354–1392–4; DOI: https://doi.org/10.1063/1.4952224)

12. Kołowrocki K, Soszyńska-Budny J (2016) Modelling critical infrastructure operation process including operating environment threats. Journal of Polish Safety and Reliability Association, Summer Safety and Reliability Seminars 7(3):81–88
13. Kołowrocki K Soszyńska-Budny J (2018) Critical Infrastructure Safety Indicators. International Conference on Industrial Engineering and Engineering Management IEEM 2018
14. Kołowrocki K Soszyńska-Budny J (2018) Critical Infrastructure Impacted by Operation Safety and Resilience Indicators. International Conference on Industrial Engineering and Engineering Management IEEM 2018
15. Kołowrocki K, Soszyńska-Budny J, Torbicki M (2017) Critical infrastructure operating area climate-weather change process including extreme weather hazards. J Pol Saf Reliab Assoc, Summer Safety and Reliability Seminars 8(2):15–24
16. Kołowrocki K, Soszyńska-Budny J, Torbicki M (2017) Critical infrastructure operation proces related to climate-weather change process including extreme weather hazard. J Pol Saf Reliab Assoc, Summer Safety and Reliability Seminars 8(2):25–40
17. Kołowrocki K, Soszyńska-Budny J, Torbicki M (2017) Identification methods and procedures of climate-weather change process including extreme weather hazards. J Pol Saf Reliab Assoc, Summer Safety and Reliability Seminars 8(2):85–106
18. Kołowrocki K, Soszyńska-Budny J, Torbicki M (2017) Identification and prediction of port oil piping transportation system operation process related to climate-weather change. J Pol Saf Reliab Assoc, Summer Safety and Reliability Seminars 8(2):113–120
19. Kołowrocki K, Soszyńska-Budny J, Torbicki M (2018) Critical Infrastructure Impacted by Operation and Climate Change Safety and Resilience Indicators. International Conference on Industrial Engineering and Engineering Management IEEM 2018
20. Limnios N, Oprisan G (2005) Semi-Markov Processes and Reliability. Birkhauser, Boston
21. Mercier S (2008) Numerical bounds for semi-Markovian quantities and application to reliability. Methodol and Comput in Appl Probab 10(2):179–198
22. Torbicki M (2019) Safety of a critical network infrastructure exposed to operation and weather condition changes. PhD Thesis, Polish Academy of Science, System Research Institute, Warsaw

The manufacturer's authorised representative in the EU is Springer Nature Customer Service Centre GmbH, Europaplatz 3, 69115 Heidelberg, Germany. If you have any concerns regarding our products, please contact ProductSafety@springernature.com

Printed and bound by CPI Group (UK) Ltd, Croydon, CR0 4YY

30/09/2026

02225506-0005